THE OVARY OF EVE

Clara Pinto-Correia

THE
OVARY
OF
EVE

Egg and Sperm and Preformation

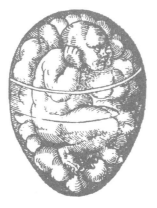

THE UNIVERSITY OF CHICAGO PRESS

Chicago and London

Clara Pinto-Correia is professor of developmental biology, director of the Master's Course of Developmental Biology at Universidade Lusofona, in Lisbon, Portugal, and adjunct professor at the University of Massachusetts at Amherst. She is the author of several nonfiction books, several books of poetry, and six novels.

Foreword © 1997 by Stephen Jay Gould
The University of Chicago Press, Chicago 60637
The University of Chicago Press, Ltd., London
© 1997 by The University of Chicago
All rights reserved. Published 1997
Printed in the United States of America
06   05   04   03   02   01   00   99   98   97   1   2   3   4   5
ISBN 0-226-66952-1 (cloth)

Library of Congress Cataloging-in-Publication Data

Correia, Clara Pinto.
    The ovary of Eve: egg and sperm and preformation / Clara Pinto-Correia.
      p.   cm.
    Includes bibliographical references and index.
    ISBN 0-226-66952-1 (alk. paper)
    1. Science—Europe—History—17th century.   2. Science—Europe—
History—18th century.   3. Embryology—History.   4. Reproduction—
Research—History.   I. Title.
    Q127.E8C67   1997
    509.4'09'032—DC21   97-14011
                                                                                            CIP

∞ The paper used in this publication meets the minimum requirements of the American National Standard for Information Sciences—Permanence of Paper for Printed Library Materials, ANSI Z39.48-1984.

TO DICK

whom God was kind enough
to ascribe to the same
generation as me.

CONTENTS

ILLUSTRATIONS

FOREWORD

STEPHEN JAY GOULD

When scientists become stymied at a frustrating point in a difficult argument, they generally assume that some unfortunate lack of information, enjoined by technological limits of inquiry, must be causing the impasse—a "factual lack," if you will. In other words, make me a better microscope or telescope and I will see further into the nature of things, thus removing the roadblock to progress.

Often, of course, solutions do follow a new technology that permits a view of the formerly unperceivable—just as the invention of the microscope surely answered some questions, while inspiring even more mystery and debate, about the subject treated in this book: the sources and causes of embryonic development. But impasses in the history of science occur just as frequently, and often cut far more deeply, when their reasons reside in limitations to the modes of thought, the conceivable range of solutions, that scientists bring to their studies—a "conceptual lock," if you will.

Some of these conceptual locks merely record the limited compass permitted by any surrounding cultural tradition. But other restrictions may run as deep as the evolutionary construction of mind. I have often wondered, for example, whether the main impediments to understanding such important, influential, and yes, beautiful theories as the embryological notion of preformation—impediments better broken by the composition of this book than by any other document ever written in the history of science—might not be imposed by some universal and deeply rooted propensities in our thinking.

We have, at least for the last three centuries of Western culture, been driven to view the history of thought and society—if not the history of all things—as a tale of general progress toward an enlightenment that must be the ultimate goal of human consciousness. At the same time, we seem compelled, perhaps by an even more ancient and universal trick of mental construction, to view the pathway of this progress as a set of victories within dichotomies—forces of good against evil, light against darkness, flexibility against dogma, cowboys against Indians (now subject to inversion for understandable reasons of political correctness).

xiii

I cannot imagine any greater conceptual lock upon the understanding of our scientific past than this tendency to view history as a path marked by gladiatorial combats, pair by pair, with victory always gained by the swordsman who holds a truer view of nature's objective factuality. I certainly learned my textbook history of science in this manner. I was taught that each of the three subjects anchoring my training as a student of macroevolution—embryology, geology, and evolution itself—had been born in the resolution of a dichotomy in favor of nature's truth against society's prejudice. (I realize, of course, that professional historians of science would never present so naive a tale of heroic triumph—though even this field featured such "Whiggish" reconstructions in generations not long past. I am, rather, recalling the genre, still so prevalent, of introductory pages in basic textbooks on scientific subjects.)

Evolution had triumphed over creation in Darwin's day. In geology, uniformitarianism had beaten catastrophism to destroy the miracle-mongering theological apologists who still wanted to restrict the history of the earth to a few thousand years filled with cataclysmic, divinely ordained paroxysms. In embryology, the epigeneticists—the good guys in the white hats who acknowledged what they saw, and knew that the egg developed from formless beginnings to final complexities—had routed the dark forces of preformationism, the tradition-blinded ideologues who actually proposed that a set of perfectly formed homunculi must have been created all coiled up, Russian doll within doll, either in the sperm of Adam or the ovaries of Eve. Preformationism, I learned, was a nonsensical doctrine espoused by men who could not bear to give up the dream of a static world order ruled by an omnipotent God—and who therefore could not admit the plain evidence of their senses when watching the development of the chick in the egg. The triumph of epigenesis marked the beginning of modern embryology. As for defining the entire field by this primary dichotomy, Joseph Needham had written, in the 1959 edition of his definitive *History of Embryology,* that epigenesis and preformationism marked "an antithesis which Aristotle was the first to perceive, and the subsequent history of which is almost synonymous with the history of embryology." And how could the preformationists be judged as anything but blind stalwarts of ancient prejudice (against the good-guy epigeneticist empiricists) when the truly blind Charles Bonnet, one of their greatest heroes, had stated as a virtual motto and mantra that "this hypothesis is one of the greatest victories that pure understanding has won over the senses?"

For two major reasons, I had reversed this silly student's view, learned in the early 1960s, by the time I wrote my first book, *Ontogeny and Phylogeny,* in 1977. First, the revolution inspired by Thomas Kuhn's work (and that of many other historians) on scientific revolutions had swept the field and made the old Whiggish view entirely unfashionable. We were now enjoined (quite rightly, I still think) to approach any rich world view with sympathy for the reasons it had gained attention, and the power of its influence, in its own time—and not to judge from the irrelevant and anachronistic standpoint of modern knowledge. Second, I finally read the major works of preformationism in its heyday during the seventeenth and eighteenth centuries—and I greatly admired the major arguments and their rationales. I understood, first of all, that the most "absurd" notions (from current perspectives) made reasonable sense under the "factual lacks" of eighteenth-century science. With no cell theory to set a lower limit on organic size, no irrationality attended the idea of innumerable generations encapsulated, one inside the other, within Eve's ovaries. Moreover, the accepted idea of an earth only a few thousand years old made the number of encapsulated generations quite conceivable, whereas a 4-billion-year-old earth might have strained preformationistic credulity even then.

But, more importantly, I came to understand that the leading preformationists had been, ironically, defenders of the general mechanistic attitude that modern science now honors, whereas the epigeneticists had tended to favor the vitalistic views now anathematized. (In fact, Bonnet's famous dictum expressed his defense of mechanism, for epigenesis seemed to require an external and immaterial shaping force: how else could complex form be unerringly imposed upon the simple homogeneous beginnings of an ovum? Logic demanded that something be preformed in the fertilized egg—for nothing can come from nothing.) The microscopes of Bonnet's day were too poor to resolve the tiny preformed creature that logic required but technology could not yet perceive. In this sense, preformationism represented the triumph of reason over our limited sensation.

Besides, and finally, who can say that the epigeneticists triumphed, while preformationism went down to flaming defeat? Yes, the visual appearance of morphogenesis is epigenetic; but the preformationists were right to insist that some system of guidance must exist in the egg. Shall we blame them because the metaphoric apparatus of eighteenth-century life did not include the "correct" concept of programmed instructions rather than preformed parts? Bonnet knew the music box, but perhaps society needed the Jacquard loom, the player piano, and the Hollerith computer punch card to bring the concept of coded instructions into general conceivability.

For all these reasons, and many others, modern historians of science treat preformationism with more respect. But, curiously, this defining world view of a major field during a crucial century has still not won its independence—for even sympathetic historians continue to treat preformationism primarily in contrast to its dichotomous partner, epigenesis! Clara Pinto-Correia's remarkable book, for all its marvelous idiosyncrasies, may be the first historical treatise devoted exclusively to preformationism as a worthy subject in its own right, and not as a foil for its contrasts. Dr. Pinto-Correia also uses the technique of dichotomy to define the history of her subject—but the dichotomy now lies *within* preformationism as the central contrast between ovists and spermists, or those who located the preformed embryos in the egg or in the sperm.

I had developed both sympathy and strong interest for preformationism before I met Clara Pinto-Correia and worked with her on this book. But I had no concept of the remarkable reach of the tentacles of this theory into virtually every important corner of life in the seventeenth and eighteenth centuries, and into so many roots of previous times. I now realize that the history of preformationism is coextensive with the history of most important intellectual subjects in science, religion, and general culture as well. I can now truly cash out an attitude that has long defined my intellectual preferences, but that I had never been able to illustrate with such a striking example—the interdisciplinary nature of all substantial knowledge (indeed, the meaninglessness of conventional academic boundaries), and the roadblocks placed before real understanding by restricting the attentions of most scholars to the certified subject matter of their own callings.

This unconventional book eschews the standard formats of chronology or biography, and opts instead for a series of Shandyesque excursions along the tentacles of preformationism into an astonishing range of cultural and scientific highways and byways. We follow the mysteries of infinite divisibility backward from the series of encapsulations in the ovary of Eve to medieval debates about angels on pins, and forward to modern doctrines of homeopathic medicine, where some dilutions of potency are so great that not a single molecule of the active agent can exist in most dosages—and yet many people claim they work! We trace the central notions of preformationism into such philosophical and social conundrums as a preference for ovism based on the egg's close approach to an ideal sphere, matched with distaste that such a preferred vessel should exist in the inferior female—while the superior male produces only a liquid filled with wriggling "spermatic worms." We watch the battle between ovists and spermists unfold in areas that would now be labeled as moral or religious—as in the even greater opprobrium heaped upon masturbation if each "wasted" sperm truly contains a preformed individual yearning for life.

I doubt that this remarkable, enlightening, and sublimely entertaining book could have arisen from the cloistered center of American academic life. But Clara Pinto-Correia possesses a most uncommon range of talents and backgrounds. She holds a Ph.D. in developmental biology and has published many distinguished technical papers on the subject. But she is also one of the best-known novelists of her native Portugal. Thus, she is a true scholar who always knows how (and when) to be funny and irreverent. She is also, perhaps above all, a "European intellectual" in the most admirable sense—a person who knows several languages and cultural traditions by living them, while most American scholars learn them only academically, if at all.

This book restores the notion of "science" to its original meaning and etymological root—that is, all knowledge, not just the restricted sort of verifiable natural data that Whewell imbued with this word in the early nineteenth century. Clara Pinto-Correia, like Bacon (whom she probably does not admire), has taken all knowledge for her province—and has shown us why we need such a full range if we wish to understand so rich a theory as the "beautiful" notion of preformationism.

PREFACE

I was born in Lisbon, Portugal (hence my unpronounceable last name)[1], in 1960—fourteen years before the fascist dictatorship that then ruled our country was finally overthrown by the democratic revolution of the most unforgettable of all Aprils, and in the exact year when our absurd colonial war broke out in Africa. Wars are horrible by definition, and ours was stupid on top of that; but I owe to it the privilege of having spent my early childhood in Angola. It followed, quite logically for a child growing up amid elephants and antelopes and fabulous translucent-bellied geckos, that I would immediately decide to study biology. And so I did, although I never became the intrepid park ranger of my early dreams. Much to the contrary, I spent the beginning of my professional life among the pipettes and petri dishes of a long succession of laboratory benches. At the beginning of my graduate studies, I learned electron microscopy and taught embryology for three years at the Lisbon Medical School. I displayed the countless marvels of countless sea urchin fertilizations to my students, all the while slicing even more countless mouse oocytes for my own research project, my lab coat full of chemical stains and my heart full of dreams.

Dreams are dangerous. They take you places. It happened to me, just as it has long been happening to so many others: slowly but steadily, the straitjacket of this quiet routine started hammering away at my nerve endings. And so I did what those so many others have long been doing. I packed a few things, shed a few tears, and came to America. I worked at the State University of New York at Buffalo, where I learned much more than how to deal with snowstorms: I gained my first solid insights into the mysteries of mammalian fertilization. Nothing special, really: I was just going through the motions of a new routine. Mission accomplished, I got my Ph.D. The wheels of routine kept spinning, and I moved to the University of Massachusetts for postdoctoral training in mammalian cloning.

Then, finally, I took the routine—and everybody around me, for that matter—by surprise when I asked Harvard University to let me come and learn a thing or two about the history of science. Having dedicated my scientific career to the arcana of fertilization, my mind was now set on studying and writing about the history of theories of reproduction.

This move may have been surprising to my laboratory colleagues, but nothing could have seemed more obvious to me. First, the early readings that had delighted me the most were nineteenth-century narratives of naturalists crossing the uncharted African wilderness and registering their exciting discoveries as they went along. Second, when I was already deep into my biological studies, I had devoured Stephen Jay Gould's *Ontogeny and Phylogeny* and had become truly enchanted by what a scientist can do when he analyzes the historical background underlying his field of expertise. And third, in a twist that seems to have emerged by spontaneous generation, for I had not really planned it and did not see it coming, in the meantime I had also become a novelist. I had been writing stories ever since I had learned to write. I had worked as a journalist while I was studying biology, and had learned to listen, especially to what is being said between the lines. I published my first book, then a second, then a third. All of a sudden I looked back and saw more than twenty publications behind me. Moving to the field of history of science was the perfect means of combining my two great professional passions, literature and scientific research.

But then, of course, if you want to do history of science, you first need to learn how to do it; and that is something that no laboratory training will ever teach you (after all, we rarely bother to read papers published before 1980 . . .). So I took a deep breath, shrugged my shoulders, repeated to myself the old refrain that freedom's just another word for nothing left to lose, and wrote a letter to Gould: I want to learn and work with you, can I? I got the answer in the mail exactly one week later: sure, come on over, let's talk about it. Ah, America.

Even after I was established in my new position, I still did not fully trust my abilities. Sure, I told Gould, I can learn. But should I dare to *write?* And in *English,* at that? I can write a scientific paper, but could I ever write a history of science book? He just said, give it a try, I'll help you. But you could be wasting your time, I said. If you're not good enough, he answered without even looking at me, I'll know it in two weeks and won't waste any more time with you. Ah, yes. America. I believe I remember having been told about the survival of the fittest.

So, three years later, here is the net result of my beginner's effort.[2] Of all the possible aspects of reproduction, I chose to take a closer look at preformation. I had been fascinated by the beauty of this doctrine ever since I first read about it, in one single hasty paragraph buried in the introduction to an embryology textbook. Moreover, I wondered how a theory deemed ridiculous and obscurantist by so many modern writers had dominated the ideas on reproduction developed during the Scientific Revolution—the Age of Enlightenment par excellence.

I should warn you that my approach to preformation entails the choice of a method whose heterodoxy may surprise some readers. Do not expect to find precise chronological accounts or detailed biographies. If the grandmother of any of the preformationists ever drowned in a well when the scholar-to-be was six years old, I did not try to find out about it. I am trying to discover the full structure of the fabric of preformation by pulling threads and tying or untying knots. I am aware that in such a process the order of factors is not as arbitrary as in a simple multiplication; but we never know for certain where each thread will come from, or what other threads it may be connected with. As in any unwinding process, expect flashbacks and flashforwards. Expect revelations that we would never have anticipated. Brace yourself for finding Plato, Panspermatism, Paracelsus, Peter and Paul, Pangenesis, and Ptolemy all in one sentence, very likely in reverse order. Please accept surprises. I like to be surprised. That is why I enjoyed writing this book so much.

You may also notice that, for the sake of simplicity, I did not try to solve the vexatious question of exactly what we mean by "Scientific Revolution" and what is the proper period to be considered the "Enlightenment." Also, I have opted not to dwell upon the distinction often drawn by historians of science between *preformation* (the assumption that the primordial organism already contains inside itself all other organisms of the same species, perfectly preformed, minuscule though they might be) and *pre-existence* (the more sophisticated version of the model, in which the primordial organism contains only the basic blueprints of all the related organisms to come).[3] I have also chosen to use modern terminology for the structures that the naturalists of the period were analyzing (gonad, spermatozoon, cell, gamete, etc.), even though these terms—or the precise modern decoding of their meaning—did not exist during their lifetimes. Again for the sake of simplicity, and bearing in mind that we are facing an endlessly complex subject, I have chosen to focus mainly upon animal reproduction. There will be some references to plants, but only when the case in point could not be illustrated otherwise. Just one last technical detail: unless otherwise noted, all the translations are my own. As for quotes of archaic English, I often modernized the text so that it would be easier to read—much to my chagrin, I would like to add, since archaisms are truly lovely.

At this point I would like to pronounce that standard and emphatic last sentence used by rule-breaking criminals in their own defense, "I rest my case." But first I must thank all those who made my work possible, for this is a book that I truly never would have written were it not for the help of a huge number of fine people who guided me through my initial foolish boldness and helped me to give it shape and sense. The whole crew of the Department of History of Science at Harvard, where I spent two years as a visiting scholar learning the tools of a trade that was totally new to me, should definitely come at the top of the list. This project was developed in 1994–95 at the Department of History of Science at Harvard, under the sponsorship of Professor Everett I. Mendelsohn. Research funds were partly provided by Fundacão Luso-Americana para o Desenvolvimento (FLAD, Portugal) and by Programa Praxis XXI (E.C., Portugal). For want of a better logic, I shall now list all the individuals who contributed to the process in alphabetical order:

Many thanks

—to Susan Abrams, for everything.

—to Martha Baldwin, who literally took me by the hand into the marvelous world of libraries.

—to my dear friend, unrepentant communist, sometime lawyer, and laureate Portuguese novelist Mário de Carvalho (and his Latinist daughter Rita Taborda) for helping me with Latin translations that were far above my deciphering skills.

—to Paula Findlen, for her enthusiastic and totally unsolicited support from the first day we met.

—to Michael Fisher, for pointing me in the right direction.

—to Scott Gilbert, for his warmth, his knowledge, and all of his countless stimulating remarks and suggestions—and also, of course, for having provided us since 1986 with *the* textbook that every developmental biology teacher and student dreamed of having.

—to Lisbet Koerner, for her unconditional support, enthusiasm, patient reading of my early drafts, and willingness to make me see how one should approach issues as a historian.

—to João Pedro Leão, for his relentless peeking into my drafts and his even more relentless illuminating and provocative suggestions.

—to Christophe Luthy, for his contagious pleasure in learning and his helpful insights into the world of early microscopy.

—to Madalena Machado, for her warm, gorgeous, and gregarious home that was always at my disposal at the end of long days of solitary work.

—to Everett Mendelsohn, for always reminding me of the question I really wanted to ask.

—to Javier Moscoso, for all of his solid and substantial teratological contributions.

—to John Murdoch, who has no idea how much I learned from him.

—to William Newman, for showing me so many things that I needed to know and would never have even guessed were there to be known.

—to my parents.

—to Norma Roche, for her marvelous editing skills.

—to Vernon Rosario, for his wit, his solidarity, and his bottomless erudition on masturbation (I was very pleased to discover that his family is originally from Goa, which in turn was a Portuguese province until the 1950s. *Les bons esprits se rencontrent,* if you know what I mean).

—to Mr. and Mrs. Wolf, at the Countway Library room of the Rare Book collection of Harvard Medical School, for introducing me to the unspeakable joys of digging hidden jewels from the cryptic contents of ancient books.

And finally, deliberately out of alphabetical order because I like to save the best for last:

—to Stephen Jay Gould, for being the absolutely stellar person that I had imagined him to be when I first read his books during my early youth—and then some.[4]

That should be all. Enjoy the ride.

Dare to Know

God is not a puppeteer
SPINOZA

Generation has always been a fundamental question, arguably one of the primordial roots of religious experience. Who are we? What are we doing here? Who PUT us here? And so on. We know the questions. If we want a word that sums them up better than any other, "birth" is certainly a powerful candidate.

Wondrous as it is, generation has never ceased to tease the imagination of those curious people who take up the task of thinking about life. One would have to be extremely narrow to assume that this interest derives mainly from the somewhat forbidden pleasures that we have long come to associate with reproduction. Above and beyond such prosaic approaches, a big part of generation's appeal relies on the fact that it concerns not only the reproduction of organisms, but also the recurrently observed phenomenon of regrowth of body parts. Moreover, any theory brought forward in this area needs to explain the constant morphology of each species and the diversity in the composition of each organism. The combination of all these pieces must have formed an irresistible puzzle for the stubbornness of human curiosity.

Ex Ovo Omnia

Within the impressive range of his contributions to the philosophical reconstructions of the Scientific Revolution, the ubiquitous French geometer René Descartes (1596–1650) brought a modern ring to the concept of generation de novo by assigning it to the work of his beloved mechanical causes. Male and female particles would mix at reproduction, and therefore start to ferment. This fermentation led in turn to the formation of the heart, and all the other embryonic parts followed, proving that even the greatest complexities of the animate world could be explained through motion and matter alone. In Descartes's own words, "it is no less natural for a clock, made of a certain number of wheels, to indicate the hours, than for a tree from a certain seed to produce a particular fruit." [1] This, obviously, was Descartes's main error: not even his

1

Fig. 1. *Ex ovo omnia:* Zeus liberating all living things from an egg. (From the frontispiece of Harvey's *Exercitationes de generatione animalium,* 1651.)

most fervent admirers were willing to accept an answer of such frustrating simplicity for the overwhelming complexity of generation, and Descartes himself was never totally satisfied with his own explanation.

Surmounting Descartes's failure with a much more sophisticated approach, the English physician William Harvey (1578–1657) opened the real debate on generation for modernity by postulating that "all that is alive comes from the egg" in his famous book *Exercitationes de generatione animalium,* published in 1651. The author was then over seventy years old, and therefore *Exercitationes* emerged as his ultimate piece of wisdom (or, as Harvey's opponents would have said, as his ultimate act of senility). Harvey based his daring statement on observations made in eggs from chickens and from deer. In reality, these "eggs" were most likely early embryos, and backed Harvey's Aristotelian idea that animals begin their existence as a homogeneous mass, from which the organs derive one after another by the process of formation de novo, or *epigenesis.* His observations of embryonic development led him to conclude that "the generation of the chick is the result of epigenesis," since "all parts are not fashioned simultaneously, but emerge in their due succession and order."

It was Harvey's misfortune, however, to still consider spontaneous generation a plausible reproductive model for "lower organisms" at a time when spontaneous generation had begun its slow fall from grace. Moreover, and certainly more damaging yet, Harvey's concept of epigenesis was perfectly consistent with the classical idea that the unformed substance eventually takes up a form that is potentially, but not actually, in it. His enlightened explanation just happened to come at a time when the fashion was to repudiate Aristotelian postulates and to look for immediate mechanical causes. Nothing is ever simple, but now everything was bound to ascend to an even more daring level of complexity.

GOD AND MECHANISM

In fact, with the onset of the seventeenth century, when something closer to what we would now call technology, and something closer to what we would now call science, started to develop, the full grasp of knowledge was reaching unexpected heights. With the ongoing change of attitudes, mechanical philosophy was beginning to have its day in the sun—carrying along the postulate that life was nothing more than an interesting cluster of behavioral properties inherent to matter itself.

The main problem with the new mechanical reasoning was that, even assuming an equivalence between living and nonliving things, mechanism could not provide satisfactory explanations for the intricate specificities of reproduction, such as the inheritance of traits from parents by children, the continuity of species, or regeneration—or, in what would become one of the most heated centers of the debate, the birth of monsters.

And then preformation surfaced.

It was the middle of the seventeenth century. The theory that was to rule the scene for the next hundred years was a resounding success in its uncanny clarity. It proposed, very simply, that all living beings existed preformed inside their forebears in the manner of a Russian doll, put there by God at the beginning of Creation with a precise moment established for each one to unfold and come to life. As one of the most creative alternatives to the sterility of spontaneous generation and the confusing demands of the initial versions of epigenesis, preformation is also an exemplary illustration of how the chemistry of ideas is catalyzed by the mental frame of a given period. The idea made such a successful entrance into the realm of natural philosophy that, after the mid-seventeenth century, another entire century had to pass before other reproductive theories took up the challenge of offering believable alternatives.

Against the background of the Scientific Revolution, preformation was certainly a logical path to follow. It added vital explanations to the mystery of reproduction without seriously challenging the prevalence of mechanistic causes. Moreover, the whole concept agreed reasonably well with some of the cutting-edge philosophies of the time, particularly Descartes's principle of the infinite divisibility of a mechanical Nature initiated, but not interfered with, by God. The theory also offered convenient religious and social backing for the status quo. Preformation "scientifically" established that all men were in fact brothers, since they all came from the same gonad. Thus it confirmed the teachings of Jesus with a new discourse in natural philosophy, and in this the theory became even more effective since it could finally explain the irrevocability of original sin—we had all been soiled by it, since we had all been encased within the first sinner. Moreover, it could now be seen as inevitable that servants would always originate from servants, just as kings would always originate from kings. By putting lineages inside each other, preformation could function as a "politically correct" antidemocratic doctrine, implicitly legitimizing the dynastic system—and, of course, the leading natural philosophers of the Scientific Revolution certainly were *not* servants.

DRIVING FORCES

Caspar Wolff (1734–1794) was the main figure to come to the rescue of epigenesis with the publication, in 1759, of his provocative doctoral thesis, "Theoria Generationes." Working with chicken eggs, Wolff verified that the adult organism developed from tissues having no counterpart in the embryo. Heart and blood vessels developed anew in each embryo, and the same happened to the intestine, which arose from the folding of an originally flat tissue. Wolff firmly believed that when everyone had grasped the significance of intestinal formation, epigenesis would no longer be questioned. To sustain this theory of creation de novo, he postulated that the embryo was created by an invisible force, the *vis essentialis,* inherent in living matter.

At this point the debate seemed deadlocked. Epigenesis was better at explaining variations and the direct observations of organ formation, but preformation was better at explaining continuity between generations—not to mention that Wolff's model relied on invisible driving forces, and such entities were by no means the toast of the moment. But the evolution of concepts on both sides had been so pronounced that the time was now ripe for Immanuel Kant (1724–1804), with the help of his biologist colleague J. F. Blumenbach (1752–1840), to come up with some sort of reconciliatory theory. This model involved a mechanical goal-directed force, the *Bildungstrieb,* or "developing drive," defined as a property of the organism itself, inherited through the germ cells. In other words, development could proceed epigenetically through a predetermined force inherent in the matter of the embryo—the primordia of modern developmental biology.

THE RUSSIAN DOLL

It is a beautiful historical coincidence that Kant should be the figure to take the next step forward and leave the old debate behind. In the launching of their model, the creators of preformation were anticipating the German philosopher's famous aphorism, the two words destined to become the central slogan of the Enlightenment: *"Sapere aude."* Dare to know. Dare to ask questions. Dare to inquire about the world around you, and bring to your inquiry the only tools your mind has at that moment to address the universal complexity. Dare to insist. Dare to imagine. If your imagination is framed by the limitations of your incipient knowledge, so much the better. You can think even more freely. You can build scenarios that further knowledge will soon take away from you. It is in the ingenuity of those early constructions that the process of knowing takes its most decisive turns.

Please play a little game with me.

Name one idea that has arisen by spontaneous generation.

The search is bound to be frustrating, for human ideas are always the children of pre-existing ideas. Thus, even if you were persistent enough to dig out that one exceptional example, chances are you would produce nothing but exactly that: the interesting exception that proves the rule—the rule being, of course, that whenever a new thought surfaces in someone's mind, it has to be a metamorphic form: a result of the continual rearranging and reconfiguring always going on deep within our mental subsoil, under layer upon layer of sediment.

Now name one idea that was as truly metamorphic as all the others, but which emerged totally out of synchrony with the social, technical, and philosophical reality of its time.

Here we could perhaps help ourselves to a small handful of visionaries and other dreamers to make our search more fruitful. But these potential examples would still be exceptional. They would still provide nothing but the confirmation of yet another, even broader rule. Ideas are born from ideas, and the length of their required gestation, and the precise time of their birth, are regulated by the calendar of cultural steps.

Now name one myth, just one, from any country or ethnic group or any civilization in any period you like, that in one way or another deals with the story of the magic box with a smaller box inside. And inside that box, there is yet another box. And then yet another box. And then yet another myriad of boxes, smaller and smaller, harder and harder to reach and to open, until the hero does or does not get to the site where some sort of precious key is hidden.

This one was easy. Countless civilizations, numerous major or minor folklores, at several different times, have produced an example of such a story under the guise of endless variations. As humans, we have much more in common than our celebrated bipedal position, or our famous ten fingers with their equally famous prehensile thumbs. All over the world, we tell our children similar bedtime stories. And we seem to be perpetually attracted to the idea of boxes inside boxes inside boxes.

And we share numerous everyday biological observations.

We have always seen animals hatching from eggs. In addition, myths about the world hatching from an egg abound in all cultures, and so do cosmogonies that compare the earth to an egg and the universe to its shell. Therefore, it only made sense for the first preformationists to be convinced *ovists* (as opposed to the *spermists,* who would claim later that everything started with the sperm cell). In their original scenario, since the naturalists who endorsed the theory were all Christians (albeit of different confessions) and did not challenge the authenticity of the biblical Genesis, all future human history had originally been encased inside the ovary of Eve, in the fashion of a Russian doll.

EVER SINCE MOSES

The professed Christianity of the leading preformationists adds an interesting subplot to their story, for the questions they raised concerning generation directly addressed the need experienced during the Scientific Revolution to redefine the very concept of God. We could perhaps use Voltaire's (1694–1778) musings on the central controversy between Descartes's and Newton's views of the universe as an example of the extreme depth of this problem.

Considering Voltaire's worries about freedom and the recurrent temptations of totalitarianism, it was important for him to favor a system that would emphasize human insignificance in the vastness of the universe, a conclusion extrapolatable from Newton's philosophy. Voltaire argued that the notion of what constitutes a "rational explanation" should be reexamined. If Newton had shown that gravitational attraction, and no other force, kept planets orbiting and heavy bodies falling, no necessity existed for an explanation of "the cause of this cause," which was God's secret and not for human minds to question. Voltaire's statement subtly implies that dissatisfaction with Newton's explanations would require that all rational explanations rely on a *complete* account of nature, which would lead to the temptation of the *"esprit de systeme,"* therefore to dogmatism and eventually to atheism. The French polemicist considered that Descartes's claim *"Give me matter and motion and I shall build a universe"*—already constituted a step in this direction, in contrast to Newton's humility in assuming that human reason could not reconstruct even a "likely story" about the way God had put together the universe. Descartes thought of God as reason, presiding over a universe expressing rational necessity. Newton regarded the universe as the result of God's sovereign and voluntary choice. Thus Voltaire concluded, in his defense of Newton, that man could only try to understand the motions and processes ruling the phenomena resulting from God's choice, but could never attain the complete system of knowledge of ultimate causes.

All the Enlightenment's finest minds were busy pondering God's place in the new world under construction, and during this period the theory of preformation ruled generation—God's first and final work, the alpha and the omega of God's will—for an entire century. God was a crucial player in the logic of preformation, since the theory assumed, as a basic tenet, that He programmed the entire world, and the entire sequence of events destined to take place in that world, all at once during Creation. And the concept could be stretched far enough to admit, for instance, that He foresaw (better yet, programmed) all the wicked events to come, including the Reformation and countless wars, and marked each one of those occurrences with a set of monstrous births. Preformationists certainly did not seem to have a problem with this type of idea. The "minister of the Holy Gospel" Jean Sénébier wrote, in his preface to Spallanzani's *Opuscules de physique, animale et végétale,* that in preformation we see "L'OUVRAGE D'UN DIEU TOUT SAGE" ("the work of an Almighty God"; Senebier's capitals). And he went on to explain how God still oversaw current events in the following terms:

> We can even imagine an INTELLIGENCE clairvoyant enough to see all the history of the Universe relating to the same object by the same means; maybe it would only suffice that such an INTELLIGENCE would be to Abbé SPALLANZANI what Abbé SPALLANZANI, with his Microscope and his genius, is to an ordinary man, who has only his eyes and never trained them in these kinds of observations.

Obviously, this is *the* classic philosophical formulation for the problem of free will: Does free will exist if God foreknows what will happen? In other words, does this concept of preformation, taken to its ultimate consequences by assuming that God programmed everything, rob people of their free will? Does preformation make human beings mere passive extras in a complex and immutable play choreographed long ago, just as it makes parents nothing but the passive recipients and transmitters of already predestined children? Could this view, cleverly taken to a strategic extreme, propel humans toward total irresponsibility, allowing, for example, the public relations department of any modern major chemical manufacturer to say casually, "Ah, who cares if we are polluting the Arctic Ocean and raping Siberia. This is just a part of God's grandiose plan"? Certainly it could, and this issue might have inspired a raucous debate. However, to my knowledge, during the heyday of preformation, nobody—not even the most rabid and vitriolic detractors of the idea—denounced the theory on these grounds. We should consider this failure an interesting detail in itself, in what it reveals about the relationship existing at this time between Christians (the preformationists were all Catholics or Protestants) and their God.

This side effect of the preformation doctrine brings us back to one of our recurrent shortcomings, the fact that we are systematically reduced to playing charades whenever we try to discuss what God can possibly have in mind (well, does God *have* a mind? Does He *change* His mind? *Can* He change His mind? Etc.?). The medieval Scholastics spent endless centuries debating such urgent topics as "Can God lie?" or "Can God be deceived?" Obviously enough, they never came to a conclusion. Therefore, during the days of preformation, God was a rather flexible agent—basically, He did whatever suited your agenda. Just like today, and, for that matter, ever since Moses.

JUST ONE STEP

The belief in the preprogrammed encasement of successive generations was the centerpiece of preformation. However, the organization of that encasement underwent several revisions. In the course of a century (from the mid-1600s to the mid-1700s), preformationist theories evolved from totally preformed persons to pre-existing fundamental parts. In its final forms, the theory came one step shy of our current models in developmental biology. But that step, in the context of the knowledge of the time, was impossible to take. And so the days of preformation came to an end, for want of desperately needed supporting data that would allow the concept to keep perfecting itself. Never mind. It was a beautiful idea. And it was a wonderful trip down to the depths of our conscience.

BEAUTIFUL LOSERS

I hope that by the end of this book I will have convinced the reader that the theory of preformation, as it existed during the seventeenth and eighteenth centuries, provides an entertaining interlude in the study of reproduction. But why write about it? Is it not dead? Certainly it was powerful in its day, but doesn't the full tale resemble the story of the dinosaurs? Basically, it emerged from building blocks that evolution had made available, took over land and sea, expanded, multiplied in form, ruled for a while—and then became extinct.

Yes, but just like the dinosaurs, it left behind a precious legacy.

Yet, unlike the story of the dinosaurs, this tale is usually not told on a note of respect and wonder. As a rule, in our day, the story of preformation is generally told with dismissive laughter.

Secondary sources took a strange liking to being particularly caustic about the preformationists of old. The only work so far published dealing strictly with the preformation of the Scientific Revolution is F. J. Cole's *Early Theories of Sexual Generation.* Despite his stated intent to "place the complete details of the Preformation Doctrine before the reader, and to avoid the common mistake of ignoring all but the most salient features," Cole starts his third chapter ("First Statements of the Preformation Doctrine") with the following quote:

> The old evolution [preformation] was the greatest error that ever obstructed the progress of our knowledge of development.—Whitman 1894

This quote speaks volumes to me, for I believe it clearly expresses Cole's own position: he wants to write about preformation, but starts from the view that preformation was a "great error" and a patent absurdity. Reading through his pages, one gets the feeling that he is implicitly saying, "now see how dumb this one was," and "just wait until you hear how much dumber this other one was." Yet numerous subsequent sources refer to Cole's work as the "commanding authority" in the field.

Another monumental authority in the history of embryology was the celebrated Joseph Needham. Needham also did not think highly of preformation. His *A History of Embryology* is full of such expressions as "the cloven hoof of preformation" or "[earlier hints at preformation] did not begin to bear their malignant fruits till the time of Swammerdam and Malpighi."

Consider just one more modern quote (there will be many more in the book), from Arthur William Meyer's *The Rise of Embryology,* to exemplify the sources that seem to have drunk straight out of Cole's and Needham's spring:

> It is impossible to convey an adequate idea of [Spallanzani's] many important experiments on generation in a few paragraphs; but it is illuminating, as well as regrettable, that so assiduous and thoughtful an experimentalist should have thought that he had found experimental proof for such a wrong theory [preformation].

My present goal is to demonstrate that this castigated theory of preformation was not a "cloven hoof," but a beautiful construction, one of the mental buildings made possible by those rare intervals when enough knowledge has been amassed to elicit them, but the limitations of further knowledge have not yet undermined their potential as explanatory models. Also, although modern authors such as Elizabeth B. Gasking and Shirley A. Roe show much more respect for the preformationists than do Cole or Needham, these authors are not analyzing preformation as such. They are analyzing the battle between preformation and epigenesis, and in the end preformation loses. It is a sad fact that history is written by the victors. In this book, I want to challenge this trend and tell the marvelous story of the losers. By addressing the deepest roots of preformation as a quest for a fundamental explanation fueled by old convictions and new technologies, I hope to inquire into the nature of our minds. For I believe that much can be said for systems of knowledge that we had to abandon along the way in one dead end or another. As discarded attempts to reach the ever-elusive Truth, they tell us about our anxieties, our deepest convictions, the secret corners of our history, and, above all, about the way we organize our thoughts within a given situation when no boundaries have yet been established for the explanations we may choose. In summary, they tell us more about ourselves.

SMALLNESS AND ENCASEMENT

I had the feeling, from page one, that straight biography in precise chronology would not do justice to my subject matter. The ramifications and implications of preformation are, to say the least, impressive. Let me try to uncover some of them:

The doctrine deals with the most fundamental of all questions, since understanding our origins is tantamount to defining our place in the universe—which, in turn, leads to our very perspective on the universe proper.

The philosophical origins of the concept can be traced as far back as the pre-Socratic thinkers. The input of ideas into the seventeenth-century model came in equal parts from alchemy; from Medieval and Renaissance Scholasticism, medicine, magic, and the old ideas of microcosmos and macrocosmos, which prepared the way for the idea that future events can be hidden inside what already is; and from the intertwining of folk beliefs with the evolution of natural history toward its modern redefinition as biology.

The emergence of preformation coincides with the development of crucial technical scientific advances such as the microscope, at a time when the precise meanings of the revelations of microscopy were far from clear. Therefore, the story of preformation helps us to understand how the learned men of the time conducted experiments and observations, and provides a very interesting case study of the roots of science as we know it today.

When analyzing the most imbricate details of preformation, we come across objets trouvés as diverse as Paracelsus's homunculus and the meaning behind the classical symbols for Venus and Mars.

The century-long debate over preformation and its continuous rephrasing encompassed the turmoil of the Scientific Revolution, and thus the need to understand matter, the nature of the soul, the mechanism of organizing principles (and even the truth behind the existence of such principles), and the acceptance of mathematics as an explanatory hypothesis for all phenomena.

And, during this entire process, along the paths followed by the theory, two main ideas were always present that seemed to reflect a strange recurrent fascination in the quest for a way to organize the human mind. One is Smallness, and how small objects can actually be. The other is Encasement, or how much matter you can actually place inside matter. We had long postulated the existence of entities that were visible only to Lynceus's eyes: microscopes turned this postulate into a tentative scientific outcome. We had eagerly debated how many angels could stand upon the head of a pin: the quest for the number of people encased inside the ovary of Eve can be seen as a naive scientific version of the same obsession.

There must be in our brains a peculiar fascination with the vision of structures contained within structures contained within still other structures. Think of all those ever-growing memory palaces, with rooms and alleys and gardens filled with ever-growing numbers of figures and pictures, that constituted the mnemonic systems of the Renaissance. And then think of how the idea lives on. Think of homeopathy, for instance.

Homeopathy is a medical system developed in the nineteenth century in reaction to the brutality of conventional "heroic" medicine, searching for kinder, gentler treatments in which patients would not have to be bled or sucked by leeches. Most importantly for the present study, homeopathy embodies a number of aspects that again have to do with smallness and encasement. Homeopathy's core concept is based on extreme dilution of the chemicals used, organized according to a system of potencies in which things should function best when diluted to the third potency. These potencies are supposed be a consequence of the special properties of water, which "becomes energized" when exposed to chemicals diluted in it. The more you dilute and the more you shake, the more energized the water is supposed to become.

Moreover, from the homeopathic perspective, your own body is a result of the generations that preceded it, and therefore your diseases were contained in previous diseases of your family. The patient's history is not sufficient for diagnosis and decisions on type of therapy. You have to know the history of all of the previous generations on both parents' sides, since you carry within yourself all the afflictions of your ancestors.

In a more recent twist on this same idea, a new school of homeopathy is now emerging in which one of the basic tools of diagnosis is the use of "regressional hypnosis" to follow your body through a number of your more recent former lives. The assumption is made that your body transports in itself all the other bodies in which your soul was previously located. If you have gastritis, maybe you were shot in the stomach in your former cowboy incarnation.

In order to verify the accuracy of this method, I went to one of these sessions myself. As it turned out, in some previous life I was the spinster chief maid in a big Victorian manor, where I was taken to be a servant by my parents when I was child because our family was very poor. There I had slowly climbed the hierarchical ladder because I was such a sensible hard worker, and had peacefully lived out my days overseeing the perfect functioning of domestic life. A friend, who protested from the beginning that she could not possibly believe any of this nonsense, heard me read my session report out loud. When I came to this passage, she immediately snapped, "Ah! THAT explains your neurotic obsession with always keeping your house so clean!" I just smiled. We do not believe; and yet, we really like to believe.

All in all, preformation addressed the core of our main sources of spiritual restlessness. From the beginning, the theory aggregated a wide variety of threads in the making of its fabric. In its configurations, preformation borrowed from mythologies, from the visions that humankind seems to have carried wherever it went, from all the factions that became involved in the remarkable war of ideas staged during the Scientific Revolution. Preformation questioned all our beliefs, our hidebound assumptions, our perception of ourselves and of our place in the world. And, in the end, it came up with a scenario that was not as far removed from our present beliefs as might have been thought before we knew what we know today—that is, before genes were introduced into our current knowledge; before the structure and replication of DNA were described; and a long, long time before we came to realize that the mechanisms through which the sperm activates the egg are so complex and hard to mimic that they sometimes seem to modern developmental biologists almost as elusive as the seminal spirit of the past.

A Road Map

Choosing to look at this specific historical period in a nonchronological fashion, as I explained in the Preface, poses a complex question: what alternative organization should one follow? As I have already stated above, I did not want to set preformation on a battlefield against epigenesis or other contemporary theories of reproduction, but hoped rather to focus on preformation per se. I decided to organize my text as a battle, yes—do we not enjoy a good fight?—but a battle among preformationists alone. Or, more precisely, between the two entities chosen by preformationists as the places where God had encased all organisms destined to come to life on earth: the egg and the spermatozoon. In a colorful display of dialectics, preformation did not allow for compromises: either you believed that God had encased all life within the testes, or you held that God had encased all life within the ovaries. In other words, you were either a *spermist* or an *ovist;* you could not possibly be both. The preformationists knew this well: the two factions fought each other with creative ardor. This book is thus constructed as an "egg versus sperm" conflict, with different chapters analyzing the advantages and disadvantages of each camp. I shall consider the subject in eight chapters:

1. "All About Eve": in other words, meet the ovists. This is my main concession to chronology: they come before the spermists because preformation was postulated, and started gathering momentum, before sperm cells were discovered.

2. "All About Adam." Shortly after the birth of preformation, sperm cells were discovered in male semen. So, now, meet the spermists. They were the children of a radical novelty, and consequently, the logical victims of meager knowledge about that novelty—and of all the unexpected moral dilemmas that that novelty entailed.

3. "One Does Not See the Wind." Why were spermatozoa treated so badly right after their discovery? Abhorrence of change? Because they *looked* like worms, and it was hard to believe that God had encased us inside worms? Because soon afterward parthenogenesis was discovered, which seemed to confirm that only the egg was necessary in reproduction? Or because the alternative explanation for the role of semen in fertilization, one that downplayed the cells in favor of a spirit, was more consistent with the ideas of spiritual insemination that permeated so many cultures—and, in seventeenth-century Europe, would make any human even more similar to Jesus? Did this not also apply to parthenogenesis, as a "scientific" explanation of virgin birth?

4. "Hopeful Monsters." If preformation had a problem with the birth of monsters (which, in my view, is highly debatable), the spermists were better positioned than the ovists to explain this complex phenomenon. The same argument applied to other events discovered at that time, such as regeneration of lost body parts or reproduction through budding. Interestingly, the spermists did not seize the opportunity. More interestingly yet, modern sources maintain that they did—only to use this claim as a means to deride them.

5. "Frogs with Boxer Shorts." Talk about fate being ironic. Spallanzani performed all his fabulous experiments in reproduction as a means of confirming the final truth of ovism. Yet he is now remembered as the man who established for good that reproduction cannot occur without the contribution of semen.

6. "The H Word." What's in a name! Why do we think that the spermists

called their "little man inside the sperm" a "homunculus?" They did not. Considering all the threatening connotations of the term "homunculus" in their day (and in our own time), the spermists should have known better than to use that word—and they did. But modern revisionism equated the homunculus with the little man in the sperm, and the end result was yet another joke on the foolishness of spermism.

7. "The Music of the Spheres." The ovists had a strong philosophical construct working in their favor, and perhaps an even stronger philosophical construct working against them. On the sunny side, eggs are spherical, and the sphere had long been considered the ultimate shape of perfection: this made even more sense if one considered that the entire scheme of life had been engendered by a perfect God. But eggs come from females. And females have traditionally been considered nothing but imperfect forms of malehood. What was God trying to achieve through this mixed message? Why would He encase us inside the shape of perfection only to lock that shape within imperfect bodies?

8. "Magical Numbers." Both ovists and spermists indulged themselves in a feast of quantification, although the ovists were favored on this front due to the monumental numbers of sperm cells in any ejaculate. At a time when life on earth was still supposed to be finite, circumscribed within a period of six thousand years, this was an advantage preformation held over any other model of reproduction: its supporters could calculate generations, global numbers, partial fractions, final sizes. Preformation was perfectly in tune with the world of infinite smallness opened up by the microscope; with the contemporary birth of infinitesimal calculus and statistics; with the *furor mathematicus* that marked the Scientific Revolution; and certainly with the magical spell that numerology consistently casts over human minds.

Another feature that may at first seem disconcerting is the recurrent mention of worldwide mythologies and religious beliefs whenever a specific aspect of preformation is discussed. I have noticed that, by doing this, I am breaking some kind of golden rule. Colleagues, friends, mentors, reviewers, have all asked me, "Why do you bring in all these different cultures and mythologies? Do you really believe that they influenced the thought of the European natural philosophers of the seventeenth and eighteenth centuries?" And I always answer—as I want to state clearly at this point—that no, of course I do not think so. But I *do* think that a transhistorical property of humankind influences all peoples' train of thought, and this is a path I am trying to trace as well.

Why did I choose to write my book in this way? Since I could not have phrased it any better, let me plagiarize M. Vandermonde in his *Essai sur la manière de perfectionner l'espèce humaine,* published in Paris in 1751:

J'ai pensé qu'en arretant d'abord le lecteur par des objets nouveaux & amusants, qu'en passant par le sentier de la beauté, il arriveroit sans peine à celui de la science.

Yes. I thought that by interesting the reader first with new and interesting objects, that by passing through the path of beauty, he would arrive without effort on the path of science. *Now* I rest my case.

All About Eve

*I*T IS THE MIDDLE of the seventeenth century, and a highly mystical Dutchman, blessed with the most refined microscopic talent, is present at a convention of learned people in the country home of the Parisian gentleman Melchisedec Thévenot, a former representative of the French king in Genoa and a well-traveled citizen of the world. The Dutchman's name is Jan Swammerdam. At some point, the generation of animals arises in the discussion, and Swammerdam's opinion is requested. Swammerdam is a shy man by nature, and not inclined to the art of oratory. He prefers to remain silent and to return during another gathering at the same place, this time bringing with him a silkworm. The intellectuals assembled in the room examine the specimen carefully, and are unable to find any trace of a moth in it. Without a word, Swammerdam peels off the skin with his customary and amazing skill. One by one, the wings, the proboscis, the antennae, all the appendages of the future moth—or at least what were presumed to be so—appear in full view of the astonished audience.[1] Swammerdam has made his point: hidden under the skin, all of these parts have been quiescent for a long time, awaiting their moment to unfold and come to life. Each creature that is going to live is already preformed inside a creature of the same species.

That night the theory of preformation was already under way, and was destined to travel a long road.

THE SEED OF ALL TREES

Preformation, as a comprehensive theoretical body, was initially shaped by a Cartesian priest. This priest was not particularly preoccupied with reproduction: his main interest was the discussion of morals and sin under the guiding lights of physics and mathematics. Pursuing this goal, he wrote a long treatise, organized in six books, titled *De la recherche de la verité ou l'on traite de la nature de l'esprit de l'homme et de l'usage qu'il doit en faire pour éviter l'er-*

16

Fig. 2. Jan Swammerdam.

reur dans les sciences. The work, first published in 1674, was extremely successful, and was quickly translated into English (under the title *Search after Truth*), Latin, and modern Greek. Its author immediately acquired fervent disciples in Germany, Italy, and even China. This wildfire expansion, together with the ensuing passionate polemics by other luminaries of the time, illustrates the need sensed at the end of the seventeenth century for a comprehensive synthesis of the true nature of Cartesianism and the true nature of God. In a certain sense, the work of the French priest was an act of intellectual charity.

Nothing was left unscrutinized in this fervent *Search after Truth*. Book 1 deals with *The senses;* book 2 with *The imagination;* book 3 with *The understanding; or, pure mind; and pure understanding: the nature of ideas;* book 4 with *The inclinations; or, the mind's natural impulses;* book 5 with *The passion;* and, finally, book 6 examined *The method.* Almost every aspect of our intellect was analyzed within this vast context, and it may even be argued that we find here the primordial sketches of what became known much later as psychology.[2] One might say that the text concerning reproduction is so short as to be almost nonexistent. And yet, the work's relatively few lines on this subject entered the memory of all embryologists and historians of biology as the cornerstone of preformation—a term the author never used—and are often quoted as having contained for the first time the expression *emboîtement*—meaning something like "encasement" and possibly derived from the German *Einschachtelungstheorie,* a term, again, that the author never used. So what did he really say that made such an impression on the minds of present-day scholars?

He said, among other things, that "one of the main errors we fall into in physics is to imagine that there is more substance in bodies that are perceptible than in those that are hardly perceptible at all"—a simple sentence, in the discussion of *The Senses,* that opens the door to the possibility of the existence of highly organized invisible things, a point to which the author returned over and over concerning all kinds of aspects of our psyche. But he said more, and more clearly; and he did so right at the beginning of book 1, when analyzing the probability of error in optics. He stated his real concern in the opening paragraph: "Of all our senses, vision is the first, the most noble, the most extensive; accordingly, if [our eyes] were given to us for discovering truth, [vision] would have a greater role by itself than all the others combined." The following development was meant to make the reader realize that "our eyes generally deceive us in everything they represent to us: in the size of bodies, in their figure and motion," and that, therefore, "things are not as they appear to us . . . everyone errs regarding them, and . . . as a result we are plunged into an infinite number of other errors."

It is at this point that what can be perceived as the road to preformation surfaces in the text:

> With magnifying glasses, we can easily see animals much smaller than an almost invisible grain of sand; we have seen some even a thousand times smaller. These living atoms walk as well as other animals. Thus, they have legs and feet, and bones in their legs to support them . . . They have muscles to move them, as well as tendons and an infinity of fibers in each muscle; finally, they have blood or very subtle and delicate animal spirits to fill or move these muscles . . . The imagination boggles at the sight of such an extreme smallness . . . and although reason convinces us of what has just been said, the senses and the imagination oppose it and often make us doubt it.

But these sensorial doubts should by no means interfere with our reason:

> This small section of matter, which is hidden from our eyes, can contain an entire world in which would be found as many things, though proportionately smaller, as are found in this larger world we live in.

Indeed, the argument followed,

> for the tiny animals of which we have just spoken, there are perhaps other ani-
> mals that prey upon them and that, on account of their awesome smallness, are to
> them as imperceptible as they themselves are to us . . . We have clear mathemati-
> cal demonstrations of the infinite divisibility of matter, and . . . this leads us to
> believe that there might be smaller and smaller animals to infinity . . . Experi-
> mentation has already partially rectified our errors by enabling us to see animals
> a thousand times smaller than a mite—why would we have them to be the last
> and smallest of all? For my part, I see no reason to imagine it so. On the contrary,
> it is much more plausible to believe that there are many things yet smaller than
> those already discovered, for . . . there are always tiny animals to be found with
> microscopes, but not always microscopes to find them.

Then come the famous lines credited with having marked the birthplace of
preformation:

> We may say that all plants are in a smaller form in their germs. By examining the
> germ of a tulip bulb with a simple magnifying glass or even with the naked eye,
> we discover very easily the different parts of the tulip. It does not seem unreason-
> able to say that there are infinite trees inside one single germ, since the germ con-
> tains not only the tree but also its seed, that is to say, another germ, and Nature
> only makes these little trees develop. We can also think of animals in this way. We
> can see in the germ of a fresh egg that has not yet been incubated a small chick
> that may be entirely formed. We can see frogs inside the frog's eggs, and still
> other animals will be seen in their seed when we have sufficient skill and experi-
> ence to discover them . . . Perhaps all the bodies of men and animals born until
> the end of times were created at the creation of the world, which is to say that the
> females of the first animals may have been created containing all the animals of
> the same species that they have begotten and that are to be begotten in the future.

These are, in essence, the true works of God; and "in these works, nothing
but infinities are found everywhere."

From here, the text proceeded to discuss the optical illusion embedded in
the range of discernment of our eyes, and to explain the mode of function of
the crystalline lens and the retina. The idea of a Russian doll initially created
by God for each species, and the claim that "nature's role is only to unfold
these tiny seeds by providing perceptible growth," were quickly left behind as
the author dived into extensive geometric and moral considerations. One last
sentence, though, loomed at the end as a visionary warning: "God never deceives us, but we often deceive ourselves by judging things too hastily."

The name of this priest was Nicolas Malebranche.

THE GEOMETRY OF GOD

Nicolas Malebranche (1638–1715) became a student of theology at the Sorbonne in 1656, and was ordained a priest in 1664. But, in the same year, he happened to notice a little book for sale at a local *bouquiniste: L'Homme de René Descartes.* Malebranche bought it, and it is said that the reading took him to such heights of enthusiasm that he was obliged to stop from time to time due to bouts of cardiac arrhythmia.[3] Under this inspiration, the young priest underwent five years of new studies, learning mathematics, physics, and physiology, and emerging in the end as a Cartesian philosopher—so Cartesian, in fact, that he easily proclaimed that "God is a geometer," conceiving of a mode of divine action similar to a mathematical function.

This divine action, according to the Cartesian priest, readily applied to generation: "Children are almost completely formed before the action by which they are conceived; and their mothers only give them the necessary growth during the time of pregnancy." While admitting that "some may ask, if God organized in such an exact and regular manner all the things necessary to the propagation of the species during endless centuries, how could some mothers have miscarriages, and even why were not their children always of the same size or of the same complexion," he quickly pointed out that "we cannot measure the power of God with our weak imagination, and we do not know the reasons He may have had in the construction of His work."

The idea of a preordained, prearranged, and forever unchangeable succession of generations, hatching from inside one another, was in all respects finely tuned to the author's groundbreaking theology. According to this theology, the only universal cause is the action of God. This action becomes manifest before man's eyes, particularly in the realm of science, through a regular chain of causes and effects. In Malebranche's deterministic reasoning, God loves order: He seeks the simplest pathways through which to impose His universal laws. Since God is the only real cause of all things, our actions are nothing but the stylistic development of God's unchangeable laws.

Moreover, the idea of infinity, an important component of the fabric of preformation, was also dear to the author's philosophy:

> The most beautiful proof of God's existence . . . is our idea of infinity. Our spirit senses infinity, even though we cannot understand it. Our spirit grasps the idea of what is infinite even before it grasps the idea of what is finite.

He also insisted that diversity in nature was necessary to emphasize God's power:

> If God, when creating the world, would have produced an endless matter without endowing it with any movement, all the bodies would be similar; all this vis-

ible world would be nothing but a mass of matter that could by itself testify to the power and majesty of its Creator; but there would not be that succession of forms and that variety of bodies which makes all the beauty of the Universe, and that brings all spirits to admire the infinite wisdom of He who governs them.

Malebranche was certainly much more concerned with devising a generational system compatible with the imprimatur of God on the simple order of things than with fighting any kind of ovist crusade. This is clearly manifested in his long friendship with Gottfried Wilhelm von Leibniz (1646–1716; see chapter 2), who also supported preformation but tended more toward the spermist theory. The two men first met in Paris in 1675,[4] and, for the following forty years, kept up a constant flow of epistolary exchanges, notes on books, indirect contacts through mutual friends, readings of each other's works, and reflections on philosophical and scientific systems—a precious legacy that encompasses all the main theoretical debates of a half-century of Western intellectual life. Yet, when browsing through this vast exchange, one is much more likely to find long debates over mathematical equations or the validity of the Cartesian system than to come across any heated controversy regarding generation, showing that, for both men, this biological problem held only a brief passing interest in the vaster context of order in nature and the relationship between the body and the soul.

The same emphasis on philosophical issues can be found in this passage of a letter from Leibniz to his friend Louis Bourget (incidentally, also an ovist), dated March 1714:

> I imagine that when father Malebranche says that we see everything in God he means the perception of the spirit, not only in what concerns visible qualities, such as figures and colors, but also in what concerns sounds and other sensitive qualities. You noticed accurately that this father acknowledges that all flies are enveloped in a certain fashion inside the flies from which they descend . . . And therefore I think that he could also admit the existence of certain thoughts inside the soul, which are born from each other.

It was the potential offered by preformation for a *mechanical* understanding of the soul and the spirit, and thus the relationship of God with his *mechanical* world, that really excited the philosophers who created and formulated the concept of infinitely encased lives. For them, the biological aspects of the system were less important side effects.

Several authors point out that preformation was necessary to counter the threat of atheism made possible by the endorsement of a fully mechanistic epigenetic position. The French naturalist Claude Perrault can be mentioned as a good example of such preoccupations. He was well aware of the amazing diversity of terrestrial species, and of the ability these species had to perpetuate themselves. The best illustration of this broad knowledge can be found in his book *Mémoires pour servir à l'histoire naturelle des animaux*, first published in 1688, in which he carefully detailed, with the aid of superb plates, the anatomy of both the internal and external parts of several animals, from the lion to the chameleon, from the dromedary to the shark, including seals and cormorants and even rarely heard of beasts such as the Brazilian "Coati Mondi." This insight on diversity through continuity made him readily embrace the concept of development through pre-existing germs, although he was more inclined to believe in panspermism than in encasement. His remarks from 1680 are extremely caustic toward the self-organizing abilities of matter:

> I do not know if one can comprehend how a work of this quality would be the effect of the ordinary forces of nature . . . for I find finally that it is inconceivable . . . that the world has been able to form itself from matter out of chaos.[5]

A QUIET FOUNDING FATHER

Malebranche was by no means the first man ever to speak of encasement. Numerous scholars trace the roots of the idea back to classical antiquity. In his *Questiones naturales,* Seneca (3 B.C.–A.D. 65) certainly hinted at it when he wrote,

> in the seed are enclosed all the parts of the body of the man that shall be formed. The infant that is borne in his mother's womb hath the roots of the beard and hair that he shall wear one day. In this little masse likewise are all the elements of the body and all that which posterity shall discover in him.[6]

The same idea reappeared in disseminated passages within the Judeo-Christian legacy, and was expressed clearly by Saint Augustine (354–430) during the fourth century. According to the Augustinian theory, God had not created everything at once, but rather had imparted *seminal reasons* to things. Thus Creation was the *unfolding* of the elements. This idea later allowed the Medieval Scholastics leaning toward the Platonic tradition to go outside Aristotelianism and draw upon the notion of seminal reasons or embryonic forms latent in matter, a concept that regained popularity in seventeenth-century chemical reinterpretations of form theory. Therefore, preformation was drinking random bits of inspiration straight from the earliest and noblest Christian sources.

Prior to Malebranche in the seventeenth century, Antonio Vallisnieri seems to have suggested plainly that not only the entire human race, but human parasites as well, were contained initially in the ovary of Eve.[7] Where Malebranche really broke new ground was in organizing a whole theory concerning the subject. This theory was much more efficient than the previous suggestions made by others because it was based on sound philosophical concepts; and also, undoubtedly, because it incorporated the cutting edge of the microscopic observations recorded at the time.

Malebranche's ideas on generation presented in *Search after Truth* were clearly drawn from the works of a number of leading microscopists. Where Malebranche mentioned the bird's egg, he added a footnote stating that "the germ of the egg is a small white dot that is in the yolk. See the book *De Formatione pulli in ovo* by M. Malpighi." And, when "frogs inside the frog's egg" were mentioned, another footnote added, "see *Miraculum naturae* by M. Swammerdam." Another example of this microscopic influence appears in the following passage of an essay published in 1695 by Leibniz in the *Journal des Sçavants*:

> This is where the transformations of MM. Swammerdam, Malpighi and Leeuwenhoek, who are the most excellent observers of our time, came to my rescue and made me admit more easily that the animal, and any organized substance, does not start at all when we believe it does, and that its apparent generation is nothing but a development and a kind of increase in size. I have also noticed that the author of 'Search After Truth,' M. Regis, M. Hartsoeker and other gifted men are very close to this feeling.[8]

So, who were these men who provided the raw materials for the philosophical rise of preformation?

According to several accounts of his life, the Italian microscopist Marcello Malpighi (1628–1694) was "modest, quiet and of a pacific disposition."[9] He remained remarkably calm and sober despite the continual efforts of the Sbaraglias, the eternal rivals of his family, to injure his scientific reputation and his good name. Although he lost his parents at age twenty-one, and, being the eldest, had to take care of his seven siblings, he pursued his studies with such tenacity that he received the degree of Doctor of Medicine from the University of Bologna four years later. He took temporary chairs at the universities of Pisa and Messina, but he kept returning to Bologna; and the city repaid him by erecting a monument to his memory after his death by apoplexy in Rome, where he had become Pope Innocent XII's personal physician.

At age thirty-eight, and already with a remarkable academic career behind him, Malpighi decided to retire to his villa and dedicate most of his free time to anatomical studies. Fascinated by the microscope, he became a master of the new trade. Working on frogs and boldly extrapolating to humans, he demonstrated the structure of the lungs, previously believed to be a homogeneous mass of flesh, and offered one of the first explanations of how air and blood

Fig. 3. Marcello Malpighi.

are simultaneously brought to these organs without actually coming into full contact with each other. Although most of his conclusions were wrong, he turned out so much new material on the structure of glands that his name is still associated with the Malpighian corpuscles of the kidney and the spleen. He also investigated the anatomy of silkworms and the structure of plants. But, in the debate on generation that was gathering momentum, it was the chicken that made all the difference. Trying to address the eternal question of how the life of an organism begins, and through what steps the new body is built, Malpighi turned his attention to the avian egg. He produced drawings of the development of the chicken embryo so detailed and so clear that they are still regarded as masterpieces today.

Malpighi never really elaborated a theory based on his findings, but he certainly furnished others with the materials they needed. In his 1673 work *Dissertatio epistolica de formatione pulli in ovo,* he claimed not to be able to observe anything in unfertilized eggs, but established that eggs that had been fertilized but not yet incubated already showed the rudiments of a tiny embryo. He reluctantly admitted that some form of development might be taking place in unincubated eggs, a mistake most likely due to incidental incubation of some eggs under the heat of the Italian summer; for, according to his own writings, this observation was made "at the end of August, when the weather was very hot."

I perceived the enclosed fetus, whose head . . . clearly appeared. For the thin and clear texture of the amnion was frequently transparent, so that the imprisoned animal came into view . . . Therefore I declare that the stamina of the chick already exist in the egg, and it must be admitted that an earlier origin has been discovered, just as with the eggs of plants." [10]

This was enough for Malebranche and other leading philosophers to appropriate the chick embryo as one of their main sources of evidence. By the eighteenth century, Malpighi was already regularly mentioned as a preformationist. And still is.

ONLY ONE FOUNDATION

Jan Swammerdam (1637–1680) was a completely different case, for he clearly took sides in his writings. Although he sometimes referred to William Harvey, the major voice in the epigenetic field at that time, with such adjectives as "honorable" and "the great," Swammerdam did not hesitate to repeatedly criticize the same Harvey for what he considered to be his numerous mistakes about reproduction. Condemning Harvey's idea that "things are produced as it were by the impression of a seal upon the matter of them, or by this matter being cast into a mold," Swammerdam remarked that the whole epigenetic idea of matter slowly taking up form was a true demonstration "of the egregious mistakes we are apt to commit, the moment we abandon the solid arguments furnished by experiments, to follow the false lights struck out by our weak and imperfect reason." He went so far as to write that Harvey's dissertation on bees "contains almost as many errors as words." [11]

Still, Swammerdam's worst blow to Harvey's credibility—and therefore to the credibility of epigenesis—was to link him directly with the then unfashionable Aristotle, since both men believed that "the nymphs of bees are so far from containing the parts of the future insects that they can only be looked upon as the eggs which are to produce them." Such "fancied metamorphosis," Swammerdam protested, was invented solely because earlier naturalists were unable to see what really was to be found inside the nymph. But, now, "the experiments we have made have, like the rising sun, dissipated these thick and dark clouds of imaginary metamorphosis"—and, in the author's words, one should nevermore compare metamorphosis to death and resurrection, an association that he considered both "preposterous" and "blasphemous." Neither should one again, like some alchemists had dared to do, compare insect metamorphosis to the metal transmutations, writing such "absurdities" as Theodore Mayerne's question, "If animals are transmuted, why may not metals be transmutable?" [12]

In Swammerdam's monumental *Biblia mundi,*[13] first translated into English in 1738 as *The Book of Nature, or, the Natural History of Insects,* a strong warning stood out: "there is a much greater number of miracles, and natural secrets in the Frog, than any one hath ever before thought of or discovered." In his exploration of such miracles, among frequent warnings that "the human understanding is confined within very narrow bounds," we frequently encounter passages such as

> those who investigate nature but superficially, look into that globe [the head of the tadpole] only for the young Frog's head, though it comprehends the whole body . . . The hinder legs are observed to increase in the Tadpole . . . that is, they insensibly spring out of the body, as the cups of flowers from their footstalks . . . about this time the young Frog's fore legs are insensibly increased and augmented, under the skin.

If these observations already seemed to suggest a rudimentary idea of pre-formed parts inside the egg, Swammerdam made that point even clearer in his following chapter, "Man himself compared with insects, and with the Frog:"

> It is evident from comparing the Frog . . . with the insect tribe, how the sanguineous animals, or such as have red blood in their vessels are, in respect to their changes, like these smaller creatures. Indeed this likeliness proceeds so far, that it extends under many names, even to man himself: for all the works of God seem to proclaim, only one foundation of propagation and increase.

Aligning himself more and more closely with preformation, Swammerdam then gave several reasons to support the claim stated above.

> It is clearer than the light at noon, that man is, like insects, produced by a visible egg . . . We observe that the Vermicle or Maggot of Man, as well as the Vermicles or Worms of other insects, have not completely perfect limbs; therefore it is increased in size, even from the beginning, till its limbs project at length out of the skin . . . It is very clearly observed, that these parts of Man-Vermicle grow by degrees into a head, thorax, belly and limbs. In the head, the colored eyes are very distinctly seen through the skin . . . We observe further, that the limbs of the Man-Vermicle, in time, acquire their due perfection, and are strengthened to such a degree, as to be able to break out of the uterus, and to disengage themselves from all their integuments. And hence this first state of man likewise resembles an insect.

THE HIDDEN INSECT

The origins of Swammerdam's interpretation concerning the development of the frog embryo are relatively easy for the average reader to grasp, due to their similarity to the general picture of all vertebrate development, in which our own species participates.[14] However, the author's main subject in *Biblia mundi* was the development of insects, then a major source of perplexity due to their bizarre processes of metamorphosis. The frog seemed to appear more as a supporting argument to his claim that "all the sanguiniferous animals" develop like insects, since the frog "is changed into a true Nymph, called a Tadpole."

Indeed, insect reproduction had been a frustrating puzzle for earlier researchers of all times. The phenomenon seemed so inexplicable by the means available before the seventeenth century that many authors, including Harvey, had resorted to the simple explanation of spontaneous generation; and it is often said that one of Swammerdam's main goals in undertaking the formidable task described in this book was to dismiss that concept once and for all. He focused his attention mostly on ants, flies, and bees, and came to the conclusion that these animals were perfect examples of encasement of the future adult parts inside the temporary skin, ideal models "to attain a perfect idea of that most remarkable property, by which they perfectly agree with each other. This property we affirm to consist in an exact representation of the future animal, and of all his parts."[15]

Scrutinizing the secrets of insect infancy with his impeccable dissecting skills, Swammerdam came to the conclusion that

> the legs, wings, trunk, horns and every other part of the animal are covered with a membrane of equal thickness, in every place where they do not lie upon each other. This is the reason why, in the nymph of insects . . . the members . . . neither touch nor can adhere to one another . . . The free space produces a slight shade between some of the parts, affording the curious eye an opportunity of determining exactly the figure of the insect's little body, and all its limbs.

However,

> their legs, wings and the rest are folded up, and as it were packed in a most intricate manner; and this difficulty has been the cause . . . of the principal mistakes of writers on this subject.

And, as for the mechanism responsible for bringing about the real insect,

> every insect . . . has a certain and determined time appointed for it by the omnipotent God to expand its wings.

Swammerdam's clearest suggestion of a preformationist position appeared soon afterward as a conclusion based upon all of these observations:

It seems probable that in the whole nature of things there is no generation that can be properly so called, nor can any thing else be observed in this process than the continuation, as it were, of generation already preformed, or an increase of, or addition to, the limbs, which totally excludes the doctrine of fortuitous propagation. Having established this principle, it is easy to explain the reason that a man, deprived of his hands and feet, may have a sound and perfect offspring. Hence, also, we may determine that famous question, whether, in order to produce a complete issue, a seminal particle drawn from every member of the body be absolutely necessary. Moreover, the reason is evident, why Levi, being yet in his father's loins, paid tithes long before he was born; for he was in his father's loins when Melchizedek met Abraham. Lastly, even original sin may stand on this principle as on a firm foundation, since all mankind have been laid up originally in the loins of their first parents.[16]

An Inspired Misinterpretation

Two aspects of the passages quoted above are of crucial importance for understanding both how Swammerdam's observations fit so well with Malebranche's search for truth and why such a careful and talented microscopist endorsed so vehemently the concept of the performed parts of the embryo inside the egg. It is indeed puzzling at first how a Dutch Protestant naturalist and a French Catholic priest could have come so close in their ideas on reproduction; and it is even more puzzling that a naturalist as skilled as Swammerdam would fiercely believe that he had detected something that was not there—in this case, all the future parts of the adult insect preformed in tiny folds inside what he called the "nymph" and the "chrysalis."

These two questions (what brought Swammerdam so close to Malebranche, and why Swammerdam believed that he had seen preformed parts of the adult inside the larva) are not only intriguing: they are decisive. If Swammerdam had not made his claims with such conviction, maybe Malebranche would have felt less comfortable in postulating the multitude of encased generations. On the other hand, if Malebranche's religious standpoint had been distasteful to Swammerdam's sensibilities, maybe the microscope and God would never have combined their efforts to produce a theory that dominated the Western view of reproduction back in the days when God was a geometer.

But all the key elements in this story were in the right place at the right time. Were it not for his passionate love of insects, Swammerdam himself would also have taken religious orders, following his Calvinist father's desire[17]—and, to a large extent, his own spiritual impulses. Although he abandoned this path to explore the mysteries of nature,[18] Swammerdam remained a deeply religious man, repeating that he wished above all things to "be free to serve God." In his last seven years, he even retreated for a time to a life of religious contemplation, under the influence of the deeply mystical writings and advice of Antoinette Bourignon,[19] although the experience seems to have been useless to soothe his anguished spiritual quest.[20]

Like Malebranche, he saw everything in God. Exhortations abound in his works, as in: "meditate on the omnipotent wisdom and the superlative goodness of God in the accretion, sustenance and change of the minutest animalcules, which form a celestial host as the angels." Very precise descriptions of different stages of insect development, illustrated by equally careful and detailed drawings, are often immediately followed by statements such as

> the changes therefore, which we observe in vegetative animals, are equally observable in sensible ones, so as to afford us in all God's works the most manifest proof of God's wisdom and power, which man can neither imitate nor comprehend; for as the foundations of all created beings are few and simple, so the agreement between them is most surprisingly regular and harmonious, everything conspiring equally to fill us with sentiments of admiration and reverence for the great Author of Nature.

These sentiments, by all accounts, truly resided on Malebranche's wavelength. Across the borders and language barriers, the two men shared a common vision.

Still—and this brings us to the second question—that vision was made possible by a supposedly authentic finding: Swammerdam was positively sure that he had observed all the future parts of the insect folded and curled up inside some sort of "egg." Considering the extreme care of his experiments and the phenomenal skills he displayed in dissection and microscopic observation, and being aware of the seriousness of his research efforts, it is unthinkable that this man would have invented such findings. The truth is far more interesting, with the added twist of showing us how close to reality one can get through a misunderstanding. Swammerdam thought that all those folded structures that he was the first to detect were the beaks, horns, wings, and legs of the future insect. By our modern standards, he was wrong. But not all that wrong, in fact. What Swammerdam had discovered were what scientists now call the *imaginal disks* of the insects.

The imaginal disks are the markers of the different parts of the adult body within the body of the larva, first present as undifferentiated clusters of cells, positioned in specific regions awaiting the signal to differentiate. In the early larva of a common fruit fly, there are about a thousand imaginal cells. They are arranged in ten major nests, which, on the whole, reconstruct the entire adult, except for the abdomen. At the appropriate moment, they form the eyes and

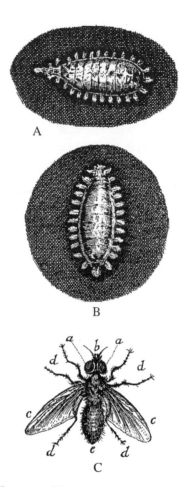

Fig. 4. Swammerdam's drawings of the "worm" (larva) of a fly "after it has cast its skin" (A and B). The darker areas that the author marked along the different segments bear a striking resemblance to our modern knowledge concerning the position and configuration of the imaginal discs, as represented in figure 4. Also, note that Swammerdam was aware of the segmentation pattern in the larva, a pattern most easily seen in the abdominal region of the adult, as shown in C. Although it is not mentioned in his text, this could have reinforced Swammerdam's conviction that the body of the adult is already preformed in the larval stages. (From *The Book of Nature*, 1738. Reprinted by courtesy of the Houghton Library, Harvard University.)

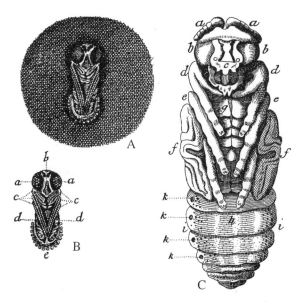

Fig. 5. Swammerdam's drawings of the "nymph" (pupa) of a fly "when cut out of the external skin, or that of the former worm, that covered it." According to our modern knowledge, this stage occurs after the larval tissue has been reabsorbed, leaving behind only the imaginal discs and other pattern-forming structures. At this point the animal indeed has all of the "preformed" features of the adult, only not expanded since they are enclosed inside the cuticle, as shown in *A*. According to the author's legends, in drawing *B*, *c* represents "the folded legs rising from the thorax, 3 in each side" and *d* represents "the folded wings, between which the extremities of the legs are closely arrayed." In *C*, depicting a later stage of pupation, *a* represents "the antennae with their joints," *b* represents "the eyes, which are now arrived at full size," and *d* represents "the first pair of legs, beautifully folded." The legend for *e* reads "behind [the first pair of legs] again another pair," and that for *f* reads "under the latter [pair of legs], again appear the wings, and their artificial convolutions and beautiful foldings." (From *The Book of Nature*, 1738. Reprinted by courtesy of the Houghton Library, Harvard University.)

antennae, the wings, or the legs. In a newly hatched larva they appear as localized thickenings of the epidermis—the famous foldings and curlings that Swammerdam saw under his microscope. Unlike the other larval cells, the imaginal disk cells divide rapidly at precisely timed intervals, increasing to sixty thousand cells in the largest disk, that of the wing. The timing of these events is commanded mainly by cellular interaction and induction, whereas the determination that a specific disk will become a leg or an eye is regulated by special sets of genes. It is still unclear what molecules cause the antenna disk to be

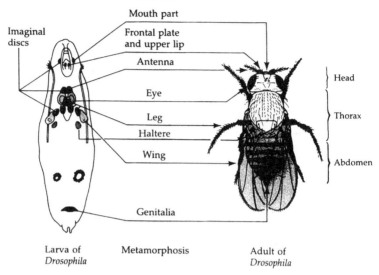

Fig. 6. A modern diagram representing the locations and developmental fates of the imaginal discs in the fruit fly. (From Fristrom et al., "In Vitro Evagination and RNA Synthesis in Imaginal Discs of *Drosophila melanogaster*," 1969. Reprinted by permission of Prof. J. W. Fristrom.)

different from the wing disk; but the discovery of these structures has already provided an extremely rich ground for understanding gene action during metamorphosis.[21] So, due to a curious coincidence, Swammerdam was not all that far off the mark. Contrary to his own belief, he had not seen the complete legs and wings of the adult folded inside the larva. But he had clearly seen, and accurately described, the clusters of cells destined to become specifically those legs and those wings—and, because of his belief system, he took those clusters for folded legs and wings.

THE WORLD AS AN EGG

There has been much debate about the origin of the term *emboitement*. Several authors cite Malebranche as the man who coined it, but the word is absent from his considerations of reproduction in *Search after Truth*. F. J. Cole, in his *Early Theories of Sexual Generation*, argues that Nicolas Andry (1658–1742; see chapter 2) was the first preformationist to write that the germs of life were, so to speak, *emboités* inside one another. But another man, a century before the surge of preformation, had already used the term—in a different context, but still directly linked to the concept of the egg.

In "Paracelse et sa posterité," Georges Cattaui quotes a brief passage from

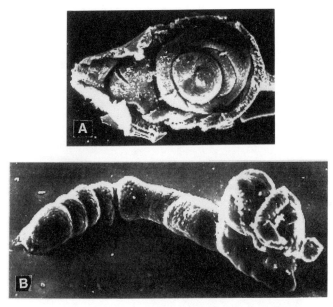

Fig. 7. Scanning electron micrograph of a leg disc in the fruit fly, still folded *(A)* and after unfolding *(B)*. In modern science, the terms "folding" and "unfolding," used by Swammerdam, have been replaced by "before and after elongation" or "evagination." If you knew nothing about imaginal disk function, would you not assume that *B* is the straightforward unfolding of the structure coiled in *A?* (From Fristrom et al., in "The Mechanism of Evagination of Imaginal Disc of *Drosophila melanogaster,*" 1977. Reprinted by courtesy of Prof. J. W. Fristrom.)

Philosophia ad Atheniensis, by the Swiss doctor and chemist Paracelsus (ca. 1493–1541):

> The Nature encompassing the Universe is One, and its origin has to be the eternal unity. It is a vast organism in which the natural things harmonize and sympathize in reciprocal form. Such is the Macrocosm. All things are the product of a single effort of universal creation. The macrocosm and the microcosm are one. They form one constellation, one breath, one influence, one harmony, one time, one metal, one fruit.

Following this argument, Paracelsus then states that Man has a special position at the very center of the world; he alone can be the perfect microcosm. In Man one can find the three divisions of the cosmos: the divine or ideal world, the astral or sidereal world, and the terrestrial world. These three worlds are superimposed; better yet, in the writer's own words, they are *Einschachtelung* (*emboités* in the French translation of the text).

The surfacing of preformation's key term in Paracelsus's text is particularly

Fig. 8. Paracelsus.

interesting when bearing in mind that the concept expressed above echoes a classical Stoic idea: the visible universe is compared to an egg, with the earth as the yolk and the firmament as the shell—and from this cosmic egg comes a progressive cosmic development of the original undifferentiated unity into the multitude of individual forms, making Creation an act of imagination. And this Imagination is of divine origin, acting through a *spiritus vitae* that regulates the balance of internal forces. This idea of Creation uses the very old concept of an undifferentiated egg from which all things, initially enclosed in the primordial matter as germs, develop and start to act. It leads us to two important concepts that were to become paramount in preformation's initial reasoning: the venerable antiquity of the idea that all life emerges from an egg, and the assumption that this egg becomes vitally organized as the result of a divine stimulus. As we shall discuss in this section, this ancient elevated status held by the egg provided a strong backbone to the belief that all humanity had been initially contained inside the ovary of Eve.

Some eggs, through their mythological power, became associated with magic rituals (and, in this respect, humans organized their mythologies and Creations in a similar way, elevating the egg to the status of *source of life,* a concept rather than a structure), and migrated from there to the realm of folktales. A very entertaining illustration appears in Pliny's (23–79) *Natural History,* concerning "the egg of the serpents, which the Latins call Anguinum"— a bizarre concoction that the author lists side by side with his descriptions of "Eggs of Hens and their medicinal properties" or "Eggs diverse in color," as if the Anguinum truly existed:

> I will not overpass one kind of eggs, which is in great name and request in
> France, and whereof the Greek authors have not written a word; and this is the
> serpents egg which the Latins call Anguinum. For in summer time, you shall see
> an infinite number of snakes gather around together into a heap, entangled and
> enwrapped one within another so artificially, as I am not able to express the
> manner thereof; by the means therefore, of the froth or salivation which they
> yield from their mouths, and the humor that comes from their bodies, there is
> engendered the egg aforesaid. The priests of France, called Druids, are of opin-
> ion, and so they deliver it, that these serpents when they have thus engendered
> this egg do cast it up on high into the air by the force of their hissing . . . They
> add moreover and say that the only mark to know this egg whether be right or
> not, is this, that it will swim aloft above the water even against the stream, as
> though it were bound and encased with a plate of gold.

As Jean Chevalier and Alain Gheerbrant point out in *Dictionnaire des symboles,* the egg is "a self-explanatory universal symbol." Throughout the world, it comes after the age of chaos as the first principle of organization. Overcoming the undifferentiated magma of the origins, the egg symbolizes the germ of the first differentiations, encompassing the totality of all differences. The world hatches from this egg in the myths of the Celts, the Greeks, the Egyptians, the Phoenicians, the Tibetans, the Hindus, the Vietnamese, the Chinese, the Japanese, the Indonesians, the Siberians, and a number of other peoples. For the Celts, it is the egg of a serpent. For the Egyptians, it is an egg spit out by a creature called the *Kneph,* while the Chinese attribute this delivery to a dragon—and, in both cases, as in the Book of Genesis, the episode represents the materialization of things through the power of the Verb. Sometimes, a primordial man is born from an egg; the Chinese consider a number of events subsequent to the original Creation in which heroes are born from eggs fertilized by the sun or from bird's eggs ingested by their mothers. Even more frequent is the concept of the cosmic egg, born from the primordial waters and incubated at their surface,[22] then breaking in two halves, destined to become the earth and the sky.[23]

The Egyptian cosmogony of Hermopolis depicts the egg as the patron of the vital forces in the human species. Even the sun was born from this all-encompassing mother-egg. In India, another tale describes the egg emerging from nothingness:

> At the beginning, there was nothing but a non-Being. It became the Being. This being grew bigger, and became transmuted into an egg. This egg rested for an entire year, and then cracked open. Its two halves released the earth and the sky. The external membrane became the mountains; the internal membrane became the clouds and the fogs; the veins became the rivers; the water inside the membranes became the oceans.

In several cosmological tales recorded in Peru by the first Spanish conquistadors, the hero of Creation asks his father, the sun, to create man so that they can populate the earth. The sun sends down three eggs. The first one is made of gold, and men hatch from it. The second is made of silver, and gives birth to women. The third is made of copper, and produces the common people. Another interesting tale of the separation between the feminine and the masculine (although this tale does not consider a separate third category for "common people") is found in the myths of the ancient Congo. Here, the egg is a living metaphor for the world and its perfection. The yolk represents the feminine humidity and the white the masculine semen. The shell, enveloped by a membrane, symbolizes the sun, which would have burnt the earth if the Creator had not transformed the membrane into a moist atmosphere. Therefore, these peoples held, man must always try to resemble an egg—and thus attain perfect balance.

This story may seem to indicate an ecological awareness too good to be true, but other models come even closer to our modern knowledge. The connection between the egg and the earth is explicit among the peoples of Mali. In their cosmogony, the first egg is the pure Spirit, and is produced by the sonic vibrations of primordial universal thunder. These vibrations cause the Spirit to spin around itself in the fashion of a tornado gaining momentum, finally causing it to condense into a rounded shape. Thus formed, the egg becomes independent of the vibration, solidifies, acquires the ability to trace its own path in space, and finally issues from its entrails all known species. Move over, Descartes's vortices; roll over, modern astronomers. We cannot invent anything that has not already been invented.

One of the Egyptian mythologies offers a scenario strikingly close to the basic ideas of preformation. Here we find a primordial ocean made of pure waters that contain all the germs of Creation waiting to awaken. From this ocean an egg emerges, and from this egg a god is hatched, bearing the name of Khnoum and charged with the task of creating order out of chaos. To fulfill his mission, Khnoum, the sculptor of the flesh, models all the eggs and embryos, all the germs of life, of all the living creatures to come. An impressive similarity to preformation is also found in Tibetan doctrines, according to which the egg is the origin of a long genealogy: it is formed by the essences of the five primordial elements, and, once opened, releases a white lake, ten different categories of beings, and a multitude of other eggs, which in turn release the limbs, the five senses, men, women . . . and a long lineage of ancestors. Moreover, what exactly does our Easter egg stand for? According to the *Dictionnaire des symboles,* it symbolizes the constant renewal of periodic generation. These remains of old myths no longer speak of Creation, but of endless re-creation: the egg confirms and promises a resurrection that is not really a birth, but rather a periodic repetition of the initial birth. Ideas are never born by spontaneous generation—even if they are the enlightened ideas of the Scientific Revolution.

"And How Foully They Hatch Them I Am Ashamed to Tell You"

Before such an overwhelming symbolic background, betting on the egg appears to be an extremely wise move. But, for the learned Europeans of the seventeenth century, this bet came with an additional safety net. In choosing the egg as the logical site for the encasement of animal generations, the first preformationists had natural evidence to back the belief that made them ovists: the hatching of animals from eggs had been recorded since antiquity and was readily visible to anyone.

A clear connection between eggs and reproduction appears in the ancient sacred books of India, from the sixth through the fourth century B.C. One passage states that generation encompasses three different classes: "that which springs from an egg, that which springs from a living being, and that which springs from a germ." Another, later passage suggests that generation is rather fourfold: "Those born from eggs, those born from germs, those born from perspiration and those born from wombs . . . Now there are the inferior beings and likewise those moving in the air. These should be understood to be born from eggs, as also all reptiles."[24]

The old Egyptians knew how to incubate the eggs of birds artificially by burying them in decomposing excrement, and used this method extensively with their hens ("and how foully they hatch them I am ashamed to tell you," wrote the Roman emperor Hadrian to his brother-in-law in A.D. 130). Some authors contend that they were already doing this around 3000 B.C., although it is more likely that the roots of the practice started around 1400 B.C.[25] In any event, the Egyptians performed their incubations with a precision and a success rate that was matched in the West only during the nineteenth century. They also engaged in a rudimentary preview of the major debates on the animation of matter by the soul so familiar during the Scientific Revolution, finally establishing that the *ruh,* or "life," entered the egg at the eleventh day of incubation.[26]

Likewise, the Chinese seem to have successfully carried out artificial incubation ever since remote antiquity. Here, eggs were put in wicker baskets and heated with charcoal pans, while an attendant slept in the incubator itself, placing the blunt ends of the eggs against his eyelids in order to keep the temperature constant.

The Romans never colonized China, but they became aware of these customs as soon as they got to Egypt, as this passage in Pliny's writings makes clear:

> Livia Augusta, wife sometime of Nero, when she was conceived by him and went with that child, being very desirous . . . to have a jolly boy, practiced this girlish feat to foreknow what she would have in the end; she took an egg, and ever carried it about her in her warm bosom; and if at any time she had occasion to lay it away, she would convey it closely out of her own warm lap into her nurses for fear it should chill. And verily this presage proved true, the egg became a cock chicken, and she was delivered of a son. And hereof it may well be came the device of late, to lay eggs in some warm place and to make a soft fire underneath of small straw or light chafe to give a kind of moderate heat; but evermore the eggs must be turned with a man or woman's hand, both night and day, and so at the set time they looked for chickens and had them.[27]

The influence of contact with the Egyptians is even clearer from another passage of Pliny, in which he states that "eggs will come to birds without sitting of the hen, as a man may see . . . in the dunghills of Egypt," immediately proceeding to describe "a notable drunkard of Syracuse, whose manner was when we went to a Tavern to drink to lay certain eggs in the earth, and cover them with mold, and he would not rise nor give over bibbing until they were hatched," finally concluding that "a man or a woman may hatch eggs with the very heat only of their bodies."[28]

The story of the drunkard also occurs in Aristotle's writings, showing that the Greeks were equally acquainted with artificial incubation. Long before Aristotle, the pre-Socratic philosophers from the sixth century B.C. turned their attention to the egg, speculating that the white was equivalent to the mammalian milk, while others contended that the yolk played this role, serving as the food of the developing embryo. Hippocrates (ca. 450–ca. 370 B.C.), and a host of pseudo-Hippocratic followers, mentioned the incubation of hen eggs as a means of studying the development of the chicken, and even hinted that this development could be compared to that of humans, as in this passage written by a Hippocratic embryologist:

> Take 20 eggs or more, and give them to 2 or 3 hens to incubate, then each day from the second onwards till the time of hatching, take out an egg, break it, and examine it. You will find everything I say [regarding the human embryo] in so far as a bird can resemble a man . . . The bird grows inside the egg and articulates itself exactly like the child . . . when there is no more food for the young one in the egg and it has nothing on which to live, it makes violent movements, searches for foods, and breaks the membrane . . . In just the same way, when the child has grown big and the mother cannot continue to provide him with enough nourishment, he becomes agitated, breaks the membrane, and incontinently passes out into the external world free of any bonds . . . That is all I have to say on this subject.[29]

Then came Aristotle (384–322 B.C.), who systematically cracked open chicken eggs and registered their development during the full three weeks.[30] In *The Generation of Animals,* Aristotle also discussed the distinction between perfect and imperfect eggs; described eggs in general; postulated (wrongly) that the embryo is formed from the white exclusively and gets its nourishment from the yolk; held that the bird embryo always develops at the pointed end (certainly because he always cracked the eggs open here, and the yolk always swims embryo uppermost); noticed that the yolk liquifies during the first week of development and grows larger; explained that the caterpillar is nothing but an egg laid too soon; and even speculated as to what extent the hen's egg is alive if it is infertile.[31] Finally, to weave Aristotle's dispersed observations even more tightly with the ovist's cause, he wrote, in his *History of Animals,* "In a certain district of Persia when a female mouse is dissected the female embryos appear to be pregnant," a sentence that could be read as an early confirmation of the preformationist doctrine.

It is impossible to establish when the idea that viviparous animals also hatched from eggs started to emerge, but several passages of a pseudo-Aristotelian book entitled *The Complete Master-Piece: Displaying the Secrets of Nature in the Generation of Man* are extremely clear on this matter. As to why women's "stones" are located "in the Hollowness of the Abdomen, resting upon the Muscles of the Loins," the anonymous author explained that this location is ideal "by contracting the greater Heat": this heat allows the "stones" to "be the more fruitful, their Office being to contain the Ovum, or Egg, which being impregnated by the seed of the Man is that from which the embryo is engendered." The writer comes back to the same point when debating "the use and action of the womb": "it has many Properties attributed

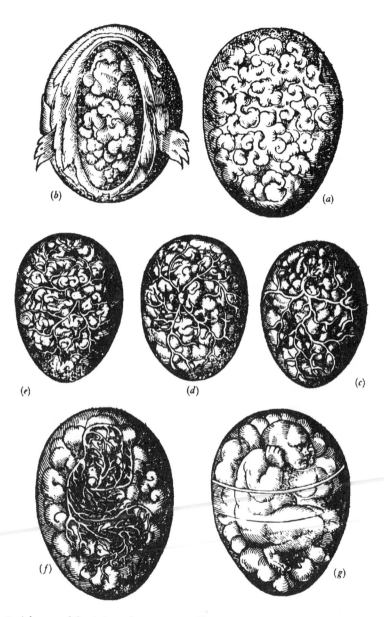

Fig. 9. A legacy of the Aristotelian concept of human development proceeding from a coagulum of blood and seed in the uterus: illustrations from Jacob Rueff's *De Conceptus Generatione Hominis*, 1554 (as reprinted by Needham in *A History of Embryology*, 1934).

to it. First, the Retention of the fecundated Egg, and this is properly called Conception." He then adds that

> the Testicles in Women are very useful; for where they are defective, Generation work is quite spoiled; for those two little Bladders in their outward superficies contain nothing of Seed, as the Followers of Galen etc. erroneously imagine, yet they contain several Eggs (about the number of 20 in each Testicle), one of which being impregnated by the most spirituous part of the Man's Seed in the Act of Coition, descends through the oviducts into the womb where it is cherished till it becomes a living child.

Never mind that Galen (129–ca. 200) lived after Aristotle: the author is not all that concerned with making his book appear to be a true Aristotelian production. According to the title page, this is the twenty-second edition, and it was for sale in London in 1741, just when reproduction had become one of the hottest topics in the natural sciences. This serves as an indication that even obscure popularizers had lost all doubt about the importance of the egg in generation.

They Have Always Hatched

From the beginning of Western civilization, the hatching of eggs was common knowledge, even for the most illiterate Europeans. After all, the poorer you are, the greater your chances of being a peasant, and if you are a peasant, you can scarcely avoid witnessing the hatching of an egg. This knowledge, mixed with all kinds of folk beliefs and superstitions, was strong enough to prompt Leonardo da Vinci (1452–1519) to scribble in one of his notebooks, "eggs which have a round form produce males, those which have a long form produce females."[32] Considerations of gender bias aside (the round shape was consistently regarded as the shape of perfection, as we shall discuss in chapter 7), casual footnotes of this sort, as well as brief references to eggs in the writings of the vast majority of medieval and Renaissance scholars who became interested in natural history, clearly show that the idea of new animals being born from eggs was beyond anybody's doubt. And then, right before the first preformationist postulates, a series of new data sprang forth from more academic circles, reinforcing the concept that life started with the egg.

The Venetian doctor Joseph of Aromati wrote a three-page letter in 1625, apparently aiming to discredit spontaneous generation, in which he noted that the chick exists in the egg even before the egg is incubated by the hen, and then grows under the stimulus of maternally provided heat, with the support of nutrients present in the egg and vital principles derived from the atmosphere.[33] William Harvey may have been a grand pioneer of epigenesis, stating,

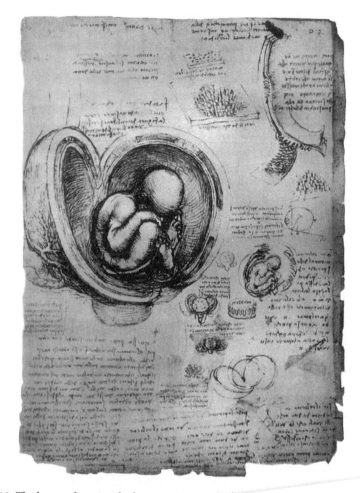

Fig. 10. The human fetus inside the egg in Leonardo da Vinci's notebooks, ca. 1400 (as reprinted by Needham in *A History of Embryology*, 1934).

in his *Exercitationes de Generatione Animalium*: "all that is alive comes from the egg." But, just by saying this much, he added yet another degree of plausibility to the egg-based model of those men who, by choosing preformation over epigenesis, were bound to become his main adversaries in the battleground of generation theories. Nathaniel Highmore, who could claim to his credit that he had provided most of the microscopic observations upon which Harvey based his epigenetic postulates, also looked at avian eggs, and asserted that the "seminal atoms" were already in the proper place in the cicatricula of the egg, displaying the same relative disposition as in the older and visible embryo. This disposition, he added, was already present before incubation, and was never altered afterward.[34] In 1664, yet another early microscopist, the Englishman H. Power, stated in *Experimental Philosophy* that, as soon as a pulsating particle appears in the incubated chicken egg, any magnifying glass will show it to be a complete heart, with both auricles and ventricles.[35]

A fundamental contribution to the credibility of the egg as the undisputed source of life came in 1672, when Reinier de Graaf (1641–1673) published *De mulierum organis generationi inservientibus* ("The Generative Organs of Women"), containing a detailed study of development in the rabbit. In these pages, the Dutch naturalist fell prey to a predictable misunderstanding, but extracted what we now consider a correct conclusion from it. He assumed that the large egg-containing follicles in the ovary (now called Graafian follicles), which are as round as the egg but much larger, and hence much easier to detect by the rudimentary techniques of the time, were the true mammalian ova. This finding firmly established that viviparous animals came from eggs. He was the first author to describe the ovarian corpus luteum ("yellow body"), a small scar left behind in the ovarian tissue when the mature egg erupts from its surface. In the course of his investigations, while trying to develop better tools for examination, he also ended up inventing the syringe. Like Harvey, de Graaf did not side with the preformationists; he even declared that he had not found any trace of a tiny embryo inside his "eggs," and that such embryos became visible

Fig. 11. Reinier de Graaf.

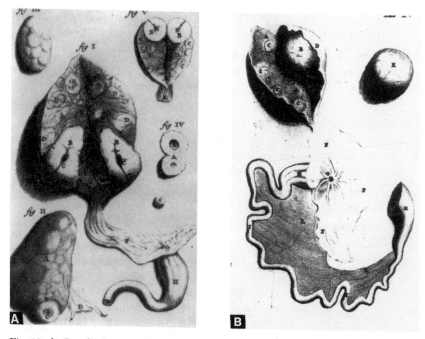

Fig. 12. de Graaf's drawings of "eggs" in bovine and ovine ovaries, after *(A)* and before *(B)* coitus. In *A,* fig. I "displays a cow's 'testicle' opened along its length," revealing "eggs of different sizes contained in the ovary." Fig. II "displays a 'testicle' not yet opened." Fig. III "displays a ewe's 'testicle' with eggs not yet irrigated by male semen showing through." Fig. IV "displays the glandulous substance of a globule removed from a ewe's 'testicle' while it still contained an egg," with the inner cavity representing "the place from which the egg was taken," and the round body marked *c* representing "the egg taken from this place." Fig. V "displays a ewe's 'testicle' from which an egg was expelled some days ago," with *c* representing "eggs of different sizes embedded in the surface of the 'testicle.'" *B* "displays a cow's 'testicle,' or ovary, with the usual appearance before coitus, after being opened," with *b* representing "a very large, i.e. mature, egg still contained in the 'testicle,'" *c* representing "small, i.e. immature, eggs embedded in the 'testicle,'" and *e* representing "a very large egg taken from the 'testicle.'" (From *De mulierum organis generationi inservientibus,* 1672.)

only at the tenth day of gestation, after roughly one-third of the total time of embryonic development in the rabbit. But he did believe that the egg contained the germ of the future organism, which was certainly a bonus to preformation. Moreover, by including mammals in the list of the egg-derived animals, he involuntarily made ovism even more plausible.

Swammerdam and de Graaf had been friends for a relatively short period when they became engaged in a race to explain the nature of the ovary and the production of eggs. As a medical student, Swammerdam carried out studies on female organs at the house of his friend Van Horne, starting in January 1667 and publishing the results in 1672. The choice of the winter season for undertaking the dissection of cadavers was most likely due to the fact that they decomposed more slowly in colder weather, at a time when methods of preserving specimens had not yet been developed. In *Biblia mundi,* Swammerdam claims to have first discovered the human ovum during this time, but de Graaf, who published a study on male sexual organs in 1668 and his famous book on female ovaries in 1672, seems to have been faster, and was certainly more widely quoted as the pioneer in these discoveries—something that his former friend considered so outrageous that he applied to the Royal Society of London to settle the dispute. This prompted de Graaf to write *Partium genitalium defensio* in response, using extremely coarse language and including such darts aimed at Swammerdam as "your little book [*Miraculum Naturae*] isn't fit for your arse, let alone your nose." These quarrels were so bitter that some said

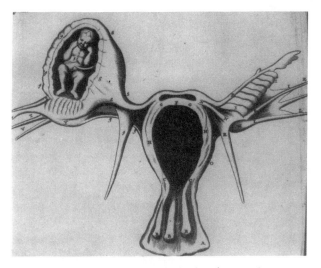

Fig. 13. de Graaf's illustration of a fetus inside the female reproductive organs, providing yet another decisive proof that humans develop from eggs. Note that this illustration, barely described in the original book, is a quagmire for modern developmental biologists, since the fetus seems to be located inside the ovary, rather than in the womb. This would suggest an ectopic pregnancy, but under such conditions, no human embryo can possibly develop this far. This illustration had to be either a symbolic representation of the site where embryos originate or a gross misinterpretation of what de Graaf actually saw. (From *De mulierum organis generationi inservientibus,* 1672.)

they ended up killing the inventor of the syringe, although it is more likely that he was a victim of a plague.[36]

This dispute was probably the consequence of a number of highly skilled men deciding to investigate the same important subject at exactly the same time, rather than any particular malice on either side. But, apart from the contributions he made on this front, and the attention he brought to the subject through his fight with de Graaf, Swammerdam enhanced the importance of the egg through yet another path. Choosing to conduct his main investigations on frogs rather than mammals, he correctly concluded that ovaries were always present in pairs of two; and he also correctly inferred that not all the eggs in the ovary were at the same state of maturation, and that their release to the uterus depended on how mature they were.[37] The fact that Swammerdam ended up being right in all these assertions, combined with the undeniable inclination that he—unlike de Graaf—repeatedly showed toward preformationist beliefs, gave "the system of the Egg" added support.

Also, painstakingly describing the day-to-day development of the frog embryo, Swammerdam stated that on the fifth day he could already "most distinctly discern the division of the young Frog into head, thorax, belly, and tail," and that on the twentieth day he could see the stomach, the liver, and the gallbladder. His subsequent claim that, at this point in development, "it is manifest that the heart was formed in the little Frog, in much the same manner with that of Chickens, according to the account given of them, by that illustrious anatomist, Marcellus Malpighi," was certainly influential in associating Malpighi with preformation. And, since Malpighi's work on the development of chick embryos was so precise and flawless, it was impossible not to respect a theory based on findings by men of such astonishing skills, as it was impossible not to respect the object of their tenacious scrutiny, the marvelous egg.

In 1703, the Flemish surgeon Jean Palfyn stated very clearly in his book *Description anatomique des parties de la femme qui servent à la génération*:

> Today it is established among the philosophers that all the Animals in general, regardless of their species, come from Eggs. This has always been obvious to the naked eye among the Birds, the majority of the fishes, and in some Insects, and likewise it is now incontestable and above all doubt . . . that the same happens among the Animals that give birth to their young, although a great diversity can be found among those Eggs . . . As for those Animals in which the matter of the Egg is not suited for reproduction, they are sterile, such as old women and mules.

Eggs did not lose their crucial importance when the troubled waters of the generational debate started to settle. According to M. Rusconi in his 1821

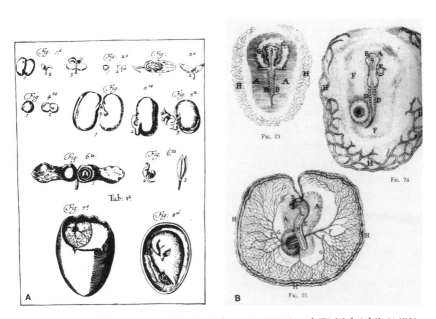

Fig. 14. "Eggs had always hatched": (A) Highmore's (1651) and (B) Malpighi's (1673) microscopic illustrations of development in chicken eggs (as reprinted by Meyer in *The Rise of Embryology,* 1939).

Amours des salamandres aquatiques, Blumenbach was able to see for the first time the development of the external gills in young salamanders by keeping females in a glass of water, waiting for them to lay eggs, and following the events occurring right after the hatching of those eggs. To this day, eggs are a major source of material for developmental investigations. In light of our modern knowledge, the ovists may have been wrong in their assumptions and extrapolations, but when they chose the egg as their banner, they at least had selected one aspect of reproduction that was hard to miss. Eggs, after all, had always hatched.

So, during all the philosophical turmoil that marked the century of preformation's domination, spermism came to life and eventually faded away (see chapter 2). Yet ovism did not die with Malebranche and Swammerdam. If anything, it became increasingly sophisticated—so much so that the theory experienced a dramatic rebirth during the eighteenth century. Three formidable contenders held firm in the frontlines until the end of their lives. Inspired by their new findings and instigated by the eventual counterattack of epigenesis, they perfected and polished their model to a point their predecessors could never have imagined.

THE MIGHTY BARON

Albrecht von Haller (1708–1777), the man who proclaimed that "we think right when we dare to think freely,"[38] seems to have had a curious effect both on his biographers and on the scholars who analyze his work. We frequently encounter the name of Bonnet adorned by such adjectives as "great naturalist" or "remarkable man." It is equally common to read that Buffon sometimes got carried away in his "ardent" proselytizing but that he was a true visionary nonetheless—and a truly exciting character at that. As for Spallanzani, everyone seems to agree that his contribution was nothing short of "marvelous." However, when we come to Haller, even those who have dedicated their lives to the study of his works will say only that he was "highly influential" or "extremely respected"—or, to put it in a nutshell, "very powerful." Somehow, nobody seems to be able to fully empathize with this man.

This ambivalence certainly has a lot to do with the fact that Haller, as a person, was full of contradictions. To say that he had a short temper is an understatement;[39] indeed, he engaged in endless polemics with his peers. However, he confessed, in a letter to a friend, "I would like to have that unchangeable serenity which makes the perfection of the saint and of the philosopher; however, that is difficult to my sensitive temperament."[40] But this

Fig. 15. Albrecht von Haller.

ambivalence is only the tip of the iceberg into which all of Haller's moral dilemmas crystallized.

Although he believed that, through the will of God, man is endowed with
very definite impulses that lead him to cultivate goodness, his general idea
of mankind was rather cynical and bitter, as expressed in passages from his
poems, such as "Here are Men! Nobody surpasses Alexander in his value, yet
thousands surpass him in madness"; or "if [Nature] raised the Alps to keep us
apart from the world it is because men are the biggest plague of themselves";
or yet, in what reads like a barely disguised reference to his own fate, "and
even when glory leads us to pleasure, is this glory worth all of our efforts? We
sacrifice to glory the better days of our lives and the greatest forces of our
spirits, yet we only attain glory after we are dead."

Other poetic passages illustrate another crucial contradiction: although
Haller was a staunch defender of the immortality of the soul, he sadly wrote
that "Achilles . . . is not less dead than the others," and indulged in multiple
tirades against mortal splendor such as "Build, vain monarchs of the South,
those eternal Pyramids, glued by the blood of your subjects," leading to such
bottom lines as "but, destined to become grazing fields for the worms, remember that, under the most precious masses, rest is not any sweeter than under
the grass."

The poems mentioned above are yet another reflection of Haller's multiple
personalities. He was an extraordinary physiologist,[41] a respected professor of
anatomy, surgery, and botany, and a fine natural philosopher. He was also, in
his day, a famous novelist and poet; and, to make the bridge between the
"humanistic" and the "scientific" even more complicated, he chose as a vehicle
for the latter skills the challenging format of poems-in-prose, which certainly
sounded awkward to Southern European ears.[42] Reflecting yet another duality,
Haller was a well-traveled urbanite, but his poems are full of praise for the
simple bucolic joys of uncorrupted nature and of awe before God's majesty
inscribed in such scenes.[43]

His private life was equally split between the two sides of the mirror. He
wrote passionately about his wife Marianne, and detailed, in long pages full of
sorrow, all of his sufferings in watching her die. Referring to himself as "a
faithful spouse," he laments, after his wife's death, "the world wants me to
forget you; it is an insult to my heart." Behaving along these lines of absolute
marital love, Haller spent his entire life crusading for restraint, religion, and
social decency, insisting that moral evil was responsible for all human suffering. And yet, most likely, he had affairs with his female servants and fathered
at least one bastard child.[44] In the preface of his letters to his daughter, we are
informed that "like the savior of the world, he went about doing good to the

souls and the bodies of men"; and that, on those occasions, he "eagerly seized the opportunities which his profession as a physician gave him, of convincing those with whom he conversed of the truth, and of converting them to the practice of the Christian religion."[45] However, it is no secret that, in his later years, suffering from numerous medical ailments, Haller became addicted to opium.[46] Here, however, we should bear in mind that this type of dependency did not have then the undertones that it carries in our days. Therefore, regardless of his addiction, Haller always stood out as one of the leading medical authorities of the eighteenth century.

His professional career was similarly marked by antagonistic facts and feelings. Accepting the offer of George II, king of Great Britain and Ireland, he took a chair at the University of Göttingen, founded by the British monarch. He did an enormous amount of work to put this young institution on the scientific map. The Hanoverian government did all it could to embellish his tenure, providing him with a botanical garden, an anatomical theater, and several honorific titles, finally buying him, at a considerable price, a patent of nobility from the imperial court at Vienna. But the Baron von Haller, a member of virtually every scientific academy in Europe, courted by Prussia's Frederick the Great to come to the University of Berlin and become a member of the illustrious Round Table, never ceased to worry about money or social status. After a seventeen-year stay at Göttingen, he could no longer bear his homesickness and returned to his native Switzerland, working in Bern from then on.

Haller's religious positions were another source of torment, reflecting the period's continual attempts to reconcile the teachings of orthodox theology with conclusions drawn from scientific investigations of natural phenomena. Throughout his writings, his scientific research, aimed at attaining a better understanding of the character and significance of nature's behavior, always seems to support in the end his preconceived belief in the infinite wisdom of the divine Creator. He even thought he held a privileged position to discuss divinity, for, as he wrote to his daughter, "what the churchmen have written on religious matters, has, in general, gained but inferior credit. Their arguments have lost much of their weight, from the consideration of their having been urged by persons who were bound, both by profession and interest, to defend the profession in which they were engaged." Since Haller was not one of those "churchmen" himself, he was accordingly much better positioned to proselytize for his faith.

According to the sources cited by Margarete Hochdoerfer in her 1932 *The Conflicts between the Religious and the Scientific Views of Albrecht von Haller,* Haller was led by the spirit of scientific research to test the truths of religion.

Yet, although he showed an earnest interest in religion from his early child-hood, Haller appears to have been unable to find happiness in the orthodox Protestant church, which demanded so much and promised so little comfort in return. Thus he vacillated continually between this inherited belief and his in-dividual opinions. And, although he took his religious creed seriously enough to wage constant war on the iconoclastic pronouncements of Voltaire, candidly admitting that he found no joy in this because he regarded the Frenchman as largely his superior in wit, his own poems were censured more than once by the Swiss religious establishment, whose leaders considered Haller's descrip-tions of God as excessively "free," deviating dangerously from the strict dog-mas of his church.

Three Different Postures

Haller's views on reproduction mirror all of his contradictory attitudes. He believed in God as the absolute Creator of a preordained universe. This Creator was able to foresee man's troubles to such an extent that He pre-arranged the coming of Christ as a mediator between our sin and His glory. Such convictions were likely to provide a solid background for the endorse-ment of preformation. Yet Haller, like Saint Peter, reneged three times.

As Roe explains in *Matter, Life and Generation,*

> during his lifetime Haller held three distinct theories concerning embryological development. As a student of Boerhaave, he adopted his mentor's belief in sper-matic preformation. Then, in the mid-1740s, he switched his support to epigene-sis. Finally, in the 1750s, Haller converted back to preformation, presenting a detailed account of pre-existence in the maternal egg in his first major embryo-logical work, "Sur la formation du coeur dans le poulet," 1758.

Haller's initial endorsement of Boerhaave's views (see chapter 2) is easy to understand on the grounds of both his respect for his much-admired master and his own devotion to the wise Creator who, as he wrote in a poem, had seen to it that

> indeed, in the semen already, before life breathes
> The ducts are already formed, that first the animal uses.

Haller's short-lived conversion to epigenesis was prompted by Abraham Trem-bley's discovery of regeneration in the freshwater polyp (see chapter 4), which led him to write in 1746 that

> eyes more attentive and minds free from systems are beginning to be persuaded that the most perfect animals are born in approximately the same manner [as the

polyp], that their formation is successive and that there was never a plan in which their parts are designed in miniature.[47]

The events that Haller considered to indicate a general mechanism for the development of the fetus parallel to that of the polyp were the formation of the heart in the chicken, which first appears as a simple tube bearing no resemblance to the complex four-chambered organ of the adult, and the apparent production of the fibers and membranes of the embryo through the coagulation of a viscous fluid. These observations, combined with the expression of traits from both parents in the offspring, including hybrids, seemed to deny any kind of preformed plans.

> I am persuaded that after several more years of observation, one will find that animals and thus plants are engendered from a fluid that thickens and organizes itself little by little, following laws that are unknown to us but which the eternal Wisdom has rendered invariable.[48]

This persuasion, however, did not last long—and, interestingly, it was the extensive series of experiments that Haller undertook to gain a better understanding of the development of the heart in chicken embryos through epigenesis that propelled him to change sides again. In 1758, he stated in *Sur la formation du coeur dans le poulet* that

> I have shown sufficiently in my works that I leaned towards epigenesis and that I regarded it as the opinion that conformed the most with experience. But these matters are so difficult, and my experiments on the egg are so numerous, that I propose with less repugnance the contrary opinion, which is beginning to appear to me as the most probable. The chicken has furnished me with reasons in favor of development [unfolding of pre-existing parts].

In 1772, he noted that

> at a time I was inclined towards the gradual formation of animals. But more mature observations, particularly those by observant eyes on the formation of chickens, have brought me back since then to evolution [preformation].[49]

As a born-again preformationist, Haller was now an ovist. His observations on the chicken had provided him with enough grounds to discredit the role of sperm in the encasement of germs. Like many of his early predecessors, he equated the eggs of animals with the seeds of plants, and resorted to preformation to explain the continuity within species. He was a keen enough observer, however, to notice the endless variety that can occur within the colors and shapes of each species, and, by bringing this plastic quality of the encased germs into the fabric of preformation, he made the whole theory much more flexible, hence more credible.

The acknowledgment of such continuous variation in nature seems to have tempted Haller to consider the possibility that the formation of animals had no model, and even that Nature herself worked without a plan. But these last hesitations came to an end when he completed his work on the development of the chick embryo. He was then fully convinced that the embryo can be found in the unfertilized egg, and that the essentials of the fetus are already contained in the ovaries of the mother. He visualized the yolk as the continuation of the intestine of the fetus, the inner membrane of the yolk being continuous with the inner membrane of the intestine, and the outer membrane of the former being an extension of the outer membrane of the latter. This concept made the fetal intestine identical to both the gut and skin (due to the inner membrane) and the mesentery and peritoneum (due to the outer membrane). The envelope that covered the yolk during the last days of incubation would later become the skin of the fetus. From these observations, Haller reasoned

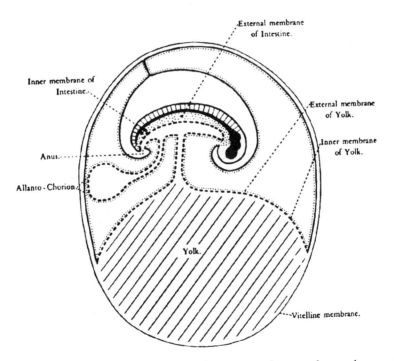

Fig. 16. F. J. Cole's oft-reprinted diagram of the continuity between the membranes of the chick embryo, illustrating Haller's ovist arguments. (From *Early Theories of Sexual Generation,* 1930.)

that, since the yolk was continuous with the fetal skin and intestine, all these structures should be contemporaneous, and truly part of the fetus—and, since the yolk was present inside the hen regardless of copulation, the fetus was also truly part of the hen, enclosed in the amnion from the beginning of her life but invisible to our eyes due to its smallness and transparency.

Once Haller had embraced the ovist system, he retained that position for the rest of his life. But, faithful to his character, he kept himself open to doubt, and did not hesitate to extend this doubt to the authenticity of his own observations. His nemesis was eventually personified by Caspar Wolff, who came to the rescue of epigenesis based on his findings of organ formation in the embryo. Whenever Wolff seemed to threaten Haller's position with new theoretical propositions and empirical results, Haller reopened his enormous database by reopening just as many eggs, noting new details of chick development and tentatively extrapolating new conclusions. Throughout this struggle, he found a firm supporter and a great rationalizer in a man, twelve years his junior, who lived just across Lake Geneva from him. Their geographic proximity notwithstanding, they met only once in person, although they exchanged long letters for more than twenty years. Haller's friend was technically blind, but his mind was full of bright visions.

LIKE THE HEAVENLY BODIES

As far as his life went, Charles Bonnet (1720–1793) was everything Albrecht von Haller was not. Due to his frail health, his early hearing loss, and later his loss of eyesight, he rarely traveled, although he corresponded with scholars from all over Europe. More than engaging in debates and polemics, he was fond of instructing and encouraging younger naturalists. He never tired of philosophical pursuits, very likely fueled, especially at his later age in blindness, by the aim of achieving a true victory of reason over the senses. Although he outlived Haller, he was mostly confined, after his early twenties, to a life of almost excruciating routine. In the end, he was the first to admit that, "As for my life, it has been nearly as uniform as the heavenly bodies." [50] He was barely twenty when, largely under the stimulus of his mentor Réaumur, he discovered parthenogenesis in aphids (see chapter 3) and, almost coincidentally, regeneration in freshwater worms (see chapter 4). After that, practically blinded by his long experiments on almost invisible creatures, he turned his energies to rationalizing and organizing the discoveries about generation that mushroomed everywhere around him. During this time, he started corresponding with Haller.

His interest in natural history started in a manner that he himself compared to the quasi-mystical impulse experienced by Malebranche after reading

All About Eve

Fig. 17. Charles Bonnet.

Descartes. He was only sixteen when he read Abbé Noel Antoine Pluche's *Spectacle de la Nature.* This book, published in eight volumes between 1732 and 1750, carried the basic message that nature was created for man to admire, thus constituting a spectacle that could not be conceived without the spectator (a view shared by the influential Swedish taxonomist Carl Linnaeus). It had an immediate and immense success and went through at least fifty-five editions in France and seventeen in England, including abridged and adapted forms. Being explicitly didactic (Pluche initially considered calling it *La physique des enfants,* or "Physics for Children"), it was often used as a textbook of natural sciences, and became customary reading for the children of wealthy families. Bonnet said that he did not read the book: he devoured it. At the core of this burst of excitement was a description of the ant lion, an animal that intrigued him so much that he searched all the surrounding countryside until he could find a specimen to observe by himself. The appetite whetted by this early revelation was fully sated only when a local librarian agreed, after being much pestered by the ardent youth, to let him borrow a copy of Réaumur's *Mémoires sur les insectes.* He read it all at once, tried some of the experiments on caterpillars described in its pages, and eventually found some discrepancies between his own data and those of the famous entomologist. He dared to write

Réaumur a letter on the subject, and thus started a correspondence that continued until Réaumur's death in 1760—and greatly boosted the young man's ego, making him start to perceive himself as a true naturalist. By 1744, when his *Traité d'insectologie* was first published in Paris, Bonnet could hardly see any longer, and had to withdraw from experiments. In compensation, he discovered the joys of philosophy, and would later acknowledge that "Insects thus led me, unaware, to the metaphysics for which I had shown until then so great a repugnance and which would soon delight me."[51]

Could the man who discovered parthenogenesis be anything other than an ovist? His mentor Réaumur tended toward that view, and so did the great Albrecht von Haller, who promptly responded to Bonnet's first letter. Although he acknowledged the existence of sperm and discussed the likelihood of spermist preformation (see chapter 2), Bonnet's final decision was to remain faithful to the egg, since "the system of the seminal Worms" presented problems that made it "dubious, to say the least." He was not afraid to address the fundamental question of "why are the Germs that entered the bodies of Females submitted to the Law of Mating unable to develop without the Liquor furnished by the Male?" His answer was simple: "The interior of the Females from those Species does not possess a Liquor subtle or active enough to open, by itself, the fibers of the Germ and thus initiate development." As for "why does the Germ continue to grow after the seminal Liquor is no longer acting," it all had to do with the "marvelous art" that presided over the construction of the "Animal machines": "[these machines] convert the food into their own substance. The preparations, combinations, and separations that those foods undergo, change them insensibly into Blood, Lymph, Flesh, Bones, etc., etc. Therefore, as soon as Circulation has started in the Germ, once it has become a living Animal, the same Metamorphosis takes place in their interior. The almost infinite diversity of Particles that make the composition of foods, their number, their smallness, their structure, the interaction between different Organs . . . convince us easily of this possibility."

THE ESSENTIAL PARTS

Thus established as an ovist, Bonnet contributed much more to the theory than his two early remarkable findings on parthenogenesis and regeneration. Led by his blindness to philosophical inquiry, he rationalized the basic grounds of preformation in the clearest fashion recorded in the history of this system. In the opening statement of the first volume of his book *Considérations sur les corps organisés,* first published in 1762, he explained that "philosophy having understood its impossibility to explain mechanically the Formation of the Organized Beings, it imagined, happily, that they existed already in small dimensions, under the Form of *Germs* or *Organic Corpuscles*. And this Idea produced two theories that are very much pleasant to Reason." The first hypothesis, he continued, "supposes that all the Organized Bodies of the same species were encased one inside the other, and from there successively developed"; although this did not necessarily mean that the future body was fully formed: Bonnet preferred to think of "essential Parts," that would "become bigger and organized through development." The second hypothesis held that those germs were spread everywhere, "and it supposes that they only manage to develop when they find convenient *Matrices,* or Bodies from the same species, disposed to retain them, to nourish them, and to make them grow."

Bonnet considered the first hypothesis, which he definitively called *Emboitement,* "one of the greatest victories that Pure Reason has ever achieved over the Senses," and added that the *Infinite Smallness* assumed by this model "overwhelms Imagination without scaring Reason": reason, he stated, "conceives with pleasure of the Seed of a Plant, or the Egg of an Animal, as a small world peopled by a multitude of Organized Beings, called to succeed each other during the course of all Centuries." In a brief personal note, he confessed that "as far as I am concerned, I like to push back the boundaries of Creation as much as possible. I am pleased imagining this magnificent string of organized Bodies, enclosed as just as many small worlds one over the other. I see them getting further away from me by degrees; to diminish according to certain proportions, and finally slide away into a night that we cannot penetrate."

As for the second hypothesis, which he called Dissemination, Bonnet believed that it coexisted with the first, taking over *emboîtement* where the smallness required for more encasements no longer made mathematical sense (see chapter 8). "This hypothesis, planting Germs everywhere, makes Air, Water, Earth, and all Solid Bodies, become vast and numerous barns, where Nature has deposited her greatest Treasures." In this case, only the Germs with the "Organic Wholes" of the species in which they found themselves could develop inside that species; but, since these Germs were "prodigiously small," they could undergo "Dissolution" if they did not land in the right species, "and they leave [that body] without alteration, to float in the Water, or in the Air, or to enter another Organized Body." Once in the right body, they could finally "grow little by little" until they achieved the adult form. Since all Organized Bodies multiply, "while the Law of Dissolutions exercises its destructive empire over the Mass of the living Beings, the Law of Generations presides over the conservation of the Species, and assures their immortality."

Although Bonnet admitted that "the fashion of growth of the Organized Bodies is a very obscure point in Physics," he was nevertheless convinced that a number of experiments could be undertaken to study this phenomenon. Unable to perform such experiments himself, he resorted to his philosophical skills to achieve "a reasonable Hypothesis." One thing he was certain of: "Nature never works through jumps," hence all growth had to proceed by "insensible degrees," starting from the "Organized Body . . . already present in small form, with all of its essential parts" that existed inside "the Seed of a Plant or the Egg of an Animal." In the end, Bonnet concluded that "Generation is a secret that Nature seems to want to keep to herself; but I believe that one day we will steal this secret from her." Generation, after all, was "nothing but the principle of the Development of Germs."

Although he did not manage to disclose Nature's big secret, Bonnet gave preformation a complete system of function that, for the first time, left almost no question unanswered. In *Considérations sur les corps organisés,* he addressed all the difficult questions that the theory had to face: Why do children express traits from both parents? How can one account for the generation of monsters? How can one explain the occurrence of hybrids such as the mule, "a sort of monster"? Rather than shying away from these issues, he insisted that "I strongly desire that experiments on Mules would be multiplied. Nothing would be more adequate to shed daylight over this tenebrous matter." Stressing that the nourishment brought to the fetus by the mother always played a part in such "accidents," he took his explanations one step further by admitting that something like a graft could occur between two germs, accounting for cases such as the hybrid and the child with mixed parental traits. He also reasoned that some germs could have more or less than the regular number of "essential parts" characteristic of the species, leading to the birth of monsters. But, most important of all, he proposed an explanation for variations in progeny that turned all of the above exceptions into a clever and original new rule: of every ten encased new germs, his argument went, one would be slightly different from the others; and this one would start a whole new lineage of slightly different beings. Thus variation, like everything else, had been encased at the beginning of life inside the primordial germ.

THE PATH TO PERFECTION

At this point, we can see that Bonnet's contribution had already established two fundamental new views that helped to make ovism viable. By proposing that the "germ" is not a fully preformed creature, but rather a loose sum of all the "fundamental parts" of the future individual,[52] he dispensed with the idea of minuscule men or minuscule horses that sooner or later would doom the entire theory to ridicule. By admitting that all generations were not precisely encased—much as he liked the concept—but that beyond a certain level of smallness they could just float around awaiting their moment, he shaped the concept of "infinite smallness" into a much more credible version. Moreover, he always left himself open to further questioning and vehemently insisted that new experiments should be done to tackle the more controversial issues, cleansing the theory of dangerously fanatical overtones. But his contributions did not stop here. He will also be forever remembered for his creative new twist on the old theory of the "chain of being."

Bonnet's reasoning assumed that God had not created only the laws of the universe: he had also created, all at once, all the structures existing in that universe. Creation was followed by a progressive unfolding of this pre-encased world, in a rigorous order meant to be undisturbed until Judgment Day. From the simplest of all structures, the *atom,* to the most elevated one, the *cherub,* stretched a portentous *Universal Chain,* which "unites all beings, connects all worlds, embraces all spheres." Only God, who created it, was outside that chain; and the chain was always ascending toward perfection, without detours and without gaps. Between plants and animals, polyps ensured the link. Between fishes and birds, flying fishes bridged the empty space. And, quite obviously, between man and apes, the savage provided the continuity. All of this emerged from one single delivery, in this same precise order; and no interchange, no miscegenation, was now possible.

Bonnet had to give up the use of the microscope, but he never forgot the lessons the instrument seemed to teach in his time. The microscope, he held, walking closely in Malebranche's footsteps, tells us that we can see only the smallest part of a colossal whole. The order of things, it follows, represents the order of our known world—but this world is by no means the only one. Other worlds have existed before, and others will come afterward; and what rules this succession is still the stairway toward perfection. Before, the world was a caterpillar. Now, it is a chrysalis. Finally, it will be the glorious butterfly. And then we can finally be resurrected. The soul of each man, according to Bonnet's order of things, was a miniature creature enclosed, for the time being, within the *callous body* of the brain: a spiritual body locked inside the animal body, as the plant is locked inside the seed. Others had called it soul. Bonnet dared to call it "the germ of restitution," an invisible entity that, after the death of the body, wandered around inside elephants or over shrubs, its whereabouts without meaning since this structure was indestructible and calmly awaited the end of time. "Resurrection," declared Bonnet, "will be the accelerated development of this germ." And resurrection would encompass all beings, including the plants, which would finally be able to walk—for it is obviously toward that goal that perfection does its work. And all of us would receive transfigured bodies, so that our souls could eventually lay inside their final homes. But the hierarchy inside the chain would meet no changes: it simply ascended, untouched, to a superior level.[53]

Before such bold speculations, it is easy to understand the subtle teasing sensed in Bonnet's letters to Haller, in which he often urged his friend to extrapolate further on his findings. Haller generally resisted these calls, remaining focused on his experiments and showing much less inclination to philosophize on their profound meanings. The two men were perfectly complementary, and the union of their efforts ruled generation theories for a time. And, whenever the confrontations became heated, a third influential voice arose from across the border to support their beliefs. In the final ovist crusade, the two Protestants from Switzerland had a firm ally in a Catholic priest in Italy. In the introduction to one of his books, this priest clearly declared that

> it is said by many, that fecundation is among the mysteries of nature; and, like many of her operations, an object of admiration, rather than of enquiry. Such an opinion is highly agreeable to the idleness of man. In times past, I acknowledge that generation, both in animals and plants, was involved in darkness, impenetrable to the human eye; but, since the appearance of Haller and Bonnet, this gloom has been rendered much less thick.[54]

ENTER THE "MAGNIFICO"

As a churchman, Lazzaro Spallanzani (1729–1799) never married, but as a naturalist, he was closely bound to two different women. Although his initial education, by the Jesuits at Reggio, focused more on topics such as grammar, logic, and the Greek classics, his devotion to natural sciences was first aroused by Laura Bassi, his professor at the University of Bologna; and his own experimental work was assisted by his sister Marianna, who learned all the tools of the trade in order to help her brother.[55] Laura Bassi herself, a sort of tourist attraction in Bologna because she managed to flawlessly combine her spousal duties and her mothering of eight children with her intensive academic work, was no stranger to the young university student—as a matter of fact, she was Spallanzani's cousin. And, most likely, she did not have to fight hard to divert her cousin's attention from the translation of Homer's verses to the exploration of the mysteries of life. As a youngster, he had already displayed a somewhat nasty side, performing those kinds of childish experiments that are so often equated with cruelty still untamed. He was described in his early years as pulling legs and wings from beetles and flies, and then trying to stick them back together again.[56]

All this promising background notwithstanding, Spallanzani's father wanted him to follow in his footsteps and take up the study of jurisprudence. The loving son did so for a while, until he confided to the famous Vallisnieri,

Fig. 18. Lazzaro Spallanzani.

professor of natural history in Padua and a close friend of the family, his total lack of interest in legal matters. Upon Vallisnieri's intervention, the father agreed to let the son follow his natural inclinations. From the Jesuits, the son then passed to the Dominicans, and received the Minor Orders from them. After completing his preparation for the priesthood, he dedicated the rest of his life to natural history while helping to support himself by saying masses. This became less of a need when, after being much requested and having long hesitated, he took up a teaching chair at Pavia. His enthusiastic lecturing style and his extraordinary ability to work the crowds made him extremely popular among his students. This admiration grew even more ardent when the *Magnifico* started undertaking long research trips, which led him as far as Turkey, in order to build a collection of specimens for the museum of his university—and also in order to visit harems and meet eunuchs, one of his long-standing interests. Spallanzani's achievements in natural sciences eventually became so impressive that he was offered chairs at more than half a dozen universities before he reached middle life; later he was made a member of academies and learned societies in several European cities, stretching from Madrid to Uppsala.

What did Spallanzani do that earned him the name "Magnifico"? To begin with, his demeanor in itself must have been quite impressive. Remarkably, in a period when disputes on natural philosophy tended to erupt with extreme acrimony and often quickly slid into personal and vicious attacks, Spallanzani, like his countryman Malpighi before him, maintained a constant gentleness and an imperturbable suavity of manners. Others could scream and shout: he would just keep on working, and let the facts speak for themselves. In the laboratory, he dedicated his attention to numerous vital phenomena, such as digestion, respiration, hibernation, and circulation. Much to his own surprise, he discovered that some microorganisms could live without oxygen, thus unveiling the existence of what we now call *anaerobic* life. He was the first scholar to understand the digestive power of saliva (until then considered necessary only to moisten food and make it easy to swallow). Using his own body as the instrument of painful experiments, he described the role of gastric juice as a food solvent, opening the way to the preparation of pepsin from animal's stomachs as an adjuvant to human digestion. We owe him that much relief in our daily lives. But we owe him even more than that. In particular, he anticipated modern experimental techniques, going well beyond simple dissection and observation to a constant invention of new manipulations meant to test and clarify his ideas.[57] In 1769, writing about the microscopic "little animals" in *Nouvelles recherches sur les découvertes microscopiques, et la génération des corps organisés,* he stated this new attitude very clearly:

> Those who made a particular study of [the animalcules], and to whom we owe
> all of our trust, were well placed to give a detailed and exact description of these
> animalcules, but they contented themselves with only announcing their existence.
> One could almost say that they made [the animalcules] a sort of mystery. Others,
> more concerned with exaggerating and embellishing their observations . . . gave
> us disguised things, that hurt the sensitivity of Philosophy.

As for himself, Spallanzani announced, the compass would be set to a totally different direction:

> I will follow this race of little animals to the end . . . with the greatest exactitude
> possible . . . and investigate . . . their shape, their laws, and the laws they observe
> among themselves.

As we shall see in chapter 5, he did so by means perfectly tuned to the modern fashion of research. He stipulated the golden rule of any current scientific paper by declaring that he did not intend to give us a full description of everything he had seen: instead, he would sort out his data, try to make the best possible sense of it, and present only the facts considered relevant to the subject under discussion. Several scientists resisted. But science was never written in the same way again.

THE HERESY OF THE HEAD

Regeneration will be discussed in greater detail in chapter 4 of this book, but it is important at this point to mention some of Spallanzani's contributions in this area, to better understand how his ideas on reproduction were shaped. Focusing mainly on animals such as tadpoles, salamanders, and snails, he demonstrated that these animals could regenerate not only severed tails and limbs, but also heads.[58] Up to that time, the spinal cord had been considered a simple bundle of nerves, running together through the canal inside the spinal vertebrae in the fashion of an electrical cable stretched through a tube; but Spallanzani's experiments on severed heads made it clear that the spinal cord formed a very important portion of the central nervous system, containing a number of reflex movement centers, able to function independently from the brain. This, of course, did not sit well with the learned community, and Spallanzani's discoveries were the subject of much criticism and an even greater share of doubt. The head was considered a sort of sacred place, the location of an imprecise entity referred to as the "intelligence" or "the soul." As the *Abbé* himself wrote in a letter to Bonnet, "I am sure that I would not have been under such a continuous attack if my experiments had to do with a body part other than the head." Regeneration of heads was felt to be a kind of profound heresy, contradicting everything that had been held as dogma about the function of animal bodies—and it was a Catholic priest who was exposing before the world's eyes just how heretical nature could be. From this point on, there was no going back. The priest's mission on earth had clearly been assigned. He was destined to dig deeper and deeper into preconceived spiritual mysteries, and to disclose their natural, material causes.

If heads could be regenerated through natural mechanisms, did it make any sense to assume that whole organisms could be created without pre-existing lives, through obscure concoctions of spontaneous generation? Confronted with his own evidence, Spallanzani obviously did not think so. But another Catholic priest, the Englishman John Turberville Needham (1713–1781; see chapters 4 and 5), disagreed. He boiled meat juices, supposedly killing any life forms existing in them, and then sealed the broth inside glass vials, assuming that if any new life forms now appeared, they would have to arise by themselves. In 1748, he reported that microorganisms had indeed regrown inside the vials, thus claiming that spontaneous generation was an indisputable reality. Spallanzani took up the challenge and repeated Needham's experiments, this time making sure that no air was left within the flasks and that the sealing of those flasks was absolutely hermetic. When nothing regrew inside, Needham objected that Spallanzani's method of sealing the containers with a flame had killed the "vegetative force" that supported the springing of life. Spallanzani just repeated all of his steps once more and *then* opened the flasks, whereupon organic forms soon started to emerge again. This experiment demonstrated that the life appearing in the broth came from pre-existing life in the air, and in one single blow weakened the defense of spontaneous generation (the *Magnifico* is now mentioned by numerous authors as the direct predecessor of Louis Pasteur). It also brought considerable attention to Spallanzani's pursuits.[59]

THE SWAN SONG

Having established that life does not arise by itself, Spallanzani had to decide which theory of generation he was going to embrace. He dedicated a great deal of attention to the function of the sperm, and strongly criticized Linnaeus's dismissive view of these cells (see chapter 3). Still, he ended up concluding that, although the idea of spermism was very ingenious, "it is unfortunate that it is not true." The publication of his *Prodromo,* in 1768, in which he claimed to have discovered a tadpole inside the egg of a frog before fertilization—the egg, in fact, being nothing but "the preformed tadpole folded up and concentrated"—marked his emergence as a firm ovist. He praised Haller's "beautiful discoveries" and held them to be too perfect to be disputed. Spallanzani's microscopic and experimental techniques were so sound and respected that Bonnet spent his last years urging him to write a treatise dealing with the systems of preformation, emphasizing their common belief that very small organized bodies existed in the female before fertilization. Coming from the revered *Abbé*'s pen, such a treatise would finally rid the world of epigenesis and other related hypotheses. Spallanzani, however, never wrote that treatise. The clock was already running out on preformation.

Spallanzani was struck by apoplexy in 1799, and then recovered a bit for a few days. He raised his head above the sheets, reciting Tasso and Homer to the delight of the friends who had come to pay him their last respects. And then he died. This beautiful swan song might be taken as a perfect metaphor for ovism. The century was drawing to a close, and so was preformation. The great minds of Haller, Bonnet, and Spallanzani had exhausted the last possibilities of the theory. Unfortunately, the sum of all these efforts was still not sufficient. New models had to be explored so that generation could be explained. Without generation, there is no life. And life had to go on.

All About Adam

Their name is the living infinite

MICHELET

*I*T IS THE MIDDLE of the seventeenth century, and a French aristocrat has sent some intriguing drawings to a Dutch naturalist. The Frenchman is François de Plantade, and often uses Dalenpatius as his nom de plume. The Dutchman is the descendant of a family of rich brewers from Delft, not a scholar, not a court figure, but already respected in the learned world for the amazing number of little animals whose existence he has been able to show through the use of that exciting novelty, the microscope. His name is Antoni van Leeuwenhoek. He has already seen little animals in the semen of several species, and has wondered aloud about their real nature. Now he has before him Dalenpatius's letter, and with it the ultimate temptation: the drawings of the Frenchman portray little men, complete with hats and beards, enclosed inside the heads of the little animals.

Why not? Preformation had already been postulated. Since the contents of the semen had hitherto been unknown, the assumption had been made that all future generations were encased inside the eggs of females. But now, in Leeuwenhoek's microscope and in those of other curious scrutinizers of nature, the existence of millions of mysterious animals in each drop of semen had been revealed. Why not replace the egg, whose true appearance in mammals was still undisclosed, with the sperm cell, whose conspicuous presence in all kinds of semen was now so obvious? Why not assume that God had encased all humans inside the testes of Adam, rather than in the ovary of Eve? The structure of the mammalian ovary was still a puzzle. The structure of these new candidates, on the contrary, could not have been more simple. They were little eels. Little worms. Little tadpoles. In any event, little animals.

> They all have the same size and the same shape, they move their tails in a way that is clearly meant to make them swim, and in consequence they are true animals. Whereas one is going to the right, another is going to the left; one comes up and

65

Chapter Two

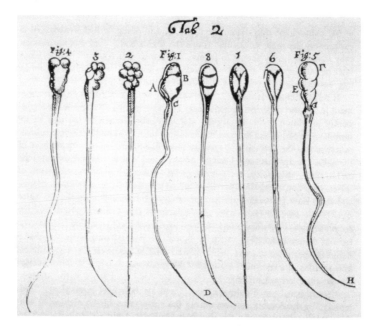

Fig. 19. Leeuwenhoek's drawings of different types of sperm cells. (From "The Observations of Mr. Antoni Leeuwenhoek, on animalcules engendered in the semen," *Philosophical Transactions of the Royal Society of London,* 1679; as reprinted by Meyer in *The Rise of Embryology,* 1939.)

another goes down. Some start moving in a certain direction, and then turn themselves around by a swing of the tail to go back the way they came. [1]

And so spermism was born.

Truly convinced of the animal nature of the sperm cells, the defenders of spermism preferred to consider themselves "animalculists." This conviction would generate numerous problems (see chapter 3), but nobody could have grasped this before the fact. As the animalculists started to put their theory together, the simplicity of their initial premise soon gave way to increasing degrees of complexity. If these animals were really animals, they had to have all the animal's organs. What did they eat? Did they have intestines? Did they have a circulatory system, and if so, how did it function? Did they copulate in order to reproduce, like many other animals? Could it be that they came in two genders, male and female, and if so, how could these two genders be distinguished? And considering those that were destined to become humans— did they have a soul? The debate over all these issues, and the fight for supremacy between the supporters of the two gametes, were bound to become one of the nerve centers of knowledge during the following century—even if often developed only on paper. As Carlo Castellani puts it, although microscopic and experimental research remained rare, "purely theoretical speculations on the matter, abundantly nourished as they were by the endless disputes between animalculists and ovists, would easily fill a whole library."[2]

As if we could never escape the mentality of Hollywood's productions, framed by a basic need to regard the world as the stage for a continual fight between heroes and villains, our modern literature on this subject tends too often to present the spermists as some sort of bizarre fundamentalists—the fanatics who produced those infamous drawings of little people in the head of the sperm cell, and came up with far-fetched theories to incorporate the entire epistemology of reproduction into their all but impossible models. However,

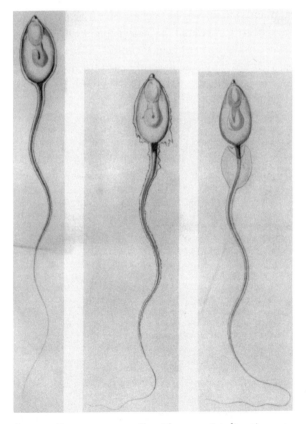

Fig. 20. Three figures of human sperm cells with a putative digestive system. (From Roujou, "Mémoires de la Société des sciences naturelles," 1878; as reprinted by Meyer in *The Rise of Embryology,* 1939.)

as we take a closer look at their writings, following their observations and their postulates and detecting their perplexities in the context of the morals of their time, a much more interesting picture starts to emerge. From the identification by the first observer of a sperm cell under the microscope to the factors really involved in spermism's early doom, each detail of this story is a mirror of the idiosyncrasies of the period—and a modest ode to human creativity.

HIS MAJESTY WAS PLEASED

Antoni van Leeuwenhoek (1632–1723) has been described as a "composed and well-balanced man," with a vigorous constitution that allowed him to live until age ninety-one and remain active as a researcher until his death. In contrast to his contemporaries Malpighi and Swammerdam, he abandoned school at age sixteen, becoming the bookkeeper and cashier of a clothing business in Amsterdam. Therefore, he lacked university training and was self-taught in natural history—a detail that may explain why he chose to reveal his discoveries mainly in the form of eclectic letters, rather than writing long and well-organized treatises. Since he ground lenses for his own use, and became famous for the quality of his work, it has sometimes been wrongly assumed

Fig. 21. Antoni van Leeuwenhoek.

that he was an optician, or a manufacturer of lenses for the market. In truth, he held the minor court office of "Chamberlain of the Sheriff," from which he drew a modest salary that he kept receiving even after having ceased his official functions after thirty-nine years of service. He was thus financially independent; and, even while on the job, the workload was light enough to ensure him a considerable amount of free time for his research.

In a letter from December 1675 he noted:

> In the past summer I have made many observations upon various waters, and in almost all discovered an abundance of very little and odd animals, whereof some were incredibly small, less even than the animalcules which others have discovered in water.[3]

This letter seems to establish that Leeuwenhoek's early observations dated from the summer of 1675, and that, as the legend has it, his fascination with the microscope started with the discovery of the myriad life forms visible through this instrument in a simple drop of water. However, he soon turned his lenses to a number of other fluids—and eventually aimed them at samples of semen.

As we shall discuss later in this book, the date and authorship of the discovery of sperm cells are hard to establish. One thing is certain, however: Leeuwenhoek was already seriously committed to the cause of spermism by April of 1679. In a letter written during this month, he fired away at detractors of the idea in an almost apocalyptic tone:

> Those who have always tried to maintain that the animalcules were the product of putrefaction and did not serve for procreation, will be defeated. Some also imagine that these animalcules do not live, but that it is only the fire that is present in the sperm. But I take it that these animalcules are composed of such a multitude of parts as, such people believe, compose our bodies.

In the following passage his voice slides from castigating to self-satisfied:

> I will bear those who reject [his observations] no grudge, the less so because, when I wrote about the number of living creatures in water, even the Royal Society would not accept it. But when I communicated my calculation and something about my method, your colleague, Mr. Robert Hooke, increased the number and wrote to tell me that His Royal Majesty,[4] having heard about it, was anxious to see it, and that . . . His Majesty seeing the little animals, contemplated them in astonishment and mentioned my name with great respect.

This royal appreciation, allied to his insatiable curiosity about the bizarre new world slowly taking shape under his magnifying glasses, led the Dutch microscopist to continue his relentless observations on several kinds of semen, from fish, birds, worms, dogs, and often humans. In this process, his own reasoning on reproduction also became increasingly sophisticated—and bolder, to the point of proclaiming, in July of 1683, "do not we see . . . that Man, even when the length does not exceed the diameter of a green pea, is already furnished with all his members?"

The "diameter of a green pea" may be a bit of poetic license; but it is true that human embryos of roughly this size are already a few weeks old, with short limbs clearly visible. Wherever Leeuwenhoek was able to observe such an embryo,[5] his faith in the truth of spermism was by then so unshakable that he found this example conclusive enough to abandon any further attempt at "finding the male animalcule in the hen's egg" and to maintain that "the foetus proceeds only from the male semen and the female only serves to feed and develop it." In making this statement he not only dismissed as ludicrous the problem raised by "a certain Doctor"[6] that "should a foetus grow from the male semen, an arm or a leg would easily get lost," but also used as a counter-argument what he took as proof of the transmission of hereditary characters through the male line, discoursing on matings between white domestic female rabbits and grey wild males. Knowing nothing about recessive and dominant genes, he triumphantly concluded that "all the young issuing from this take their father's grey color; and, indeed, it has never been seen that any such young rabbit had a single white hair or any other hair than grey. Moreover, they will never grow to the size of the mother, nor have long ears; also they will never be so tame as the mother, but will always remain rather wild."

At this point, Leeuwenhoek appeared to feel that he was sitting atop a huge pile of evidence against ovism. In a letter to the Royal Society in January of the same year, he assumed the pose of a man who holds the final truth when he stated,

> now that I have discovered that the animalcules also occur in the male seed of quadrupeds, birds and fishes, nay even in vermin, I now assume with even greater certainty than before that a human being originates not from an egg but from an animalcule that is found in the male semen, the more so since I remember having seen that in the semen of a man and also of a dog there are two sorts of animalcules. Seeing these I imagined that one sort were males and the other sort females.[7]

His belief in the sperm cell as the sole carrier of primordial new lives seemed now to be so solid that he did not even hesitate to take his case one step further and dismiss the existence of mammalian eggs as supposedly observed by the ovists, or all those who "would take their oath on having found the supposed eggs in the tuba fallopiana":

> But I need not believe that the round parts found in it, are those which are sucked off from the so-called ovary . . . But I would rather believe that what remains of the sperm (for one animalcule would be required, not many), would coagulate into a round ball or globule, and I also imagine that if a so-called egg should be found in the tuba fallopiana immediately after copulation and should be examined closely, the prevailing opinion would be abandoned.

He supported this belief by citing the size and solid consistency of the "glandular so-called egg" as opposed to the narrow diameter of the tubes it would have to travel through after being sucked from the ovary, and also the fact that "I take it for certain that these imaginary eggs would be found not occasionally but in every case after the male has served the female." He then cogently used the assumption that this "service" pulls the eggs down the fallopian tubes: "it seems rather incredible to me if we remember how little time a bull, rabbit, boar, etc., require for copulation, that in that space of time . . . eggs . . . would be sucked off the tuba fallopiana."

The next step in the dismissal of the "imaginary, so-called glandular egg" becomes frankly scatological. "Do not we see that all excrements, discharged either by human beings or animals, consist of globules . . . ? And . . . we see that fat, pus, and certain parts of a horse's urine also consist of globules." This line of reasoning had the double merit of associating the egg with the least noble of all materials and offering an interesting mechanical cause for the sperm cells' appearance in a rounded form within the female ducts: "why then should we not be allowed to assume that the remnants of the sperm of a male human being will stick together in the uterus . . . forming a globule."

The mechanism for a reproduction generated by animalcules gained further structural elaboration in later passages of the same letter. When entering these specific speculations, Leeuwenhoek was obviously not insensitive to descriptions of the shedding of the skin during insect metamorphosis, which by then had been carefully illustrated by several naturalists, most remarkably by his countryman Jan Swammerdam.

> Cannot we imagine that if such an animalcule comes to attach himself to a vein, this animalcule will, within a short time, assume an entirely different character, that is to say, that its skin will serve for afterbirth and that the inner body of the animalcule will assume the figure of a human being, already provided with a heart and other intestines, and indeed having all the perfection of a man.

Carefully anticipating the criticism of his peers, the microscopist immediately proceeded to explain how single births can occur in this manner. Using one of his favorite analogies, he presented the example of the apple tree that would in much later times prove so useful for Darwin.[8]

Against these hypotheses propounded by me, one might well ask why a woman does not produce many children at one birth, for although I recognize only one or two places in the womb, fit to feed animalcules, many of them adhering to one place might all of them live. I might answer this question in the following manner. Let us make a little hole in the earth, the thickness of a straw, and throw into it 6 or 8 seeds of an apple: they will not produce 6 or 8 trees, but the seed that has first put out the biggest root will become a tree and will crush all the others, and this is what I fancy will happen to the animalcules.

With this, spermism gained progressively complex undertones. Even through Leeuwenhoek's uncoordinated prose—constantly jumping from sperm to fleas, and from fleas to globules found in wood, and then to the blood of frogs—spermism starts to emerge as a sound technologically based theory whose explanatory possibilities for the endless variations of life are apparently unlimited. Its attraction was evident in the prompt allegiance of George Garden to the animalculist cause. Garden stated, in his "Discourse Concerning the Modern Theory of Generation," published in the *Philosophical Transactions of the Royal Society of London* in 1691, that

> it seems most probable that the fibers of all the Plants and all the Animals that have ever been, or ever shall be in the World, have been formed at the origin of the world by the Almighty Creator within the first of each respective kind.

Like Malebranche (see chapter 1), Garden was not a naturalist, but a theologian. It was his admiration for Leeuwenhoek's discoveries that propelled his endorsement of spermism, which he backed with the assertion that Marcello Malpighi's first view of the chicken in the cicatricula of the egg was similar in shape to these exciting, newly discovered "little animals." "Now at least," he exclaimed, "Leeuwenhoek has discovered an infinite number of *animalcula in semine marium* of all kinds which has made him condemn the former opinion about the propagation of all animals *ex ovo*."

Leeuwenhoek was gathering a following. Spermism was definitely taking over.

THE HEART-SEARCHER

Leeuwenhoek's main rival was another Dutchman whose last name means, literally, "heart-searcher." The young and restless Nicolas Hartsoeker (1656–1725) kept ridiculing his contender's views even after his death. He made fun of many of Leeuwenhoek's claims, such as to have dissected a flea and observed its testicles: the heart-searcher promptly noted that this task was all but impossible due to the small size of the animal. Ironically, these charges saw print only after Hartsoeker had also died, in his *Cours de Physique*.

This aversion on Hartsoeker's part to Leeuwenhoek's accounts and merits was probably based on their fight over the true authorship of the discovery of spermatozoa.[9] Reportedly, Hartsoeker first saw a microscope when visiting the old and venerable Leeuwenhoek as a timid and avidly curious youngster. But the chemistry between the two men was apparently not very successful, to say the least. According to the "Eloge de M. Hartsoeker by M. de Fontenelle," affixed by the latter to the posthumous edition of the *Cours de Physique,* Hartsoeker was the first to see "these little animals until then invisible, which must transform themselves into men, which swim in prodigious amounts in the liquor destined to carry them, which do not occur but among males, and which have the appearance of young frogs, with large heads and long tails and very vivid movements."

According to Fontenelle's version, Hartsoeker was then only eighteen years old. He had rebelled against his father, who "wanted him to learn something useful," in order to pursue his academic interests. He was studying mathematics, and had become a fervent Cartesian. Then he looked through a microscope and saw the sperm cells. It followed that he was at first so shocked by his own discovery that he did not say a word about it, fearing to have fallen prey to some strange malady. Only two years later, in 1677, did he get the courage to return to his microscopic observations in Amsterdam—and he saw the little tadpoles again. He then dared to mention them to his teacher of mathematics and to a friend of his, and the three made another round of observations. They saw the sperm cells of a dog, and then those of a pigeon and a rooster, "more similar to worms or eels." At this point, according to Fontenelle, "they practically no longer doubted that all the animals are not born but through a series of metamorphoses invisible and hidden, just as all the species of flies and butterflies come from metamorphosis visible and well known."

Was Hartsoeker the teenage genius he appears to be in these pages? If Fontenelle's description of how he built his first microscope is true, then he certainly had to rank above average.

> At the house of his mathematics teacher . . . one day, wandering around without a precise aim and presenting a string of glass to the flame of a candle, he noted that the tip of that string became rounded, and since he already knew that a ball of glass could magnify the objects placed under its scope . . . he took the small ball that had formed and become detached from the string, and with it he made a microscope that he first tried on a hair. He was delighted to find it excellent, and of having the art of manufacturing it at such lower cost.

We shall never know for certain just how romantic this version of the real events may be. In the absence of certainties, I, for one, would very much like to believe it—the idea that such a stream of observations, experiments, and ideas sprung from such a simple piece of proto-technology is somehow blissfully soothing to a modern-day scientist.

HOMO VERMICULOSUS

The French physician Nicolas Andry du Bois-Regard was probably the founding father of helminthology. His interest in worms and parasites, and his focus on how these animals could account for numerous human diseases, was so intense that some of his contemporaries took to calling him "homo vermiculosus." His book, *An account of the Breeding of Worms in Human Bodies,* first published in 1700, had numerous subsequent editions and several translations. Among all the specimens considered in this work, the "spermatic worms" stand out in the penultimate chapter [10] as the only ones that do good things for their carrier, instead of "devouring his flesh." Andry may have done some damage to the respectability of spermism by establishing such a tight bond between worms and spermatozoa (see chapter 3). But he certainly did so with the best of intentions, taking sides from the beginning with Leeuwenhoek's cause.

> Leeuwenhoek, to whom we are beholden for these Discoveries, says that he took out the Belly of a Doormouse, and, having separated the Testicles from the deferent Vessels, saw . . . an immense number of living Animals . . . some of which were wound or folded up, one upon another. Some appeared to be fully grown, and not yet come to life . . . He made the same experiment several times, and still discovered the same thing.

This opening statement clearly indicates Andry's willingness to play Paul to Leeuwenhoek's Jesus, producing the first didactic account of the revelations hidden in the midst of the almost parabolic letters of the Dutch microscopist. Andry's prose, most probably intended to transmit the brand new data to upcoming generations of medical students, has a pedagogical tone throughout. In this mode, the perspicuous text explains details such as

> When the Spermatic Worm is entered the Egg, it becomes a Foetus, that is to say, there it is fomented and nourished, its parts increase, and they unfold themselves insensibly; and when they have attained to that bigness which they ought to have . . . the Animal breaks prison and is born.

or

> The Spermatic Worms have all of them long Tails, but those disappear when they become a Foetus; just like the little Frogs which are at first only a Head and Tail, but at last they lose that Tail when they begin to take the sensible Form of Frogs.

There is even a curious passage on how the spermatic worms can fool people about the real date of conception, by staying for several weeks in the woman's "matrix" (a term then used for "womb") before one of them finally decides to enter the egg. Andry based this statement on his observations of the sperm of a dog kept in a bottle for seven days, after which "I found them alive . . . some of them having as much motion as on the first day." He reasoned that since the uterus is a much more comfortable environment for the semen than a bottle, the waiting period could be up to four months. Although "the case is difficult, because the number of Spermatic Worms is too great for none of them to enter in so long a time," it is not impossible—and "hence it may happen that a Woman, whose Husband dies some days after having conceived by him, may not be brought to bed till the 11th, 12th, or 13th month after . . . So that we have sometimes seen Women thus brought to Death without being criminals."

Having dealt with these mundane details, Andry proceeded to prove that the validity of Leeuwenhoek's conclusions was beyond doubt, describing his own observations on human and animal cadavers, and establishing in very clear terms the core of preformation's basic model:

> What are we to conclude . . . if it be not that those Spermatic Worms are the occasion of the Generation of all Animals? These Worms are not found before the Age proper for Generation. They are found dead or dying in old Men, and in those who have Gonorrhea, or Venereal Distempers. What must we infer from those Circumstances? Does not the thing seem to speak of itself, and tell to us plainly that Man, and all other Animals, come of a Worm; That that Worm is the Epitome of the Animal that is to come of it; that if the Worm be Male, it produces a Male; and if it be Female, it produces a Female; that when it is in the Matrix, there it takes its growth by means of an Egg into which it enters, and where it stays the time appointed by Nature; when it is grown to a certain measure, it forces the Membrane of the Egg, and then is born?

Unlike Leeuwenhoek and Hartsoeker, Andry was no longer wandering from one experiment to another, investigating the possibilities offered by the magnifying glasses, listing his findings and publicly hesitating on the correct interpretation of them. Nor was he seeking to be credited for an important discovery. He was a medical doctor, not a naturalist. Moreover, he was a man who had made a choice and seemed to have no doubts about it. His self-assigned task was now to spread the news, to give it a coherent shape and to proselytize on its absolute veracity—which he did in a distinctive oratorical tone, totally absent in the work of his predecessors.

> According to this System, we must necessarily suppose that the Spermatic Worm not only includes the Compendium of the Animal which is to be born, but that

it also includes the Abridgment of all those that are to be born of this animal; and not only the Abridgment of all those, but likewise of all the rest that are to come of the line of that . . . This Littleness can neither be imagined nor comprehended . . . but . . . must necessarily be admitted . . . We must admit that there are Animals a thousand times less than a Grain of Dust, which we can scarcely see. This is not enough: those Animals . . . have a motion like other Animals; they have Muscles then to move, Tendons, and an infinite number of Fibers in each Muscle; and also have Blood or Animal Spirits, very subtle and fine, to fill or move those Muscles, without which they could not transport their Bodies into different places . . . Our imagination loses itself in this thought, it is amazed at such a strange littleness; but to what purpose should it deny it? Reason convinces us of the existence of that which we cannot conceive.

Appealing to reason to explain the seemingly impossible, Andry closes his chapter using a rhetorical technique similar to that of Malebranche: he asks us to imagine ourselves reduced to the size of a worm, and to try to imagine how unlikely the gigantic size of men would seem to us under those circumstances; or, inversely, to imagine that we are giants created by God on another planet, and, through the use of telescopes, have just discovered the existence of human beings on earth and cannot believe how small they are. But, before this suggestive ending, he attempts the tour de force of appropriating for the sake of spermism the fundamental preformationist theory laid down by Malebranche in *Search after Truth,* casually ignoring the fact that the Cartesian priest had based his doctrine solely on the findings of ovists.

According to this thought, which cannot appear extravagant to any but those who measure the Wonders of the infinite power of God according to Ideas of their own sense and imagination, we may say that in one only Spermatic Worm there may be organized Bodies proper to bring forth Foetuses and Infants for an infinite number of Ages, always in proportion lesser to lesser . . . that all the Bodies of Animals that have been born since the beginning of the World, or that shall be born to the Consummation of Ages, were created in the first individual Male of each Species. We might carry this thought further, were we not afraid, with the author of the *Search After Truth,* to dive too far into the Works of God. Let us hold this great Principle: That nothing is great or little in itself . . . and let us consider that no littleness, how inconceivable soever, ought to occasion the least scruple in us; and if there be no other difficulty in the System which I have just now proposed, nothing ought to hinder our embracing it.

MULTIPLE VIEWS

Exhortations of this sort, which are frequent in the preformationist literature, may sound a bit too militant; but we often overlook the fact that, for all the vibrant conviction expressed in some of their writings, most of the leading spermists remained, throughout their careers, remarkably open to different views. Leeuwenhoek is an excellent example of this attitude, as he made room for a multiplicity of hypotheses, even while staunchly adhering to the crucial role played by his "animalcules" in the process of generation. Among other things, he was the first naturalist to describe parthenogenesis (see chapter 3), which would naturally appear to make a strong case for ovism. In his letters to the Royal Society, he provided a description of how beetles and silkworms lay their eggs without any mention of sperm, as if the egg were everything in the process—as if the author himself were the ultimate ovist.

Moreover, in his repeated attempts to explain how sperm cells originate, Leeuwenhoek's observation of spermatocytes (the much larger, round cells that precede the mature spermatozoon during the course of spermatogenesis, whose real nature was obviously impossible to understand at the time of the earliest microscopic discoveries) led him to consider that the little animals might hatch from eggs—a conclusion that, in its ultimate implications, would both give a higher status to ovism and pay homage to the "ex ovo omnia" credo of epigenesis.[11] In fact, when he describes very clearly the beak, horns, and claws of "weevils" appearing in a temporal sequence, and then gives a similar account of the formation of color patterns, we could just as well be reading the words of a convinced epigeneticist.

These fluctuations, with all their apparent contradictions, reflect a curious spirit at work before a mystery larger than life, and the writer of the letters to the Royal Society remains all throughout the process remarkably frank, often admitting his own perplexities and presenting points "that I cannot satisfactorily solve." One of these problems concerned how long the little animals in the sperm could wait before they released the little creature within: "For if we assume that they have been in our bodies, from our birth or even from the moment of procreation, the seeds, in my opinion, could not remain in our bodies for sixteen years or more, without producing life, for I certainly believe that, when there are in the testicles animalcules that have received life, there must be a desire for coition."

As a result of his relentless observations, Leeuwenhoek also ended up on what we now regard as the right side of an old debate, thanks to his repeated dismissal of "what is called equivocal or spontaneous generation, that is to say, from inanimate substances without any parent." Echoing Francesco Redi's (1626–1697) suspicions (see chapter 3), he suggested that animals reported to reproduce in this manner, such as the corn beetle, could simply have been carried from one granary to another in the clothes of the workers or in cars of corn. He attacked "a book published in Rome by a learned Jesuit, named Philippo Bonani, wherein he maintains that the animalcules, or small living creatures, can be produced out of inanimate substances, such as mud and sand." To discredit this claim, he referred to the eggs laid by "a fish called mussel" and discussed how these eggs appear in boats, which led him to describe eggs attached by tendons to the outer shell of the mussels, with creatures inside "with the shape of a mussel." Any ovist would have been proud to sign such a statement.

Likewise, Nicolas Hartsoeker, when writing his *Essai de dioptrique,* first published in 1694 and so often associated with the drawing of a little man inside the sperm head (see chapter 6), was for the most part much more concerned with mathematics and physics than with reproduction—attempting, among other things, to prove that the universe was infinite and that Newton's system was not valid ("nothing contributes more to the advancement of the sciences than this innocent war between authors," noted the circumspect Fontenelle). Hartsoeker's reflections on the possibility of spermism as the answer for generation appear only in the tenth chapter of the book, "On the observations made with approaching glasses and with microscopes." Here, it is true, is a long entry under the title, "That the small bodies of the water, the air, etc., are in turn composed by even smaller others," in which he estimated that "the small bodies that constitute the mercury, the water, the air, and all kinds of solid bodies are each formed by even smaller bodies." In this passage, we glimpse a first theoretical model for the dynamics of preformation, formed without regard for spermism or ovism. But even this entry comes only after a number of other quite distinct sections, including "I will produce the chart of the moon, as soon as I find the way of using this glass for it"; "There are rocks on the moon"; "There are several bands, light and dark, in the planet Jupiter"; "This planet is accompanied by four satellites and how they do their revolution"; "Saturn projects its shadow over the ring, and the ring projects its shadow on Saturn"; "The planet Mars seems to be red, and why"; or "The observations that can be made with the microscope are without number"—all of them, judging by their order of appearance, ranking much higher in the author's priorities than the problem of reproduction.

Furthermore, when he finally discussed the point that later history made so famous, his explanation of how the "little man inside the sperm" comes into being is close to how we would now describe mammalian pregnancy:

Each little animal actually encloses and hides an even smaller being under a tender and delicate skin . . . and when an animal enters the egg that the female, *tempore congressus,* has expelled from her testicles or ovaries . . . this animal becomes united through the softest part of his body to the egg and the egg to the matrix . . . these three bodies, the female, the egg, and the small animal should therefore not be considered any longer but as one single body, the blood passing through circulation from the woman to the egg, from the egg to the small animal, from the small animal to the egg, and from the egg to the female.

Dwelling further on this scenario, Hartsoeker added that the "softest part" of the "small animal" is "the tip of the tail, said tail encloses the umbilical vessels, and perhaps if we could see the small animal through the skin that hides it we would see something as this figure represents."

The figure, needless to say, is the famous drawing of the little person with the huge head curled up in the fetal position, which is now commonplace in any publication dealing with the history of embryology (see chapter 6). Yet Hartsoeker did not say any more than that "perhaps" we could see this creature if we had the appropriate tools; his only added comment stated solely that "it is difficult to maintain the correct proportions," and that maybe the head was larger than what the drawing showed. Centuries of studies have made an issue of this drawing. Yet, when Hartsoeker first presented it to the world, he was much more preoccupied with a series of experiments to test the resilience of the sperm, carefully submitting it to cold, heat, alcohol, and aging, noting the comparative results and further reporting that he was unable to find a single one of these animals "in the seed of a man whom I examined after he had met a woman several times in a row."

Hartsoeker's convoluted line of reasoning is a good demonstration that these men were much more obsessed with establishing some sort of fundamental empirical truth than with constructing a theory and then fanatically sticking to it. Throughout his writings, we see him changing his opinions several times, pursuing different models for the interaction between the sperm and the egg, hesitating over how to explain the incredible numbers of animals in the semen, proposing a model for the birth of monsters and then retracting it—and finally, toward the end of his career (in his "Critique of the Letters of Mr. Leeuwenhoek" appended to the *Cours de Physique*), abandoning preformation altogether (see chapter 4). After having agonized over "what can we do with a thing that is enveloped by such a thick darkness," he admitted the impossibility of explaining many phenomena through this model, and left us with no answer for the mystery of reproduction.

GOD'S WASTE

All of the above difficulties, however, paled in comparison with the moral side of the coin. In the context of their Christian environment, the spermists of the seventeenth century had more than one barrier interposed between their ideas and general acceptance. For starters, they were assuming that God had placed all of our lives inside the testicles, the residence of semen with all of its little animals. And Augustine, the powerful early organizer of Christian thought, had made it very clear in his correspondence that God had given men testicles as a reminder of the irrevocability of original sin.[12] Pushing this reasoning to the extreme in his blatant, unorthodox expressions of radical faith, Paracelsus would later suggest, in his brief tract "De Homunculis," published only a century before Leeuwenhoek made his observations public, that God had given men testicles just so that they could castrate themselves in order to become impeccable Christians: "Christ does not want to have virgins whom He has not chosen, because they are unsteady . . . rather He wants to have his own chosen, who remain faithful to Him. But if man wants to hold himself chaste by force, he should castrate himself or have himself castrated."[13] Paracelsus's reasoning was that lust, by itself, had the inevitable effect of generating seed. Thus it was for the convenience of enacting their own self-mutilation that God had blessed men with external genitalia. Such a cursed place might appear to be a distasteful choice for localizing all generations to come.

Moreover, of course, one had to address the very ancient problem of spilling of seed.

Leeuwenhoek was among the first to assume that one animalcule, not many, would be enough to bring about the new human being.[14] But this was bound to be a risky assumption, since it inevitably implied that millions of potential lives were wasted with each copulation. Such waste, it was well known, would certainly not please God, who had delivered a manifesto of sorts against this profligacy right in the book of Genesis (38:4–10), taking Onan's life because "whenever he slept with his brother's wife, he spilled his seed on the ground so as not to raise up issue for his brother," which "was wicked in the Lord's sight." Whether Onan's real sin had been masturbation or *coitus interruptus* is open to controversy;[15] but either choice implied that God definitely stood against the wastage of whatever precious, or even sacred, element was hidden in the semen. If the learned world was now to assume that this sacred essence constituted millions of minuscule people ready to initiate existence, was it not perturbing to accept in the same breath that the vast majority of these creatures were meant to die before even being born? Moreover, many a theologian had seen in Paul's first letter to the Corinthians[16] a warning against masturbation as a practice that would close the gates of Heaven to those who indulged in such secret vices. Thus both the Old and the New Testament seemed to agree on God's profound distaste for any hint of sperm waste. By the seventeenth century these brief passages from Scripture had echoed in Western morals in numerous invasive and influential ways, so the issue was certainly not to be treated lightheartedly.

These imprecations crystallized mainly in the criticism and demonization of masturbation, the most obvious of all ways to waste lives in the sperm. Writing during the sixteenth century, the Franciscan monk Benedicti had clearly characterized and condemned the practice in his book *Somme des péchés:* "Those who seek voluntary pollution outside marriage . . . sin against natural order . . . The voluntary pollution, willingly obtained, either by cogitation . . . by conversation with women or men, by the reading of shameless books, or by any other possible means, is a mortal sin."[17]

In issuing this statement, Benedicti simply followed an old tradition inside the Church. All the Medieval prescriptions of penitence for sin contain specific penances for masturbation, of variable severity but constantly present.[18] In the thirteenth century, Thomas de Cantipré relates, in his *Bonum universale de apibus,* two cases in which God punishes the guilty with death. One of them dies crying "The vengeance of God on me!" The other one feels an irresistible desire to stroke his virile member and suddenly senses a snake crawling in his hand. Cantipré insists repeatedly on the "abomination of this sin." Two centuries later, Gerson wondered how to extract from young sinners the confession of "the most abominable of this detestable sin which is called *mollities,*" and suggested a method of interrogation that should start with a seemingly innocent question such as "friend, do you recall when in your childhood, at the age of ten or twelve years, your member first came into erection?"[19] Both men were following a line of thought initiated by Thomas Aquinas's statement that "provoking pollution without any carnal union to obtain sensuous pleasure is a sin against nature," a concept firmly endorsed afterward by Saint Antonin.[20]

At the close of the sixteenth century, Benedicti's work stood as a reconfirmation of these traditional views, organized for the first time in a systematic way. Here, three main aspects frame the status of sperm waste. First, masturbation is unarguably a mortal sin—with its seriousness increasing along a sliding scale encompassing the different kinds of thoughts that masturbators summon up during the act, since thinking of a married woman is to commit adultery, thinking of a virgin is rape, thinking of a parent is incest, and, as later Monseigneur Bouvier would add, thinking of the Virgin Mary is a "horrendous sacrilege."[21] Second, masturbation should be condemned not only for being an act against nature, but also for having been described as the object of God's wrath in Scripture. Finally, even to this rule some exceptions apply, the most complex ones concerning how to classify "nocturnal pollutions"—which, depending on whether they are caused by lascivious dreams, states of drunken stupor, agitations due to the ingestion of certain spicy foods, or numerous other causes, have different degrees of severity. In analyzing this issue Benedicti paid tribute once more to Aquinas, fully agreeing with his predecessor's view that many nocturnal pollutions are caused by the Devil, who inspires thoughts in the sleeping victim that lead to the shameful outcome. For both writers, the goal of Satan was unequivocal: with this malicious little trick, he makes the victim feel soiled upon awakening, and unable to receive Communion during the following day.[22]

Leeuwenhoek, however, had his own answer ready, even before anyone confronted him with the difficulty of proposing an explanatory model for reproduction that made the God of all these theologians appear remarkably wasteful. "People will ask me," he wrote to the Royal Society in January of 1683,

> why are there so many animalcules in one drop, if one little animal in human sperm suffices for the procreation of a human being. My answer is as follows. Do not we see that an apple-tree, though it can become a hundred years old and even older, will produce thousands of blossoms annually, and that each blossom may become an apple, and that each apple will contain 6 or 8 seeds, and that each seed may become a tree? Suppose now that under such a tree grass and weeds will grow wild and that all the apples this tree produces will fall into the grass. Now will all the seeds the tree produces grow into a tree? You will say "no," for the shade of the tree, and even more the weeds and grass will smother the seeds, if any should sprout, and deprive them of food. But let us put one good seed in the earth and keep it carefully to prevent its being smothered by weeds, and it will grow into a tree.

Voila! Now all that was left for the naturalist to do was to strengthen the analogy between the tyranny of shade and wild weeds and the meager support offered to the animalcules by the body of the female. It made perfect sense, especially when "seeing that the womb or tuba fallopiana is as it were an entire world in comparison with an animalcule in the sperm, and that, besides this, there are very few little veins in them or recesses fit to receive and feed an animalcule."

Judging from Fontenelle's "Éloge," Hartsoeker had also come up with a solid defense against the accusations of God's wastefulness. "He imagined," wrote the secretary of the Academie des Sciences,

> that [the animalcules] must be diffused in the air, where they flew around, that all the visible animals took them all in a confused manner, either through breathing or through feeding, that from these the ones corresponding to the correct species sought refuge in the male parts proper to enclose them, or to feed them, and that they passed afterward to the females, where they could find eggs, which they would seize in order to develop. According to this idea, what a prodigious num-

ber for primitive animals of all species? All that breathes, all that feeds, breathes
only them, eats only them. It appears, nevertheless, that eventually the difference
among them would necessarily start to diminish, so that all species would not be
equally fertile. Maybe this difficulty did help to make Mr. Leibnitz believe that
the primitive animals never died, and that after having shed their rough envelope,
that sort of mask, they would remain alive in their original form and start again to
fly in the air, until a favorable accident could make them become human again.

This citation of Leibniz probably refers to the German mathematician's
statement, in "Principes de la nature et de la grace," that "[the souls . . . and
the animals] are ungenerated and imperishable: they are only developed, enveloped, reclothed, stripped, transformed."[23] Leibniz (see below) held that
souls had been created when God created the world, but that they were attached to material bodies undergoing continuous transformations—which, in
short, implied that generation and death were only appearances. If a sperm
(Leibniz's vehicle of choice for such souls) failed to develop into an animal,
then, it did not have to die: it just continued to live as a simpler organism. The
soul, and the animal to which it was connected, never ceased to be alive, although the visible forms of their existence might change. This concept derived
directly from the idea of continuity embodied by the great chain of being,
according to which all creation, both organized and apparently unorganized,
was connected by imperceptible degrees, everywhere alive and nowhere inter-
rupted by gaps (see Bonnet's account of the same concept in chapter 1). All of
the above made for a beautiful rationale, but its depth could be hard for the
layperson to grasp. This is where Fontenelle made a difference.

In the last sentences from the "Éloge" quoted above, the flowery prose
of Fontenelle contributes one added possibility that renders the "no-waste
model" more credible. It combines Hartsoeker's and Leibniz's philosophical
efforts in a synthesis that suggests a continuous recycling of the animals, which
are metabolized and re-expelled into the air in degrees of species-specificity
less and less prominent. Through this simple physiological touch, the original
number of such creatures required to make generation possible becomes less
astonishing, hence more believable. We are left with a much more sober numerical estimate, and it is smartly combined with a brand new version of an
old, familiar theory—that ever-seductive chain of being. In fact, the assumption that the animals not used in one generation would not die, but would
rather resume their airbound cycle, is immediately reminiscent of the idea of
"seeds of life carried by the West wind" endorsed by the proponents of spontaneous generation, from Aristotle to William Harvey. With its appealing
combination of factors, this idea, known as *panspermism,* gained considerable

popularity among the spermists of the seventeenth century. Leeuwenhoek himself toyed at times in his letters with the idea of seeds disseminated through the air. Most interestingly, the fact that, as early as 1698, texts mocking pan-spermism had started to circulate clearly indicates how widely the concept had been endorsed.[24]

Much in accordance with his errant, constantly dissatisfied nature, Hart-soeker initiated this view and then eventually abandoned it. In the commentar-ies appended to the *Cours de Physique,* he stated that his detractors were right in condemning "such a bizarre thought," which he himself now repudiated. Instead, he proposed the existence of some sort of "Intelligence" residing in males that "fabricates, as in a laboratory equipped for that process, the sper-matic animals in the testicles." This "Intelligence," he added, could just as well be "a single soul that makes everything, maybe a small portion of the Soul of the Universe . . . disseminated throughout the whole body . . . that needs noth-ing more, to move a muscle, than, so to speak, to open a faucet so that the animal spirits necessary for that movement may move in."

But, even with this grandiose scheme involving the Soul of the Universe in the fabrication of the animals, Hartsoeker still had to address "the scruples raised by their infinite number." Ironically, his explanation came very close to that of his archrival Leeuwenhoek: "Since Nature is prodigal in everything she does, don't we often see a thousand flowers for one single fruit, a thousand eggs for a single fish, and an infinity of animals born by dawn and dead by dusk? Besides, this prodigious number of spermatic animals serves Nature to attain her own goal, which is to fertilize the egg, which without it would occur but very rarely, indeed it would be a miracle." But, exalted spirits, beware: God is not operating "these continuous marvels" to fill men with admiration for "His majesty, His power, and His immensity"—"as if God had any need to follow the example of men, who do many things only for the sake of osten-tation and vain glory!" Such a thought, concluded the anxious microscop-ist, is not only absurd: "it is impious."

Writing after these two contenders, with no doubts in his mind, but with a pedagogical task to fulfill, Andry offered, in a short paragraph, his own re-interpretation of Leeuwenhoek's position, proposing a scenario that would not put any real lives at risk: the sperm, he argued, holds nothing but a *potential* organism, which becomes expressed only when entering the egg. "We must not conclude from this System, that the Spermatic humor of Dogs includes little Dogs, that of the Cock contains little Pullets, or that of Man little Chil-dren . . . We ought not to say that the Spermatic Worms that are still in the Body of Man, are little Children, though they are to become such as soon as they enter the Matrix, or rather the Egg contained in the Matrix." If this sounds a bit like flirtation with the epigenetic idea of matter taking up a form that was there potentially but needed a stimulus to become expressed, it only proves that, for all the heat of the ongoing debate on reproduction, preformation and epigenesis were never really as far apart as we now tend to view them.

A CHANGE OF PACE

If, in the seventeenth century, the spermists got away with vague theoretical models for excluding waste of seed from the reproductive picture, they were bound to encounter greater difficulties in the following century. It should be noted that, all these charges of mortal sin notwithstanding, masturbation was still taken quite lightly during the opening decades of the Scientific Revolution. Although punishments were prescribed for those caught in the act, most of the writers who analyzed the penal aspects of the practice agreed that court sentences would necessarily be rarely enforced, since the sin rarely occurred in public. The Flemish jurist Josse Damhouder implicitly accepted this fact when stating that the crime was "horrible and execrable" and would be punished by God, but, since it generally took place in a clandestine setting, human justice would not stand much chance of directly dealing with it.[25] Moreover, the medical texts from this period never question the physical occurrence of the act—and some go as far as prescribing it as a remedy. Judging from passages from the Spanish Jesuit Toledo[26] and the Portuguese Jesuit Rebellus,[27] contemporary doctors considered masturbation a perfectly acceptable way of expelling from the body degraded sperm that could become detrimental to the organism. "We must condemn," wrote Rebellus, "the assertion that it could be permitted to use the hand and the repeated strokes to expel a corrupted and poisonous seed, with the aim of improving the health." Following this argument to its ultimate logic, even if a cure could be obtained through such means, it was better to die than to masturbate. The urgent tone of these texts seems to indicate clearly that many people were being prescribed masturbation as a treatment.

This relative laxity must have been connected to the confusion surrounding the true nature of the semen, as nicely illustrated by the *Theologia moralis fundamentalis* of the Spanish monk Juan Caramuel.[28] In his pages we realize that, as Leeuwenhoek and Hartsoeker were putting semen under their lenses and discovering little animals, most savants were still debating whether the semen was sweat, saliva, milk, or blood. A reference to milk is present in a brief passage of the pseudo-Aristotelian text *The Complete Master-Piece,* wondering "by what secret Power of Nature it was that coagulated Milk (as a Divine author calls it) came to be transubstantiated into a human Body." But blood was certainly the most favored option, according to the Galenic idea of the "ebullition of the blood" during coitus. Even Descartes reasoned along these lines, postulating that the embryo was formed by the fermentation in the womb of seminal particles, produced by the blood of both the male and the female (see Prologue). This idea is beautifully expressed in a passage from *The secrets of Nature Revealed, or the Mystery of Procreation and Copulation considered and Explained,* a booklet attributed to the astronomer, alchemist, and sometime magician and court entertainer Michael Scott, and supposedly first written in the thirteenth century "to the use of the Emperor of Germany." This brochure was printed and widely circulated in the 1730s, when it was for sale in London at the price of one shilling. Semen, stated the author, was "the blood of man or woman, which is, by agitation, converted into a white, clammy substance . . . more light and more active than the part remaining doth thereof." Further connecting semen to blood, pseudo-Scott recommends that those who have too much desire for sex should "forbear phlebotomy and live upon the most simple diet."

Caramuel drew directly from these common beliefs. Everyone, he noted, agrees that relieving a body of some unhealthy excess of any of these fluids is a sound medical practice. Even if you argue that semen is different because it has to do with procreation, that argument does not stand if you recall that woman's milk is also made for survival of the progeny, and the blood for proper feeding of the whole body. Thus, he concluded, voluntary pollution is not bad because it goes against Nature, but solely because it has been forbidden by God. Had God not touched this issue, the spilling of seed would never even have come into question.

Caramuel was asking for trouble—and he got it. In March of 1679, his propositions were condemned by Pope Innocent XI. But it would be misleading to read this anecdote as a case of one man's lunacy before the strict censorship of the Church. Caramuel was writing in the days of a still mainly Galenic medicine, and Galen had stressed the dangers of semen retention, especially if the fluid was abnormally abundant. He even praised Diogenes the Cynic for his reported practice of masturbating in public to rid himself of the discomfort of having his testicles too full.[29] Galen's legacy had produced all kinds of results, including the recommendation issued by the Italian anatomist Faloppio, to whom we owe the discovery of the fallopian tubes connecting the ovaries to the uterus in mammals, to stimulate the penises of young boys through repeated and vigorous strokes, in order to render them more fit for their future contributions to the propagation of the human species.[30]

This was in the sixteenth century. At the same time, Vettori was advising an efficient remedy for curing men who suffered from nocturnal pollutions for four months in a row: attach the end of a string to the penis, and the other end to the neck. If they happened to have an erection during sleep, the pain they would experience would immediately awaken them—and cure them.[31] But, even in the presence of such cautionary measures, a century later the notion of sin linked to masturbation was still uncertain enough for Samuel Pepys to openly recount in his diary that he had done "the thing" on Christmas Eve in the Queen's Chapel of Saint James, during the midnight mass, just because he happened to be watching a beautiful maiden—adding nothing in his defense but the laconic "may God forgive me for having done it in the church."[32]

However, when, in 1742, the Dutch insect anatomist Pierre Lyonet (1706–1789) confronted Leeuwenhoek's and Andry's belief in the role played by the "spermatic animalcules" in reproduction,[33] he made it quite clear that the massive destruction of lives resulting from such a theory was against the laws of Nature, where there is no room for waste.[34] The spermist's line of defense, involving analogies with the seeds of plants, was not acceptable to him, since, in his opinion, seeds were never wasted, but rather used for other purposes, such as the nourishment of animals. In the case of sperm cells, on the other hand, if each one was actually a future person, then that person had to be endowed with a soul—and "can we conceive that, in order to form our body, the least noble part of our being, God was willing to create so many hundred of millions of rational souls only to destroy them?" He went on to add that, had Andry considered this fact "and the consequences which flow from it . . . the pen would have fallen from his hand and that part of his work would never have been written."[35]

Lyonet was not alone in expressing his horror at God's wastefulness. It may be more than a coincidence that he was writing in the middle of the eighteenth century, when the former relatively lighthearted attitude toward the spilling of seed experienced a radical change of pace.

MAKE IT VAGUE

Under close scrutiny, the writings of the spermists of the eighteenth century show a striking difference from those of their predecessors: their approach becomes much more guarded, often repeatedly apologetic. One of the clearest indications of this newly sensed caution is a tendency to narrow the focus of the text to specific aspects of reproduction, rather than discoursing about spermism as a whole. The work of Jean Astruc (1684–1766), physician to King Auguste II of France, is a good example of this trend.

In his 1930 work on preformation, F. J. Cole presents Astruc as the man who "enjoys the doubtful honor of being the last of the animalculists."[36] This, however, does not do justice to Astruc's work. Astruc was a pioneer gynecologist when many of the matters related to such a specialty were still social taboos. His six-volume book *A Treatise of the Venereal Disease*s was a groundbreaking investigation into the nature and mode of propagation of syphilis, the mysterious sexual affliction supposedly brought to Europe from the West Indies. Among his other important contributions to understanding the problem, here he first held that "this poison shall sometimes lie lurking in the blood for several years without any manifest prejudice, and consequently without any apparent symptom; and yet it shall be preserved there so entire, as to be capable of exerting itself when occasion offers, and frequently with so much violence, as to produce the most fatal effects."

William Barrowby, who translated this work into English in 1737, said in his preface that he undertook this task due to "the great demand which has been made from Paris for this treatise," and promptly admitted that "interest and prejudice will necessarily raise enemies to this piece," especially since the subject has been tackled by so many "empirical practitioners" and all kinds of charlatans. In his own prologue, Astruc argued that "there appears to be a wanting for a more accurate description of Venereal symptoms, a more just knowledge of Pharmacy and Physicks, a fuller exposition of causes, and . . . more certain diagnostics and prognosticks," all of which he had now set about to provide. He stiffly accused all those who "have some extravagant remedy to recommend, which they set off with such extravagant commendations, as sufficiently expose their fraudulent designs, and clearly show . . . that they wrote their dissertations with no other view, than to impose upon mankind, and to gain to themselves some private advantage"; and lamented that in most recent treatises "you will find new-devised notions, called hypotheses, in which the Physicians . . . evidently indulged themselves too much, and as they refused to submit to authority, so they did not enough consult experience and reason." Yet, in the rare modern accounts of Astruc's ideas on reproduction, the man who took such pains to explain a difficult medical question still wrapped in so much obscurantism is depicted exactly like one of those whose fallacies he sought to expose.

Here are some of Astruc's comments "to explain how this mystery of Nature is preformed," from his *Treatise on All the Diseases Incident to Women*, first published in 1765:[37]

> Conception is the Formation of an Embryo in the mother's Womb . . . A complete Embryo is defined as a Totum composed of Parts of a human Body, organized and united to an organized Placenta, both which grow inside the Mother's Womb . . . In order to conceive, the Union of the two Sexes, or the Male or Female, is necessary.

So much for rabid spermist rhetoric. As for defending the spermist model against accusations of mass murder, Astruc's line of reasoning was nothing short of astute:

Why should an infinite number of animals perish, and but only or some few be preserved? For this seems contrary to the Simplicity and Uniformity of Nature: But do we know all the secret recesses of Nature, and all the views of its Author? Moreover, why do so many thousand Ova remain useless in the Ovaria, particularly of those Women who never conceive? . . . Nor are these Animals in the Semen of Man united to so many Souls, nor are they the Rudiments of Man, only in as much as they are joined to the Ova of Women. Nor do we know in what Time afterwards the Soul is united to them; but that we conjecture in general, that this happens when Generation is perfect.

Astruc was using a good deal of wit to rescue spermism—but, mind you, he was not even all that sure that he himself was a spermist.

What is the Origin of these animals? This I answer by demanding what is the Origin of the Vesicles in Woman's Ovaria? Some, I know, say that the Rudiments of all human kind were contained in the Ovaria of Eve; and those who maintain the contrary sentiment, hold, that they were all contained in the Semen of Adam: But I do not believe either the one or the other.

His only reason for siding with the spermists—and this was Astruc's ultimate insight—was that this system entailed the better explanation for the birth of offspring with mixed parental traits.

As the Embryo originally proceeds from the Semen of his Father, it still retains its Lineaments; but as the small Animal afterwards introduces itself into the Pore or Passage of the Ovum, it is therein shaped, and, as it were, moulded; whence it is more or less stamped to the Likeness of its Mother, whose Lineaments are impressed by the Author of Nature on that Hole or Passage through which the Animal enters the Ovum; so that if this Creature exactly fills the Hole, and is obliged to force itself, the Features of the Mother will be the more impressed on it; but if it be so little, as to enter without any Resistance or Compression, it will always retain its Father's Features.[38]

If we sense in Astruc the skills of a shrewd commentator, we also sense a clear willingness to shift the focus away from the global model and closer to a single detail—in this case, the imprinting of both parents' phenotypes in the final shape of the fetus.

Other approaches bear testimony to the same trend. J. N. Lieberkuhn, for example, invested his intellectual energy in investigating minute details of the spermatozoon's anatomy. In 1751, he announced that the tail of the spermatic animal was the backbone of the future fetus. He then tried to provide added credibility for the sperm's role in reproduction by establishing that the type of spermatozoon fits with the type of adult into which it develops. The sperm of the snail, which has no backbone, he wrote, has a long sluglike structure and moves accordingly. Likewise, the tortoise has no movable backbone, since it is fused with the shell—but the turtle has a mobile neck, and hence its spermatic animal has the tail in front and not behind. Nevertheless, the animalcule does not move backward, but is pulled forward by hooking movements of the tail.[39]

PHILOSOPHICAL APPROACHES

Another trend of the later spermist literature was to approach the subject in philosophical, rather than physiological, terms. Moreover, these philosophical approaches often made theoretical room for accommodating mixed possibilities. The work of the eminent Dutch physician Hermann Boerhaave (1668–1738), professor of clinical medicine, botany, and chemistry at the University of Leiden and twice appointed its Vice-Chancellor, is an exemplary case study. In his writings, Boerhaave tended at times toward moderate versions of epigenesis, and ascribed an important role to the female in his system. According to Cole, Boerhaave, in his *Academical Letters on the Theory of Physic,* declared himself inclined to believe that the seminal animalcules did contain the rudiments of the future adult body, including all of its organs.

> Received into a fit place or nidus, and there supplied with most subtle nourishment, forwarded by a friendly warmth and motion, they grow up and unfold themselves, so as to display the latent parts of which they are composed even to the naked eye . . . The father communicates the embryo and first rudiments of life, the structure of the body being already determined and assigned in the animalcules of the male semen in all creatures, which yet receives some alteration according to the different species of animal or female from whence it is nourished . . . The mother . . . retains and nourishes [the fetus], affording therefore an habitation to the foetus, and nourishment by the liquor in which it swims.

This starts to sound like an attempt to reach some sort of compromise, wherein the sperm would not totally steal the spotlight. But the author took the path to compromise one step further, candidly admitting that he had no idea about the origin of these animals, or how they could live in places inaccessible via the external air.

Boerhaave's mild endorsement of spermism might have been inspired solely by the aesthetic appeal of the theory, for he was clearly fascinated by the beauty and harmony expressed by the imperturbable continuity of all things, animate and inanimate alike. In his "Discourse on the Achievement of Certainty in Physics," first published in Latin in 1717,[40] marveling at how "Na-

Fig. 22. Hermann Boerhaave.

ture, the artificer, sure of her aim, occupies herself with the tiniest things," he exclaims:

> And how true to themselves these seeds of all things remain! Plant, animal and mineral have this very same day the identical characteristics and form which they have had throughout, from times immemorial . . . They all carry hidden in their veins prolific seeds from which similar bodies may then be brought forth; these again, in fruitful breeding, are capable of supplying and replacing one generation of progeny after the other, by means of procreation . . . so much that in the one seed of the most simple thing are the first beginnings hidden and concealed of all that the sequence of evolving ages will propagate from the offspring of offspring . . . Sound Physical science stands amazed at these things, whenever they are pondered, yet grants that they are absolutely true.

And, to complement physical science, he quoted "the Prince of Sages," Moses, who "has in a few words explained and set forth the order of nature far more beautifully than the other philosophers in their enormous volumes." He vehemently addressed his audience: "Listen, I beg of you, listen to wisdom's summary on the subject of physics! Accept the most energetic utterance, the effective power of which has made the world fruitful! *Let the earth bring forth the green herb, which shall yield its own seed, after his kind.*"

Stunned by the infinitely small world coming into view during his time, and paying homage to the "anatomists" who were starting to uncover it (note that his list includes both Malpighi and Leeuwenhoek), Boerhaave delivered in the same address an interesting account of how nature had planned the unfolding of generations down to the minutest details of hair growth. "Bear in mind—for this is how it happens—that through the results of their artful effort in the embryo, the foetus, and the new-born child, these hiding places have already been so far finished that they need not remain idle but may form and perfect the hair that is there hidden and safely guarded, until the length of days, having accomplished the circuit of time, makes it emerge in the very moment when it is required." Absorbed in these infinite wonders, to which he often referred as "delights" or "pleasures," Boerhaave stated very firmly his admiration for Descartes—no preformationist, to be sure, but nonetheless a man "who certainly excelled all others."

Entranced by the perfection of systems and the clarity of philosophical approaches, Boerhaave never seemed too anxious to defend the cause of the male line, and his pursuit of a global celestial scheme for the order of all existence often verged on epigenetic views, as is very clear from this passage in *An Essay on the Virtue and Efficient Cause of Magnetical Cures*:

> We may say that the Seed furnished the Matter and Heaven putteth on the Form. Also a Man doth beget a Man, the Man by virtue of the Seed procureth and brings together the Apparel . . . The Spirit and the Warmth which is the Seed prepareth the Matter of the Seed: first it caused a Mingling and Tempering together consonant to each final Part that is to be made thereof; and then it giveth the Structure, Conformity and Posture; lastly it diffuseth itself through all the Parts and abideth therein as a Ruler . . . which giveth a Species Form and Shape to this single prepar'd Matter, which is in Heaven, for the Species of things are cast in Heaven, and not by the Matter itself.

As above, so below, as Paracelsus would say, approvingly.

Moreover, it is to Boerhaave's efforts that we owe the preservation of Jan Swammerdam's *Biblia mundi*: after Swammerdam's death, Boerhaave bought, edited, and published the manuscript, completing it with an extensive preface in which he repeatedly praised the work of the great Dutch naturalist. Translated into English as *The Book of Nature, or, the Natural History of Insects,* this was the work that provided ovism with most of its main foundations. And it was a gentle spermist who made it possible.

Among these philosophical endorsements of spermism by prominent scholars of the time, one early case is particularly striking. It was most likely sheer philosophical convenience, and certainly not enthusiasm springing from personal observations made with the microscope, that led the German mathematician and man of affairs Gottfried Wilhelm von Leibniz to align himself with the spermists' ranks. Although he made his choice and stood by it, as is clear in his letters written in 1714 and 1715, shortly before his death,[41] he did not really consider it particularly important to make a final decision in favor of either ovism or spermism, his reluctant preference for the latter being due mainly to his great respect for Leeuwenhoek. It was the *basic concept* of preformation, rather than the *specific gamete* to which the concept should be allocated, that occupied most of his writings on this subject. His dynamic philosophy required a model for generation that would encompass his prominent preoccupations with Force and Inertia acting on Matter. And, as is clear in his *Monadology and Other Philosophical Essays,* preformation was a perfect vehicle for Leibniz's cause.

Defending his concept that "there is no soul of the universe," Leibniz added, in the chapter titled "What is Nature? Reflections on the Force Inherent in Created Things and on their Actions":

> I grant that the admirable events which occur every day and of which we are
> wont to say, with reason, that the work of nature is the work of an intelligence,
> are not to be ascribed to some created intelligences . . . The whole of *nature . . .* is
> *a perfect work of God's making,* and this so much so that every natural machine—
> this is the true but rarely observed *difference between art and nature*—consists in
> turn of an infinity of organs, therefore evincing the infinite wisdom and power of
> its creator and ruler . . . I am satisfied with thinking that the machine of the uni-
> verse is constructed with so much wisdom that all those admirable events are pro-
> duced by its very operation and that the living beings in particular develop, I
> would hold, from a certain preformation.

Preformation offered the advantage of being easily adaptable to Leibniz's idea of Monadology. "This same substantial principle which is called *soul* in living beings and *substantial form* in other corporeal objects," forming a "truly unified substance" when joined to matter, becomes "a unity by itself," which the author called a *Monad,* and upon which he conferred extreme importance in the organization of life: "Eliminate these genuine and real units, and you will have only beings which are but aggregates and, consequently, bodies will not be real things at all." And, since "just as in matter natural *inertia* opposes itself to *motion,* so also in the body itself and even in every substance inheres a *natural constancy* which opposes itself to *change,*" it follows that "the *interaction of substances* or monads has its cause not in an influx, but in a harmony created by divine preformation." Through this reasoning, Leibniz hoped "to prevent the mechanical explanations of natural events from being abusively used to the prejudice of piety, as though they implied that matter can subsist by itself and that mechanism does not require any intelligence or spiritual substance." The idea of carefully arranged lines of preformed generations must have been extremely appealing to the philosopher who clearly expressed his belief "that God has been determined by reasons of wisdom and order to give to nature those laws which we observe in it."

Some years later, about 1728, confronting the combined powers of Haller and Bonnet on the preformationist front, a bold newcomer called Caspar Friedrich Wolff would draw heavily on Leibniz's dynamic philosophy to support the existence of the *vis essentialis,* the inherent force in matter that completed his model for the renewal of epigenesis.[42] But that, within the immense realm through which theories travel from one concept to another depending on the convictions of their carrier, is yet another, equally complex, story.

THE GREAT FEAR

Some authors have associated the new trends in spermism during the first half of the eighteenth century with the failure of the microscope to produce clear images of the miniature creatures presumably encased in the sperm head.[43] But, together with disappointment in the microscopic observations, and the resulting urge to move in new directions, an added factor must be taken into consideration. A great fear had spread over Europe, and its premises required the spermists to proceed with extreme caution.

In 1715, an anonymous pamphlet was printed in London under the title *Onania; or The heinous sin of self-pollution, and all its frightful consequences, in both sexes, considered: With spiritual and physical advice to those, who have already injur'd themselves by this abominable practice.* This booklet was an astonishing success, having reached its 22d edition by 1737 with thousands of copies sold. It also triggered a number of replies and denunciations as a piece of charlatanism—but even its critics agreed with the mysterious author that masturbation was, indeed, a "heinous vice." The appeal of the anonymous book must have relied on a trinity, first developed at this time and destined to become a dominant dictum of the nineteenth century, of sin, vice, and self-destruction reunited in a novel fashion under the newly coined term "onanism."[44] "For fornication, and even for adultery," the author stated, "one may always claim that human fragility and natural inclinations were behind it; but masturbation, that, it is not only a sin against nature but a sin that perverts and destroys nature, and those who are guilty of it work for the destruction of their own species, and strike a blow, in a certain way, against the very Creation."

The author presented a long list of diseases resulting from such habits: ulcers, convulsions, epilepsy, consumption, could all be explained at this light. "It definitely retards development, both for boys and for girls," he added. "Few of those who, during their youth, have sinned with excess and during a certain time reach the vigor and the force that they would have achieved without this vice." Impotence was also one of the logical consequences of masturbation. Even if the sinner manages to procreate, *Onania* assures us that the resulting children will be feeble and prone to all kinds of illnesses, and will never reach adulthood. Among women, this habit was likely to lead to sterility or repeated miscarriages.

The idea itself was not exactly new. The author of *The Secrets of Nature Revealed* had already pointed out that any person indulging in excessive sexual activities would look pale and become "slow in action, heavy in gait, dull in conversation, stupid in apprehension, unadvised in labor and apt to believe everything," and that the bodies of these persons would be "suddenly infected by any corruption of the air," causing them to have their lives "shortened by two-thirds"—and that if the persons in question were still virgins, then all these ailments were most certainly being caused by "that beastly practice" of self-pollution. What was new in *Onania* was the clear and systematic way in which all these pseudoscientific data were now presented to the general public.

According to Stengers and van Neck in *L'histoire d'une grande peur,* the publication of *Onania* triggered a formidable chain reaction because it was easy for readers to finally discover, in its pages, all their troubles explained. "With the brochure in hand," the authors reason, "people discover the origin of the maladies they suffer and whose nature had been until then elusive." Now they knew that it was "their unfortunate vice" that was responsible for all their agonies. And herein lies the cause of the dramatic change in attitudes toward masturbation during the transition from the seventeenth to the eighteenth century. Fear of sin is one thing, but straightforward fear of illness and death is quite another: less grand, certainly, but much more effective.

If *Onania* was nothing but the clever product of a charlatan,[45] the same cannot be said about the work of the Swiss doctor Samuel-Auguste-André-David Tissot, who, having carefully read all the (real or fake) "letters from readers" increasingly appended to each new edition of *Onania,* decided to dedicate a lifetime of work to the systematic medical characterization of all the diseases that accompanied self-pollution. His undertaking of this task is a paradigmatic mirror of the movement toward the secularization and medicalization of morality that became one of the trends of the Enlightenment. In Vernon Rosario's words,[46] Tissot's publications made it very clear that "not only was onanism the defilement of the temple of the soul, but also of the temple of Nature—the body."

The Swiss physician's work, and particularly his book *L'Onanisme,* had such a pervasive influence that, in the nineteenth century, the entry on "masturbation" in Pierre Larousse's *Grand Dictionnaire Universel* said dryly "it is not to us to describe an act unfortunately as well known as it is shameful," and then proceeded directly to an entire page on the "funereal results" of that act. The same dark tone was present in the 1819 edition of the *Dictionary of Medical Sciences,* which went one step further to remark, under "masturbation": "the diseases that are the product of the excesses of onanism become more frequent in proportion to the higher degree of civilization achieved by modern societies. This opinion, which is generally adopted by medical observers, seems to rest on numerous and well established facts." At this point, and all throughout the nineteenth century, wax museums had didactic cabinets specifically dedicated to the matter, portraying young men and women before and after they had taken to these hidden pleasures, radiant and youthful before, decadent and prematurely aged after.[47] This was also the time when, in 1853, King Leopold I of Belgium wrote to Queen Victoria explaining why he was in such a hurry to marry his elder son, then only eighteen years old, since "young men often fall into a habit destructive of health, mind, spirit, in short, everything," and asking her, should she not clearly grasp the meaning of this worry, to consult "our dear Albert."[48]

Against such a background, and being aware of its repercussions throughout the nineteenth century, it should now be easier for us to detect—and understand—the apologetic tone that Astruc chose to defend his spermist views in an unfriendly environment. "It is said with truth that death levels all distinctions," he wrote in his *Treatise on All the Diseases Incident to Women,* "but one may say with greater truth that our origin indeed humiliates us, for we are all of us only vile insects which among a thousand million other insects have been more active and more fortunate, and, having penetrated the puny vesicles of the ovary, have developed and been found worthy to receive the Divine endowment of a spiritual soul which elevates us to the dignity of man." The subtext is clear: although it is undeniable that millions of "vermiculi" are pitilessly slaughtered with each ejaculation, this should not be regarded as a problematic massacre of innocents, since those that die are not yet miniatures of men, but rather vile insects, just like those myriads of others abounding everywhere, and everywhere dying, without those deaths becoming a matter of theological concern. Later passages of the same book suggest that the author nevertheless ascribed a significant degree of complexity to the spermatic animals. At one point, he endowed them with a circulatory system different from that of the adult in that it did not involve a heart. In another section, he claimed that even the fetus proper was not yet a miniature preformed human being in the first days after conception, but rather a little formless mucus attached to a thread. These apparent contradictions seem to indicate that Astruc himself was still struggling to come up with a system compatible with the knowledge and beliefs of his time—and some of those beliefs made him defend, even in spite of his own contrary observations, a careful system that did not imply the loss of a single human life. "The good man is falling low," Haller commented to Tissot upon receiving the first edition of *Diseases Incident to Women.*

A DANGEROUS SUPPLEMENT

Tissot was not alone in launching this "great fear" that swept across Western Europe. His work coincided in time with the publication of Jean-Jacques Rousseau's *Les confessions,* with all its commentary on how self-pollution, "that dangerous supplement that deceives nature" and may "sometimes even surpass reality," was destroying civilization. Here, in 1782, the author claimed to have lost his mental virginity in Italy, when he was a catechumen in a monastery. Upon being sexually molested before the altar and watching his molester masturbating, he saw "shooting towards the chimney and falling upon the ground I don't know what sticky, white stuff that turned my stomach."[49]

"The vice which shame and timidity find so convenient," wrote this influential contributor to the tell-all book tradition,

> has an even greater attraction for lively imaginations: it allows them to dispose, so to speak, of the whole [female] sex at their will . . . Seduced by this fatal advantage, I worked at destroying the good constitution that nature had established in me . . . Add to this disposition my situation at the time, rooming with a beautiful woman whose image I caressed in the depths of my heart, seeing her constantly through the day; in the evenings surrounded by objects that reminded me of her, sleeping in a bed where I knew she had slept. How many stimuli! The reader who imagines all this must already think me half-dead.

Without first knowing each other, these two men were crucial in associating the spilling of seed with the decline of humanity. Rousseau's *Émile,* which came to print in 1762, contained passages such as "if [your student] once knows the dangerous supplement, he is lost. Thenceforth his body and heart will be enervated, he will carry to the grave the sad effects of that habit, the most mortal one to which a young man can be subjected." Upon reading this, an enthusiastic Tissot sent the French writer some of his books and begged for a visit, pleading that *"L'Onanisme* will prove to you that finally there is a physician who recognizes all the danger of that odious practice that you have so vigorously attacked, and has the courage to make that danger public knowledge." The two men were somehow united by the faith entailed by the candor of their statements.[50] Their meeting, their exchange of letters, the renewed strength springing from their conviction of not being alone in an arduous fight, eventually made them even more efficient at spreading the terrorizing news through their world. Special attention was dedicated in subsequent anti-onanism literature to the protection of children, since the flow of semen, either male or female, was coincidental with the onset of puberty. In this light, semen was praised as a precious fluid essential to maturation—and, since it was held that during childhood and adolescence the body needed to invest all its energy in self-construction, masturbation appeared as a major threat to the harmonious completion of this process.[51] In a passage of *Les conféssions,* Rousseau explained the dangers of having young people reading certain kinds of "dangerous books . . . whose inconvenience is that one can read them with one hand." He also blamed women for being the main causes of this evil. In this, for certain, he was by no means alone. More than one medical source warned of "abominable women" and nannies who used masturbation to pacify infants, or "more infamous yet, engage in simulacra of coitus with little boys."[52] Tissot, in turn, affirmed that "a whole college, by this manoeuvre [masturbation], sometimes diverts the tediousness of metaphysical scholastic lessons that are delivered by a drowsy old professor."[53]

This explosive mixture of moral and medical implications was by no means solely the concoction of conservative and Christian men. Diderot and d'Alembert's *Encyclopédie,* published between 1751 and 1772, had three entries for "pollution." One was medical, and consisted of diseases causing the involuntary ejaculation of seed. Another was legal, and had to do with the defilement of sacred places. But before these came a connotation that was frankly moral: "the effusion of seed outside the function of marriage." Readers were then referred to a separate article, dedicated to "manustupration," where they could find more information on the afflictions imposed on "those abandoning themselves without restraint to that infamous passion and sacrificing themselves to that false Venus."[54] At least on this particular horror many disparate voices finally seemed to be in accord.

Submitted to Ridicule

Incidentally, Tissot was a very close friend of Albrecht von Haller, the man who, together with Charles Bonnet and Lazzaro Spallanzani, revived preformation during the second half of the Enlightenment and brought it to new, much more sophisticated heights—this time under the banner of ovism alone. "I remember with pleasure the hours I had the satisfaction of spending with you; I am flattered by your friendship, and I offer you all my esteem and my affection," Haller wrote to Tissot in one of their first letters, sent in August 1754.[55]

In their subsequent correspondence, which continued for more than twenty years at a remarkably constant pace, we find the two men complaining about their publishers' policies and occasional shortages of money ("an exact Jew shall be preferable to the most generous man, who will not be exact," wrote Haller in a moment of desperation over the handling of his *Elements de physiologie,* considering one more time a change in publishers); constantly exchanging and recommending books and remedies; agonizing over positions sought or offered; discussing politics and religion; commenting on details and choices of forewords for translations they seem to have done for each other's works; describing their own occasional medical problems, including Haller's repeated cases of gout—and, obviously, indulging in all kinds of intrigues about their work and the fate of their peers. Here is a flavorful sampling of such sound bites: "I regret this impossible hate of Voltaire, who shall persecute Rousseau in his grave, and who makes our theologians satellites of his passion"; "We should try to prevent [Caspar Wolff] from getting that position"; "Please let me know what you thought of Harvey's book as soon as you have read it"; "In Anatomy Rondelet, in practice de la Riviere, in botanic Magnol the father, this is all that [Astruc] seems to find illustrious!"; "How verbose is John Cooke[56] on generation. One could as well be reading an author from the sixteenth century."

The two scholars were apparently close enough for Haller to ask Tissot to look after the health of his daughter Émilie, who was missing her menses and suffering from recurrent fevers, together with loss of strength and appetite. This problem appears to have been solved in a most orthodox manner. In 1757 Haller wrote to Tissot: "You know that Émilie got married. She honors Hymen with her good colors and her happiness." Reading their letters, it seems that they even trusted each other enough to voice discredit or suspicion about those who were supposed to be their best friends or allies.

"I beg you to moderate your praise of M. Bonnet, which could be read as a critic against M. de Réaumur," Haller wrote to Tissot in 1754, as if Bonnet were not his main mentor and supporter in the ovist crusade. "The drawings of the former have always seemed suspicious to me, and his experiments are even more so. It is false, for example, that one never finds a *corpus luteum* in a virgin female." Then, in 1762, Haller turned his sharp tongue on the positions of Tissot's dearest friend, Jean-Jacques Rousseau. "They say, because I have not seen it, that Rousseau writes against the Christian religion with an imprudence that surpasses all that has ever been attempted. The remains of good sense that he presents here and there only make this poison more dangerous . . . Even exile is too good for him." In reply, Tissot, while admitting that "nothing surprises me in this world any more," presents a moderate defense of a man who, for all his flaws, writes "with a strength that one does not find elsewhere." Haller replies: "I have not read his work. But it seems to me that this man wants, regardless of whether or not it makes sense, to follow a path different from that of the entire world"; and adds, later in the same year, "Rousseau loves to prostitute his own hernias."

Haller also kept Tissot informed of the progress of his work intended to establish the final truth of ovism, stressing repeatedly that Malpighi's observations had some defects, due to "a microscope too enlarging": "I am presently working on the formation of the chicken in the egg. These experiments are part of those, that I have made to shed light on the mystery of generation, or at least to correct some errors. But this is still very far away" (1755). "My experiments on the chicken are going marvelously, and are close to their end. It is now certain that the heart almost does not change at all, that all the mystery that some thought of having found there comes solely from the length of the artery . . . I shall continue my observations as soon as I have hens" (1757). And, on another note, in 1765, "I have worked again on the egg, to answer to a certain Mr. Wolff,[57] who has attacked our friend M. Bonnet a little too heatedly. I have found numerous interesting new things." From his consultations with Haller, Tissot probably derived the notion that sedentary lives produced nervous weakening and visceral congestion by slowing down the circulation of the humors, further claiming that the resulting degradation of the seminal fluid led to the well-known fact that the sons of geniuses are generally less gifted than their fathers.[58] Haller, in turn, offered continual praise for Tissot's books.

It must have been somewhat rewarding for all of these men to watch Gautier d'Agoty (1717–1785) bend himself over backward in what seems an almost desperate attempt to produce one more spermist model that would not require the waste of millions of lives. In his *Conjectures sur la génération contre les oviparistes & les vermiculistes,* published in 1750,[59] the French naturalist and illustrator rejected the views of both sides of the preformation controversy, and proposed instead that the fetus arises in the seminal vesicle of the male from the purified blood of the testes combined with humors present in the vesicles themselves, passing from here to the maternal uterus during intercourse. This bizarre sort of male parturition meant that only one embryo (or, at worst, very few) was present in the semen at each ejaculation; and, therefore, massive mortality did not occur. Once in the uterus, the little creature generated in the semen would be fed by the female semen and by the menstrual blood, condensing into a final shape. His drawings of the seminal embryo picture a little man with a long umbilical cord and a gigantic head, readily visible, the author claimed, if the semen were discharged into a glass of clear water.

These propositions may now sound almost hilarious, but once again, in his time, d'Agoty was not by any means some sort of oddball or loose canon.

Fig. 23. Gautier d'Agoty's drawing of the preformed fetus contained in the human semen, visible to the naked eye if the semen were discharged into a glass of clear water (1750; as reprinted by Cole in *Early Theories of Sexual Generation,* 1930).

He held the title of "anatomiste pensioné du roi," and his *Anatomie des parties de la génération de l'homme et de la femme, representées avec leurs couleurs naturelles, selon le nouvel art, Jointe à L'angéologie de tout le corps humain, et à ce qui concerne la grossesse et les acouchemens,* published in 1753, is a wonderful anatomical atlas, whose engravings, still pristinely preserved in their original lines and colors, speak for themselves on behalf of the art this man sought so persistently to perfect. Here, he stated again his belief in the male parturition model in very sober and clear terms.

> The tubes and ovaries receive blood through the spermatic arteries, and produce a light secretion. The testicles of men have the same spermatic vessels; but these organs, more perfect than those of women, provide the true seed that forms the embryo.

This sentence is dropped with no further comment in the middle of an extremely long and detailed description of "the parts of generation in women." His disbelief in the egg and its putative role in reproduction is then conveyed to the reader in a similarly understated fashion when he writes, again without further comment:

> [the ovaries] are composed of a very spongy tissue, extremely tight, where one can find small globules, very clear and transparent, which are called eggs . . . These globules are true glands for the filtration of the female seed . . . the male has similar glands, but less apparent, more compact and solid, which filter a more perfect seed.

The argument is pushed further with a subtle suggestion of the necessary inability of the female apparatus to play a major part in generation:

> I think that the exact anatomical description of the natural parts of the woman which we have just seen confirms . . . the conformity and parallel of the sexes in humans, just as in animals.

Bearing this in mind, we should now be ready to assume that being parallel does not mean being equal, and hence follow the author's logic without hesitation when we enter his considerations on male anatomy:

> Considering the admirable structure of the male vesicles, and their good situation for the production of the seed, should we not guess their function? On the contrary, the female parts, seem to be but a reservoir ready to dilate and retract according to the needs of the embryo. If the organic molecules were shared by man and woman, would not the two share the same organs?

They do not, and for d'Agoty there was no doubt that "the small canals coming from the prostate provide the clear and transparent liquid that surrounds the embryo [described elsewhere as "a small membranous body very fine and delicate, shaped more or less like the external orifice of the matrix in the woman"] and preserves it during the transit through the urethra toward the bottom of the matrix." In a later passage he explained that "at the moment of conception the uterus closes itself and embraces the seed that contains the embryo." To strengthen his point, he made once more a comparison between what is seen in plants and what happens in humans, with sentences such as "the root of the germ, with its hairs, expands and attaches itself to the ground; and the cord of the embryo, which entered the uterus like the seed of a plant, expands and attaches itself to the bottom of the uterus." As for the birth of twins, he offered a simple explanation: "the seed which caries the germ can carry two twins, since men have two testicles . . . and even three, in extraordinary cases."

Doubtlessly, there is some degree of fantasizing going on here. But d'Agoty was the first to admit that he would have been ready to accept other models were it not for the fact that he observed the small transparent embryo in the male semen. And not all the contents of the book were equally imaginative. After explaining conception, d'Agoty gave a lengthy and quite accurate description of pregnancy, illustrated by several superb plates. This description included long and careful accounts of the circulatory system of the fetus, of its diseases, and of its enveloping membranes. Taking a stand against the still familiar idea of absolute similarity between living and nonliving things, he wrote:

> Animal generation relies on all that is organized, moves, and is independent of the action of the earth . . . To give life, one must be alive . . . The earth has never conceived a man, and a mountain has never produced a mouse.

He also made a clear distinction between conception and delivery and the act of generation, which can be self-perpetuated ("such as in numerous insects") or achieved through coitus. Most of all, in an incisive chapter titled "On the Impressions of Pregnant Women," he strongly refuted the idea that a woman's imagination can influence her fetus (see chapter 3), and firmly concluded that "one must not overstate the power of imagination. It might have an influence on the character of the child, but not on the child's members; and never a monster was born as the result of imagination."

None of this, however, could save this man, who had tried to forge a new middle ground that would dispense with both spermism and ovism but still retain a crucial role for semen in reproduction, from becoming the central character of a long-lived joke on spermism. d'Agoty also described a small transparent creature in the semen of a donkey and in that of a rooster, stressing that these small embryos were visible to the naked eye, without any need for the use of a microscope. In 1756, with the help of his young son, he produced an engraving with the figure of a fetus discovered inside the semen of a horse— not bigger than a bean, but with distinctive equine features. Similar plates were presented for embryos of roosters and donkeys, the latter with very long ears. "I enjoy the fact that the French have again submitted themselves to ridicule," Haller wrote to Tissot in 1764. "I trust that the young generations will be wiser than their elders."

d'Agoty's evident lack of amphibian anatomical knowledge when describing similar events in the frog[60] prompted Spallanzani's patronizing criticism, with the declaration that "I will not pronounce it to be a mere fiction; I will rather suppose that some fallacious appearance has misled him, in consequence of his inexperience in observing frogs and his ignorance concerning their internal structure, which is exceedingly obvious."[61] The detractors of the heinous sin of the spilling of seed could breathe a sigh of relief. Spermism was becoming a laughing matter.

After this episode, the theory faded quickly, struggling to the surface one more time with Astruc's efforts ("This Astruc is incredibly weak," Haller commented in his letters to Tissot. "Why does he write?"), and then vanishing from the literature. The battle over generation was thenceforth left to ovists and epigeneticists alone. Spermism had started as a solid, all-encompassing model for reproduction, but its inability to address the problem of God's wastefulness probably led to its demise. In this respect, Albrecht von Haller owed his friend Tissot a big favor. With the publication of *L'Onanisme,* the physician had provided "serious" medical grounds for a final and decisive blow to the credibility of encasement inside the sperm cell. The way was now cleared for the late-eighteenth-century ovists to develop their sophisticated arguments against epigenesis—and for all of us to remember spermism as nothing but an amusing dead end.

"One Does Not See the Wind"

Blessed are those who have not seen and have believed
JESUS' LAST WORDS IN THE GOSPEL OF JOHN

ERE IS AN INTRIGUING PUZZLE.
Faute de mieux, when preformation first started taking shape, largely on the basis of Swammerdam's and Malphigi's observations, the egg was considered the only possible vehicle for all the encapsulated generations. But then, within a very short time after the formulation of the first postulates, came the discovery of spermatozoa. When the same microscopes that had shed light on developing chick embryos and cocoons of insect larvae were aimed at human sperm, a whole new world of possibilities opened before the eyes of the preformationists.

However, a bizarre paradox, whose implications are rarely explored by modern accounts, immediately followed: the microscopists' discovery of the sperm cells led almost instantaneously to a fierce denial of the very existence of these cells by several of the same microscopists. As observations and descriptions of spermatozoa from different animals multiplied and gained more and more consistency, numerous voices of respectable naturalists raised the claim that the semen contributed nothing but a spiritual essence to the process of fertilization, totally unrelated to these newly discovered "animals." An analysis of the taxonomic approaches prevailing during the seventeenth century, combined with the background of the Western traditions and religious heritage that dominated the scene in which these events unfolded, could perhaps make this paradox resolvable.

THE DISCOVERY

"I have no doubt," Leeuwenhoek wrote to the Royal Society in April of 1676, "that the testicles have been made by no other purpose than to furnish the little animals in them, but to keep them till they are ejected." He illustrated his findings with figures copiously charted by arrows and numbers, engaging in a very serious technical description of "the animalcules in the male semen":

105

Fig. 1 is an animalcule which mostly has this aspect when, living and moving, it swims with his head or frontpart in my direction. ABC is the body, CD the tail, which, when swimming, it lashes with a snakelike movement, like eels in the water. In my opinion, ten hundreds of thousands of them would not make up in size a grain of coarse sand. Figs. 2, 3, and 4 are some of the same animalcules lying dead and showing various aspects; and however frequently I beheld them I never saw any that, when dead, resembled others in appearance.

And, to his detractors, he retorted:

If certain persons had seen as much of this matter as I have, they would be able to narrate miraculous things about it.[1]

Was Leeuwenhoek the first to see them? Or was it rather his elusive student Ham, referred to by Huygens as "Hammius," who seems to have had two completely different identities, one as Johan Ham of Arnhem (eventually the Burgomaster of that city), who studied medicine at Leiden and who never published any work of his own, and another as Ludwig von Hammen of Danzig, the author of *De Herniis*?[2] Or could it have been an enlightened eighteen-year-old Hartsoeker, who certainly did not mince words in crediting this feat to himself?[3] And exactly how much did the Dutch mathematician, astronomer, and physicist Christiaan Huygens (1629–1695), who reportedly drafted Hartsoeker's first publication for the *Journal des Sçavants,* contribute to the discovery?[4] Was the momentous year 1677, 1694, or 1699?[5] Historians can debate these details forever and never come to a precise conclusion, but the real heart of the matter lies elsewhere. From the turmoil following the description of spermatozoa, we take home an important message that largely transcends dates and names. Much more interestingly, the discovery of sperm cells offers us yet another reminder that being visible does not necessarily imply being credible.

Here were very active organisms swimming in a fluid long known to have something to do with reproduction. In the absence of achromatic lenses and efficient microtomes, visualizing a mammalian egg with the same degree of clarity was all but impossible, since these eggs are enclosed within thick and blood-filled layers of ovarian tissue. The observers were naturalists, not theologians. They were working at the cutting edge of the technology of their time. Yet, when presented with samples of semen under their magnifying glasses, many chose to believe that the sperm cells played absolutely no role in the mystery they had set out to solve. They considered them artifacts, parasites, products of diseased testicles, signs of putrefaction, agents destined to incite the male to perform the sexual act, inactive fibers left behind when the viscid mass of the sperm starts to evaporate, filamentous ramifications arising and breaking away from the semen some time after its discharge, even inert oleaginous corpuscles set in motion by the heat of the semen, originating during the sexual act and perishing on its completion—and insisted that the egg was the one and only carrier of humanity.[6]

THE DENIAL

Radical novelties are rarely welcome in the context of already complicated thoughts. Even the term "spermatozoon," coined by von Baer in 1827, means "animals in the sperm," or "seed animals," showing how long the idea of the parasite persisted in embryology. During the seventeenth century, many naturalists were willing to accept the existence of some sort of seminal spirit, which would stimulate the egg to initiate the unfolding of the new human being. Jan Swammerdam first christened this "influence" with the term *aura seminalis* in 1685.[7] But, even when distinguished by a "technical" name, that evanescent property was the only role several researchers would grant to semen in the process of reproduction.

Something worth noticing is happening here. Men of reason, hard-working men of experiment, are presented with a visible entity in a fertilizing substance. Yet their first reaction is to deny that such an entity plays any role in fertilization.

Let us examine a powerful example. Swammerdam, the skilled observer behind the launching of the seminal aura, described, in *The Book of Nature,* the process of copulation among the frogs in clear and lively terms that seemed to imply a very active participation on the part of the male—and of its semen. "We find that the testicles and spermatic vessels in the male Frog fill with sperm," he wrote in the last chapter of the book, totally dedicated to the study of amphibian metamorphosis.

These animals become then so eagerly intent on the business of propagation, that they take no care in a manner of their own safety . . . I believe that they eat nothing, or very little, during this fit of lust . . . The male Frog leaps upon the female, and when seated on her back, he fastens himself to her . . . he throws his forelegs round her breast . . . he most beautifully joins his toes between one another, in the same manner as people do their fingers at prayer . . . and closes them so firmly, that I found it impossible to loosen them with my naked hands . . . and they continue sometimes on their backs, for forty days successively . . . sometimes even for a longer time . . . Such is the male's eagerness to act his part, that he is not to be parted from his mate . . . Thus these little animals swim, creep, and live together for many days successively, till the female has shed her eggs . . . and, while this is doing, it may be observed, that the male acts the part of the midwife, and promotes the expulsion of the eggs by working with his thumbs, and compressing the female body harder with his forelegs. Thus, at last, the eggs are discharged in the female's fundament in a long stream, and the male . . . immediately fecundifies, fertilizes or impregnates them, by an effusion of his semen . . . As soon as these eggs have escaped from the female body, between her's and the male's hinder legs, and have been impregnated by the male's semen, the two Frogs abandon each other. The male swims off, and works his forefeet as before, though they had continued so many days successively, without the least motion, in the most violent state of contraction.

This passage seems at first to read as the ultimate defense of the material engagement of the male. Yet the man who described it so well, and many of his contemporaries, still preferred to resort to spirits rather than to material entities. And, on top of such an apparent contradiction, in making this choice they recovered from the lexicon of epigenesis the idea that the semen, through some "irradiant faculty," acts as a spiritual agent to bring the embryo to life—in other words, they borrowed an old concept from a group they were trying very hard to undermine.

SPIRITUAL AGENTS

In his works *De formatio foetu* and *De formatione ovi et pulli,* published in 1604 and 1621, respectively, the Italian physician Fabricius d'Aquapendente (1537–1619) had defined the semen as "the agent or efficient cause of generation." The rooster's semen, he believed, never entered the egg; indeed, it would be impossible for the semen to do so. The length of the uterus and the foldings of its inner surface prevented the viscous fluid from traveling up the oviduct; and when the egg reached the lower end of this tube, it was already covered by a shell, which served as a barrier. In Fabricius's view, the semen from the rooster was stored in a little sac; and, from that place, by virtue of its "irradiant or spirituous faculty," it rendered the entire uterus and the egg fertile—just as the testicles appear to affect the vigor of the entire body. "For if you will recall that incredible transformation which affects a gelded animal when it loses its heat, vigor, and fecundity throughout its whole body," he wrote, "you will readily admit that what I have said happens merely to the hen's uterus alone. Certainly capons of the same species support this contention, for they lose all vigor and fecundity after they have been castrated and deprived of testicles and semen." The obvious conclusion was that, at least among oviparous animals, the material and the agent were distinct, and were separated from each other by a long distance.[8] Much to the author's satisfaction, this system was perfectly in tune with the Aristotelian theory that the female furnishes the passive matter for generation, whereas the male acts as the prime agent, energizing

Fig. 24. Hieronymus Fabricius d'Aquapendente.

the passive substance to take form:[9] "the semen perfects the egg; it does not, however, exist within that which is generated but endows it with form and makes it a living creature by the power residing in it."[10]

Fabricius died in 1619, before his painstaking work on the formation of the chick in the egg was published. By then, the study of generation had already been taken up by his pupil William Harvey, who published *De Generatione Animalum,* his own work on generation, as the corollary of a lifetime of research. The numerous flaws identified in his master's thesis seem to have stimulated the disciple to reinvestigate the whole field with even more urgency.

In *De Generatione Animalum,* Harvey criticized Fabricius's views on the part played by the semen in the production of the fetus, saying that "all those things he hath conjured up to guard his opinion seem for the most part false, or very suspicious." One of those "things" was Fabricius's statement that "after the cock has been admitted one single time, all of the eggs are made fertile for the entire season." When researching this possibility, however, Harvey found out that a hen laid a fertile egg on the tenth day after her separation from the rooster, and another on the twentieth, finally conceding that "it may seem possible that one or two acts of conjunction may fructify the whole cluster, and consequently all the eggs of that year." This may sound dubious; but Harvey's idea that "intercourse, once or twice repeated, suffices to impregnate the whole bunch of yolks" was later confirmed by Spallanzani, and since then fertile eggs have been secured from hens as long as 32 days after copulation.[11]

However, Harvey again adopted a spiritual factor as the sole possible ex-

Fig. 25. William Harvey.

planation for generation in viviparous animals when he encountered serious problems in dealing with the embryonic growth of mammals, based on his observations made on female deer from King Charles's Royal Park. He regarded the mammalian ovary as a kind of venous plexus, designed to concoct a fluid for lubricating the parts. In his opinion, the ovaries did not exhibit any changes during the period of sexual activity, and did not play any part in generation. After repeated dissections of both birds and mammals, Harvey could not discover any trace of male semen in the female genital ducts. Therefore, he agreed with his master Fabricius that the male semen never reaches the site of generation in any animal. He followed this path with extreme emphasis, returning to the same point over and over. Just like Fabricius, he started by quoting Aristotle, "one of nature's most diligent inquirers," according to whom "the principles of generation (are) male and female, she contributing the matter, he the form; and . . . immediately after the sexual act the vital principle and the first particle of the future foetus, viz. the heart in animals that have red blood, are formed from the menstrual blood in the uterus." Still, Harvey insisted that the male semen contributed neither form nor matter to the egg, but was only the provider of the energizing essence through which the egg became fertile and fit to engender an embryo—and that this happened by a kind of contagion, effluvium, intangible penetration, seminal aura, or essence. Having performed this subtle office, the male semen either escaped from the body, dissolved, or turned into vapor and vanished.[12]

It is ironic to find the same Harvey who caustically derided Fabricius's idea of the "irradiant, spirituous faculty" of the semen eventually writing that "the geniture of the male is not the architect of the fetus, because the first conception assumes its body from it, but because it is spirituous, and boiling, as being inspired with a fertile spirit, and turgent like a thing possessed." [13] The main difference between Harvey's point of view and those of Fabricius and Aristotle lies in his belief that the semen activates the egg by "contagion" rather than by providing the "efficient cause." This change does not, however, make the contribution of the semen to generation one bit more material.

Still, the main irony behind these philosophical tribulations is that Harvey, the founding father of modern epigenesis, used the seminal spirit to back his belief in the formation of each organism de novo at each generation. This means that, in their endorsement of the *aura seminalis,* some of the leading preformationists not only dismissed a conspicuously visible structure,[14] but also used, in this dismissal, a concept long postulated and perfected by the enemy. What could possibly be their problem?

JUST ANOTHER WORM

Considering the cultural background of the time, I would like to argue that, for the researchers of the seventeenth century, at least a substantial part of the problem must have resulted from the belief that the spermatozoon was "a worm." The use of this zoological designation for the spermatic "animalcules" was by then widespread, crossing all party lines. The ovists were prompt to dismiss any active function for the "vermiculi" or "tailed worms." Even the spermists referred to them in this way. Leeuwenhoek preferred the term "little animals"—but then again, he applied the same term randomly to all kinds of protozoans and algae he encountered. To verify how long and how deeply this association endured, it is sufficient to recall that Clifford Dobell's book *Antoni van Leeuwenhoek and His Little Animals,* published in 1932, refers solely to bacteria and parasites. Not good company for sperm, when one is trying to promote them as God's chosen guardian of human lives.

In 1694, when Hartsoeker described his observations on sperm in his *Essai de dioptrique,* he noted that the semen of animals "is full of an infinite number of animals similar to young frogs," but he felt compelled to explain that this applied only to "men and quadrupeds," since "when it comes to the semen of birds, flies, butterflies, and other flying insects, it is full of an infinity of animals that are similar to worms." He then repeated the term "worm" several times when discussing a possible model for sperm entrance into the egg.[15]

Nicolas Andry's *An account of the Breeding of Worms in Human Bodies,* translated into English in 1701, dedicated an entire chapter to "Spermatic Worms." The author started by defining an important difference between these specific "animals" and all others of the same kind: "The Worms of the Body are distinguished into *Flesh-Eaters* and *Spermatics.*" The first are "those which devour the Creature." The second, easy for a microscopist to find in numbers "above fifty thousand" in samples of "Humor contained in the Testicles and other parts of Generation" and of a size "but as much as a grain of sand," are typical of "the Spermatic Vessels of Creatures" and "do 'em no harm." Based on this difference, Andry said that "the name of *Worms* is but improperly given 'em." But he had no other term to use. He could only try to distinguish such "living Animals, resembling Eels, and all in continuous motion" from the remaining "bad" worms by affixing an adjective to the name: "I call them *Spermatic Worms.*"

Andry carried his defense of the "good" spermatic worms even further by bestowing upon them a morality of sorts. For example, he noted that "the Flesh-eating Worms are bred in the greatest part of violent fevers" whereas "the Spermatic worms for the most part die at that time." Furthermore, "you find them languishing, and for the most part dead in *Gonorrhea*'s, and in Venereal Distempers"; "there is not one, or at least not any living in those that are Impotent"; and "they who commit the greatest Excesses against Continency, generally have no Spermatic Worms."

In a later passage of the same chapter, the author proceeded to demonstrate the connection between these creatures and generation by enumerating several examples.

If you open a dead Man presently, or an Offender that has been executed, you shall discover in the Humor of the testicles, and in the *deferent Vessels,* an innumerable train of living Animals, having a large Head and a long Tail. I said that these Worms are not perceived till the Age proper for Generation. To be convinced of which, there need no more but to examine the Testicles of a Cock Chick, and you shall not find a Worm. You may do the same thing in a young Whelp, where you shall not find so much as one of those animals. If you open an Infant newly deceased, you shall not find a Worm either alive or dead, whereas in a Man you shall find millions . . . They are not found in the *Ovaries* or Eggs of Females; of which you may be convinced if you open Women that come to violent ends. After the mixture of the Sexes, the bottom of the Matrix, and Tunicles of the Matrix, are all full of Worms, whereas before there be none. To demonstrate this, we need but open a Bitch immediately after she hath been covered, for then millions of those Worms may be found in the Matrix . . . whereas if that Animal be opened before she be covered, there are none to be seen at all . . .

What are we to conclude from thence, if it be not that the Spermatic Worms are the occasion of the Generation of all Animals?

Nevertheless, worms they remained.

OR JUST ANOTHER INSECT

Ironically, the worm has come a long way since the time of the controversy over the true identity of sperm cells. These days, top-flight investigators from state-of-the-art laboratories wander around cell biology meetings sporting T-shirts with the complete cell lineage map of the fashionable worm *Caenorhabditis elegans.* The backs of those shirts display a quotation from Nietzsche: "Ye have made your way from the worm to man, and much within us is still worm." During the last decade, *C. elegans* made headlines, was granted covers on all of the best scientific journals, was featured in a video for teaching developmental biology to undergraduate classes,[16] and even made its way into the science sections of the popular press worldwide. It is repeatedly mentioned by the spin doctors of science as one of the "sexy" organisms for modern research. But one century's hype is another century's doom. During the preformationist debate, nothing could have been less sexy than being associated with a worm.

At the time of the discovery of spermatozoa, the worms were not even a precise group. In a letter addressed to the Royal Society in 1683, Leeuwenhoek discussed metamorphosis with sentences such as, "In a flea's egg there are nothing but globules floating in a watery substance . . . Finally this becomes a worm and then a pupa . . . The same is the case with many caterpillars"; or, "The frog is at first a thick worm when it has grown to a fair size."

Likewise, in René Antoine Ferchault de Réaumur's (1683–1760) *Mémoires pour servir à l'histoire des insectes,* published between 1734 and 1742,[17] it is clear that, for the highly respected French entomologist and his contemporaries, the term "insect" applied freely to many groups of invertebrate animals, such as spiders, crustaceans, worms, and polyps. This vagueness was still the rule during the thirteen years before Réaumur's death (1747–1760), when he worked on the last, and unfinished, manuscript of the series, *The Natural History of Ants,* all the while studying numerous other matters. The taxonomically pre-Linnaean "insects" discussed by Réaumur and a number of his colleagues included many other invertebrates now regarded as distinct classes or even phyla.[18] Réaumur stated in the introduction of the *Mémoires*:

> I confess that I am not in the least inclined toward a precise enumeration of every kind of insect . . . it seems to me sufficient to consider those kinds that prove to us they deserve to be distinguished, either on account of their peculiar industries or because of their unusual structure or because of other striking singularities.

Fig. 26. René Antoine Ferchault de Réaumur.

This statement leads naturally to the affirmation, almost immediately ensuing, that "the crocodile is certainly a fierce insect, but I am not in the least disturbed about calling it one."

Even the immensity of such a group did not stop the French count Buffon (1707–1788), superintendent of the Jardin du Roi in Paris and author of the monumental *Histoire naturelle, générale et particulière*,[19] from scorning those who studied "useless insects." As Buffon put it, "a bee should not occupy more space in the head of a naturalist than it does in Nature; and this marvelous republic of flies will never be more in the eyes of reason than a swarm of little beasts that have no other relation to us than to supply us with wax and honey."[20] Réaumur reacted to these statements by writing to Bonnet that "the whole misfortune of bees and other insects is that I love them and dare to admire them," implying that Buffon's disrespect for insects had solely to do with personal rivalries. But Réaumur's complaint was private, whereas Buffon's scorn was very public.

If this picture seems bad enough for "insects" in general, when we come to worms proper, "useless" seems a mild euphemism. Worms were definitely considered a lesser category of creatures, although they did not yet have a clear position within the animal kingdom. In the chapter dedicated to "The action of the egg" in *De formatione ovi et pulli*, Fabricius included a reminder that

Fig. 27. Francesco Redi.

Aristotle considered "the generation of grubs *(vermes)*" to be the product of "imperfect eggs." Several decades later, worms remained so insignificant that Buffon lumped them together with the fungi in the bag of beings reproducing by spontaneous generation,[21] even though this imaginary propagation had already been disproved by the Italian physician Francesco Redi nearly a century before.[22]

Swammerdam himself helped to make worms a lesser category, writing in his *Biblia mundi,* in the section dedicated to taxonomy:

> After giving above all the insects which really belong to the fourth order, I shall also place in this order all the nymphs of the worms in the first, second, third and fourth orders which change into worms, caterpillars, pupae, warts, leaves, growths and so forth, and take on the appearance of a nymph. I shall do this not because they belong in this fourth order, but because, as do worms of this fourth order, they change into nymphs in an obscure and altogether inaccessible manner; furthermore, their changing is so difficult to observe that it can only be distinguished by persons who are greatly experienced with these insects.

This, plainly, is a concession to his public's taste: Swammerdam placed these awkward cases in the "fourth order" to facilitate the arrangement of his system and the general understanding of the system by his readers. In doing so, he contributed to the equation of "worms" with the group of animals that could not be easily fitted into other categories—and this perception of worms as "leftovers" became a generalized trend in the years that followed. But being a leftover was only a small fraction of the philosophical equation. The really serious problem was that worms were also perceived as obnoxious.

SICKENING MIDGETS

In 1735, the Swedish naturalist and founder of modern taxonomy Carl Linnaeus (1707–1778) published the first edition of the *Systema Naturae* and finally separated insects (*Insecta,* group V) from worms (*Vermes,* group VI). "In the human intestine three species of animals occur, viz. Lumbrici, Ascarides (round worms) and Taeniae (tape-worms)," reads note 9 on the appended "Observations on the Animal Kingdom."

> That the *Lumbricus* of the intestine is one and the same species as the ordinary earthworm, is shown by the appearance of all its parts. That the *Ascaris* species are identical with those very small worms *(Lumbricus)* one finds anywhere on marshy spots, becomes very clear by close inspection . . . Worms do not take their origin from insects' eggs, flies and the like (for, if that happened, they could never multiply inside the intestinal tract, and would perish during the stages of metamorphosis); but from the eggs of the worms above-mentioned, taken in with the water by drinking; from this it is evident that medicaments detrimental to insects need not necessarily kill the worms.

This classification should have been a great breakthrough for the emancipation and dignity of worms, since they had finally been recognized as a distinct taxonomic group, with precise characteristics ascribed to it. But Linnaeus's observations on *Vermes* mention only notoriously sickening representatives, such as tapeworms and small intestinal worms. To make matters worse, Linnaeus did not think highly of the sperm cells, either; and he tied the two negative concepts together in describing his own observations of dog sperm through a microscope in Leiden, in 1737, rejoicing in the fact that "the belief in these 'seed worms' is nowadays almost completely vanished."[23]

The damaging connection was completed in 1752, when, in his "A History of Animals," "Sir" John Hill (1716–1775) classified spermatozoa as distinct animals, under the label of Infusoria or Parasites, in the group of "lesser animals called animalcules." They were assigned to the new genus *Macrocercus,* and divided into six species. Now they were worms for sure. And, as Antoine Maître-Jean had written in 1722 in his "Observations sur la formation du poulet," "as little worms have been found under the microscope in pond water, vinegar, and all kinds of liquids, there is no reason to suppose that those in the semen are in any essential way connected with generation."[24]

A STAIRWAY TO HEAVEN

The implications of this worm problem cannot be fully evaluated without considering the religious views of the time. These were still the days of strict Christian vigilance over the thoughts of the naturalists. In the middle of the eighteenth century, we still find the righteous Abbé de Lignac vehemently attacking Buffon's 1749 essay "Theory of the Earth" on the grounds of its dubious orthodoxy. Moreover, we know that the count sensed this attack as such a threat that he feared falling from grace with the Sorbonne and the Académie des Sciences—and, in order to avoid such disaster, felt compelled to shield himself behind Madame de Pompadour's influence on her lover Louis XV.[25]

Episodes of this sort are even more striking when we bear in mind that Buffon was by then one of the most prominent spokesmen of his time, putting forward a variety of scientific postulates that did not isolate science, but rather linked it intimately to all other realms of human thought and to man's basic needs. The "heresy" that so irritated some of his peers lay in his view that the earth was originally covered with water, that the most ancient of animals were fishes, and that all the fossil seashells found today in the interiors of continents and on the tops of mountains were derived from the original waters, and not from the deluge sent by God during Noah's days.[26] To avoid further trouble, the French naturalist ended up writing a letter of retraction to the Sorbonne, later printed in the fourth volume of the *Histoire naturelle*. The statement is made in a form that had been the standard for such cases for more than a century:

> I abandon whatever in my book concerns the formation of the earth, and in general all that might be contrary to the narration of Moses, having presented my hypothesis on the formation of the planets only as a pure philosophical speculation.

If a small deviation from the geophysics of Genesis could create such a stir, it is not hard to imagine the consequences of challenging the most basic premise of the same book, the creation of Adam by God.

One has only to remember the terrible tempest that swept the learned world between 1859 and 1871, when Charles Darwin (1809–1882) initiated the line of thought that finally prompted the daring suggestion that men had descended from apes. Darwin, and the path followed from *On the Origin of Species* to *The Descent of Man*, were still almost two centuries away when microscopy first revealed the existence of sperm cells. Under such circumstances, it would probably be too much to expect broad public support for the idea of a future mammal carried inside a worm—especially if that mammal was a human being, supposedly made in God's image.

The Catalan Monravà, who taught anatomy in Lisbon during the first half of the eighteenth century, summarized the problem quite bluntly:

> if we were to believe (the spermists), then each man would be a son of a beast; which is even worse than being the son of a whore. Oh, we should all feel really honored!"[27]

The same disgust resonates in the short passage dedicated to spermism in the book *Essai sur la manière de perfectionner l'especè humaine,* published in 1751 by the French doctor Vandermonde:

> just like certain insects that only become such when they stop being worms, some people [Leeuwenhoek and Hartsoeker] believed that worms were the first germs of our lives, just as after our death they are the instrument of our destruction. We were reduced to little worms, and our existence was reduced to a sad metamorphosis!

Moreover, in the context of its time, the *aura seminalis* was probably not such a farfetched concept as it may now appear. During the Scholastic period (see chapter 7), Thomas Aquinas had related the effect of the vapors of the semen on the unformed egg to that of the Spirit of God over the unformed deep. In Europe, the first description of the lesser circulation of the blood was printed in 1553 by the Catalan Miguel Serveto in the theological work *Cristianismi Restitutio,* and one of the author's main concerns was how the Holy Spirit permeated the blood to become the soul of the body.[28] And Swammerdam, after all his careful and painstaking examinations of metamorphosis in insects and frogs; after having meticulously dissected larvae of both groups and after having carried the understanding of their anatomy as far as the discovery that tadpoles simultaneously have lungs and gills, "both serving to circulate, cool, alter and purify the blood"; after having employed in his observations experiments as advanced as the use of electricity on frogs' legs more than a century before Galvani—after this immense ground-breaking work, brought all this effort to a vibrantly pious conclusion in the closing words of *The Book of Nature,* totally coherent with his initial statement that "it is not in search of glory to ourselves, but the great Creator, that we ought to survey and examine his works":

> Are not these changes admirable? And do they not lay before our eyes the omnipotent hand of God, conspicuous in his inaccessible radiancy and infinite majesty? He, in this case, forms another out of one and the same animal, which though different in appearance, yet remains one and the same creature. May not the resurrection of the dead be exemplified in this illustrious instance?

More importantly yet, it is my suggestion that the idea of the seminal aura could have been accepted as just another case of spiritual insemination—and a very flattering one at that, since Jesus himself had been conceived in the Virgin Mary by the sole power of the Holy Ghost. In the context of Western Christian culture during the seventeenth century, the seminal spirit could have been almost a bonus—a scientific creed that brought men even closer to the perfection of their Savior.

FAITH VALUE

One cannot make such a claim without allowing for a considerable amount of perfectly legitimate skepticism. Do people, deep down, really take at face value all the dogmas of the religions they profess? It is one thing to admit that the Gospels spoke of very peculiar circumstances surrounding the birth of Jesus. But it is quite another thing to suggest that Swammerdam and his peers had no doubts about the Immaculate Conception. Did the champions of the *aura seminalis* really believe in the prowess of the third member of the Trinity?

Examining this question requires a careful look at both the Scriptures and the myths converging in their composition. For preformation developed its postulates within the framework of the Christian heritage, and this heritage combines, in its teachings, a wide number of visions derived from the primeval anxieties of the human mind.[29]

Let us start with one of those spiritual essences that come closer to material entities, those ever-present angels flying subtly over the entire body of Scripture. Gabriel, the messenger sent to Mary to announce her awesome destiny, seems to have achieved impregnation by the sole means of the Annunciation, so that the Verb and the Spirit are one (see note 29). And Gabriel had a celestial host of predecessors in the Old Testament, always involved in the unfolding of an important pregnancy. Three angels announce the birth of Isaac to the barren Sarah (Gen. 18:9–16). In the Book of Judges (13:3–5), an angel comes to another barren woman, the wife of Manoah, to notify her of the future birth of Samson (the "little sun"), who will deliver Israel from the hands of the Philistines. The same angel later tells Manoah: "How can you ask my name? It is a name of wonder" (Judges 13:11–18). In an obscure passage of Genesis (6:1–4), before the Flood, the "sons of God" take to wife the most beautiful "daughters of men," and with them beget the "giants" of the primordial time. In the New Testament, it is also Gabriel who announces to Zechariah the birth of John the Baptist to his old wife Elizabeth (Luke 1:5–17). As additional signs of the magical halo that surrounds the acts of God's messengers, Zechariah remained mute as long as he did not believe what he had heard

(Luke 1:20–22); and, later, John jumped with joy in his mother's womb when he sensed the presence of Jesus in the womb of Mary (Luke 1:40–41).

MANLY TRAITS

Do all of these considerations, when combined, vaguely suggest an image of God having a certain amount of fun by playing occasional tricks of spiritual insemination? This may not be the most orthodox exegesis ever presented, but the theme certainly has been hinted at repeatedly in the sources of Christianity.

Powerful men, mighty men, have shown all through history a tendency to express their power by—among many other assorted feats—inseminating numerous women and thus bringing a multitude of children to life. God, we are told over and over, is more than powerful. He is Almighty. Insofar as we think of God as a man, it would make sense to view Him as expressing this all-powerfulness in the specific insemination of selected women of his choice. And, for that matter, although His description is supposed to be beyond the reach of our words, the God of the Scriptures is definitely a man.

For one thing, He created Adam (not Eve) in His own image. This type of discourse can certainly be viewed as strictly metaphoric; but in God's subsequent behavior, or at least in our perception of it, there is little room left for metaphors.

God begins to appear truly human through the shrewdness of the serpent right in the Garden of Eden. The now excommunicated German Catholic priest Eugen Drewermann,[30] makes this point brilliantly when he reflects on the implications of the dialogue between the Temptress and Eve.

> We hear it ask a question that sounds quite harmless, inquiring about what God had said: "Did God really tell you not to eat from any of the trees in the garden?" Could there be a more fanciful or fantastic beginning for a temptation than inquiring about something that God had said? Could there be a nicer question than one asking about the words of God? The only catch is that the serpent's question calls God into question. It foists on God a line of talk that is wholly alien to God. But it is possible to think along those lines.[31]

And we do. A single question from the serpent changes the whole image of God, hence of the world. In Eve's ensuing confused defense, God's original words start to be twisted in her perturbed speech so that now it appears that He could descend to execute the human beings who would disobey him—like an avenging angel, yes, or else like any other truculent warrior. From now on, God is playing games. Human games.

God as a man is also perceivable through the words of Jesus. When Jesus says He is the Son of God, He adds that He is the Son of Man. According to

Mark (14:62) and Matthew (26:64), He maintains this even before the High Priest, when it has become clear that such a statement will lead Him to die on the cross. And, to bind the spiritual even closer with the carnal, this Son of Man never shied away from all manner of superhuman undertakings verging on pure magic. He walked on water (Mark 6:48). He climbed a high mountain and became transfigured[32] as a voice spoke out from a bright cloud, "this is my Son, my Beloved, on whom my favors rest; listen to him" (Matthew 17:2–5). He told His disciples to lay their hands on people and drive out demons (Mark 6:7–13), after repeatedly having done so Himself from the beginning of His public life (Mark 2:23–27). Jesus behaves as if our lives have been taken over by a host of alien spirits and forces, as if we often have scarcely any control over ourselves, no matter how clear our thinking might be. In this light, the concept of an *aura seminalis* is perfectly reasonable as a reflection of the inherited beliefs of the researchers who proposed it and of the culture that produced them.

VIRGIN BIRTH DETECTED

The impact of this heritage on the views expressed during the seventeenth and eighteenth centuries gained added force through an unanticipated discovery. Among the many contributions of his frantic quest to understand microscopic life, Leeuwenhoek produced a perturbing observation. He noted that plant lice, or *aphids,* were viviparous and apparently could reproduce in the absence of males.

"I wrote that I had dissected lice, and taken from them eggs, which we call nits, and from these eggs again lice," he reported to the Royal Society in 1677. The letter contains only this one sentence, by all accounts rather cryptic, concerning the virgin birth of lice. But Leeuwenhoek, who at that point was mainly focusing on the mysteries of insect metamorphosis, had already detected something peculiar distinguishing these eggs from eggs of related animals. "From that moment I had given my mind to the flea," he adds, pointing out that in the latter he never saw "little fleas gradually increasing in size, as in the case of the lice." Moreover, the eggs of the fleas were "several times smaller than those of a louse"; and "I several times broke eggs (of fleas) but could not discover anything inside, only observing that they consisted of an incredible small quantity of exceedingly small globules hanging or sticking together."

In another letter, dating from 1683, he came back to the same subject in much more definitive terms: "We see, then, that generation does not take place in one and the same manner: the louse already has all its members when still in the egg and in his mother." Now the statement that intercourse is not necessary to produce new generations of lice appears in a firm and clear sentence. It is only briefly tossed, very matter-of-factly, into the middle of a long discussion of the relative parts played by eggs and sperm in the production of offspring; but this should strike us as even more remarkable—and reliable— since it is coming from the pen of a microscopist who dedicated strong and relentless efforts to persuading his peers that all animals were encased inside spermatozoa before their birth.

The same phenomenon was later confirmed by others—and, this time, the whole learned world was listening. Charles Bonnet was but twenty years old, and had not even finished the law studies forced upon him by his father, when, in 1740, he followed Réaumur's suggestion and raised a female louse in seclusion. The animal produced 95 young, many of which Bonnet claimed to have watched hatching before his eyes. The event was presented to the Académie des Sciences and scrutinized by several highly regarded personalities, including Bonnet's cousin Abraham Trembley (1710–1784) and Réaumur himself, and it made Bonnet famous overnight. Such fame, however, was not without a price. As naturalists became more and more interested in this phenomenon, and questions mounted on the possibility of copulation taking place among embryos in the uterus,[33] or of one single copulation accounting for the delivery of successive generations,[34] Bonnet undertook more and more complex experiments, eventually losing his eyesight in the process. With one species of louse he obtained four generations without mating. With another, the number grew to six. And finally, with yet another louse, he proceeded to increase these cycles for nine rounds, making, for three continuous months, detailed tables of the day and time of each hatching. He was still not entirely satisfied and wanted to go as far as the thirtieth generation without a father, but felt secure enough to say that a single mating capable of inseminating nine generations would be even more unthinkable than reproduction without copulation.[35]

It was now clear that, among certain organisms, the female could do without semen; but no male was known to reproduce without females. This seemed at first a definitive triumph for ovism. Those who accepted the egg as the source of all seeds rejoiced in this apparent support for their views. But they had discovered more than a simple supporting observation. The cards had been redistributed. Parthenogenesis had reared its strange head, pushing the concept of virgin birth further into the realm of science. And, in science, as in knowledge in general, there is no such thing as a one-way street—no such thing as an accidental discovery recording nothing but an innocuous incident. When the oldest myth meets the newest discovery, no postulate is too daring. The louse was now, according to Haller, an important being in physics.[36] And, in all its minuteness, it held the doors of scientific respectability wide open for evanescent spiritual explanations. In 1742, the Jesuit Abbé Pierquin was already invoking the reproduction of aphids to explain the maternity of the Virgin.[37]

This, obviously, was not meant to please Diderot[38] (1713–1784), who immediately seized the occasion to deliver one more bite of pamphletary atheism. "Here we have a Jesuit," wrote the scandalous author of *La Réligieuse,*[39]

> whose work I have under my eyes, who asks how Mary could become a mother without losing her virginity and who, through an effort of sagacity of which he is very proud, decides that the thing took place just as among the hermaphrodite lice; that she participated in copulation as man and as woman; that the birth of her son was the result of the mixing of two fluids springing from the same source and that the only difference between this woman and all the others was that she experienced double voluptuosity, without consequences on her innocence or on the innocence of her little louse.

Even without this iconoclastic truculence, the ironies of virgin birth were by then obviously an attractive source of inspiration. In his vivacious *Lucina sine concubitu,*[40] John Hill also mocked the insistence of a maiden and of her aristocratic family that she had gotten pregnant without "ever had known a Man"; and used this fictitious episode to ridicule, in the most spirited manner, the ideas of the spermists. But neither defiant attacks nor subtle humor could undo what the combination of unusual observations and theological conjectures had done. Parthenogenesis was now a fact, and the name of Jesus had been involved in its discussion.

THE MIGHTY SEA-HORSE

A belief that the Holy Spirit penetrated Mary becomes even more credible when recalling how old and persistent are the stories of spirits, good and bad alike, fathering children with particular features or destinies assigned to them. Writing only a century before the discovery of spermatozoa, the French royal surgeon Ambroise Paré (ca. 1493–1541) had no doubts about these events, whose "scientific" backing in his treatise *On Monsters and Marvels* is provided solely by repeated quotes from the Scriptures. Evil spirits, he maintained, can be found in the air, in the water, on the shores, on earth, or "in the deepest center of the latter"; and, in order to "lose and ruin the human race," they can "transform themselves into anything they please."

The ability of such spirits to interfere with everyday life is seemingly unlimited: "They howl at night and make noise as if they were in chains: they move benches, tables, trestles; rock the children, play on the chessboard, turn the pages of books, count money; and one hears them walking about in the

Fig. 28. Ambroise Paré.

chamber; they open doors and windows, cast dishes to the ground, break pots and glasses and make other racket; nevertheless in the morning one sees nothing out of its place." When they enter human bodies, they speak, "their tongues having been torn out of their mouths," through the belly or, most importantly, "through the genitals"; and they do so in "various unknown languages." To alert us to the danger of sexual encounters with such creatures, the French surgeon recalled several cases reported in his time, like "a lost woman (who) 'did her business' with an evil spirit at night, (it) having the face of a man, and suddenly her belly swelled on her, and thinking she was pregnant, she fell into a strange malady, in that all her entrails fell"; or "a butcher's helper who, being plunged deep in empty musings on lust, was astonished that he suddenly perceived before him a Devil in the form of a beautiful woman, with whom, having 'done business,' his genital parts began to burn, so that it seemed to him that he had a fire burning within his body, and he died miserably." Paré was by no means alone in holding this type of belief. In the pseudo-Aristotelian *The Complete Master-Piece,* the author agrees that "it cannot be denied that the Devils transforming themselves into Human shapes, may abuse both Men and Women, and with wicked people use Copulation."

But the Western culture was not original in bringing spiritual penetrations to its sexual configurations. Expressions of the belief in spiritual insemination are many, and they have many faces. Eventually, they always seem to find their way into the conventional cornerstones where civilizations inscribe their perceptions of themselves and the context of their relationship with the surrounding universe. Even admitting that for each example we could probably find a counterexample, such tales are so widespread that they must represent a deep building block of human nature—and, in this light, it could be argued that the seminal aura was easy for the naturalists of the seventeenth century to accept, not only because they were Christians, but, above all, because they were human. As Haller would write to Bonnet in 1758, voicing his exasperation with the constant demands of the epigeneticists that the performationists produce a visible proof of pre-existent lives-to-be, "one does not see the wind."[41] And the wind, that quintessential metaphor for all that exists but cannot be captured in perceivable images, can perform many wonders indeed.

Still accepting the possibility of spontaneous generation in certain organisms, Harvey had conjectured that the invisible "seed" of creatures was disseminated by the wind.[42] And, long before him, Pliny stated this idea quite clearly when he described the weather and soil of the floodplains along the river Tagus, in the Roman province called Lusitania, which was later to become Portugal: everything here is so fertile, he wrote, that the mares got pregnant with the wind. This seductive quote has been reproduced by several authors.[43] Jorge Luis Borges translates it as "nobody ignores that in Lusitania, in the surroundings of Olisipo (Lisbon) and of the shores of the Tajus, the mares turn their faces to the Western wind and thus become fecundated by it; the young engendered in this manner are admirably light, but they die before they are three years old." He relates this belief to a broad Greco-Latin fiction about the wind that inseminates mares, also expressed in the verses of the Third Book of Virgil's *Georgics*. Borges then reports that several ethnologists see in this myth the origin of the Islamic legend of the "Sea Horse," mentioned during the thirteenth century in al-Qazwiní's *Marvels of Creation,* and recurrent in many other stories—a magnificent wild horse who lives in the ocean and comes out only when the wind brings him the odor of mares in heat.[44] A similar horse roams the Icelandic bestiary under the name of Minir.[45]

Winds of fertility blow from all quadrants of the compass. The resulting beliefs find shelter and fertile ground in their religiousness. After all, most likely from an initial Darwinian necessity to ensure reproductive success, has not motherhood been proclaimed sacred, in one form or another, more or less throughout all our histories? It is tempting to comment that such proclamations were most likely the doing of the male sector; women, who knew better about the burdens of reproduction and child-rearing, wisely said yes to everything and then quietly proceeded to devise methods of abortion or contraception.[46] But our culture has been, whether we like it or not, the product of a male-shaped world; and women's passive resistance could only have been a side effect of the dogma. Between 500 and 200 B.C., religion promoted fertility by persuading the Romans that if they left behind no sons to tend their graves, their spirits would suffer endless misery.[47] The same belief was held by medieval Jews.[48] And if maternity becomes sacred, it is only logical for spiritual connotations to adorn our perceptions of reproduction.

In Portugal, every Christmas, at every midnight mass, we sing quite light-heartedly a carol celebrating the Immaculate Conception, in which the meeting of the Holy Ghost and the Virgin Mary is described in the most incorporeal way imaginable: the "Divine Grace," we intone from childhood without even chuckling at the thought, "entered through her and left through her/as the sun shines through the window." I would argue that we move so easily through these dubious territories not only because memorized Christmas carols are rarely the stuff of profound adult meditations, but also because we come from a lineage of a world full of spirits, as ancient as ourselves. And our traditions like to think of these spirits as able to enter and leave our bodies as freely as they please. "Let no one wake a man brusquely," says one of the Upanishads of ancient India, "for it is a matter difficult of cure if the soul finds not his way back to him."[49] Never mind that we cannot see the wind. We know it is here, with us, every minute, every breath we take.

Fishbite

Mankind has certainly been prolific in exploring the possibilities of spiritual close encounters.[50] The Trobriand Islanders, or so the legend has it, did not attribute pregnancy to sexual interchanges, but rather to the entrance of a *baloma* into the woman. *Baloma* stands loosely for any kind of ghost; but in this particular case the ghost had to be some sort of spirit of the waters, since it was supposed to enter only when the woman was bathing. The sole explanation offered by unmarried girls for their pregnancies was "a fish has bitten me." Reality or myth, this story is made even more interesting by the fact that the entire community seemed to accept this fertilizing fishbite as a fact, rather than a lame excuse conveniently endorsed to save everyone trouble. Otherwise, why would the only method of birth control practiced here by all women be to avoid bathing at high tides?[51]

The association of spiritual essences with childbirth is also quite apparent in the reasons offered by different civilizations for inflicting a death sentence on the newborn. Several peoples resorted to infanticide if the baby was deformed, or disabled, or a bastard, or if the mother had died upon giving birth. The spiritual connotations of the sacrifice become even deeper in such cases as the strangling of children who did not enter the world headfirst by the Bondei natives; the killing of babies born in stormy weather by the Kamchadals; or the exposing, drowning, or burying of children born in March or April, or on a Wednesday or a Friday, or in the last week of the month, by Madagascar tribes.[52] True, it can always be argued that such infanticides were nothing but a prosaic response to the harshness of the environment—a ritualized way to prevent famines and extra burdens—especially knowing that, in some cases, the sacrificial lambs were eaten afterward.[53] But when we encounter such cases as the torturing to death of newborn girls so that their souls will be induced to appear in the form of boys in their next incarnation,[54] it is impossible to insist that arguments about spirits are not important.

Fertility Rites

If more evidence is needed, we have only to look back at the oldest central point of most primitive mythologies, the fertile mating of earth and sky and its inevitable repercussions in multiple forms of sex worship. When we consider the ever so precious fertilization of the soil, it seems that no avenue was left unexplored—from the glorious promiscuity of spring harvest festivals to sophisticated methods of suggestion by example. Zulu medicine men fried the genitals of men who had died in full vigor, ground the mixture into a powder, and strewed it over the fields.[55] Some peoples chose a King and Queen of May, or a Whitsun bridegroom and bride, and married them publicly, so that the soil might take heed and flower forth. In certain localities the rite included public consummation of the marriage, so that Nature would unequivocally get the message.[56] The peasants of Java mated in the midst of their rice fields to ensure their fertility.[57]

Knowing nothing yet about sperm and eggs, most peoples came to deify conception and the visible external structures involved in it—and ended up worshiping sex, in some form and ritual, in Egypt and in India, in Babylonia and in Assyria, in Greece and in Rome. And, since no real ritual exists without an appropriate totem, phallus worship has prevailed in most countries. In India, it has survived from ancient times and remains fundamental to modern Hinduism. The signs of this devotion, as Gandhi pointed out in *Young India,* are visible everywhere—"in the phallic figures on the Nepalese shrines and the temples in Benares; in the gigantic *lingas* that adorn or surround the Shivaite temples of the South; in phallic processions or ceremonies, and in the phallic images worn on the arm or about the neck." And, at least according to the indignation of Augustine, when the Greek god of fertility Priapus became domiciled in Rome, maidens and matrons alike frequently sat on the male member of his statue, as a means of ensuring pregnancy. His figure adorned many gardens. Little phallic images of him were worn by simple persons to bring fertility or good luck or to avert "the evil eye."[58]

FLIGHTS OF IMAGINATION

Another variant of this connection of spiritual powers to insemination involves the long-held belief that women could imprint upon their children the effects of their states of mind—a theory later dignified in science by the name of *telegony,* or the theory of *maternal impressions.* Once again, this belief is not an exclusive conjecture of the West. The Fiji Islanders had tribes in which husband and wife were discouraged from sleeping together, for fear that the female breath would enfeeble the man.[59] Such is the dimension of the hidden powers bestowed upon the feminine mind.

As reported by Joseph Needham in his 1934 *A History of Embryology,* the pre-Socratic Greek philosopher Empedocles wrote about monsters and twins in the fourth century B.C., asserting that the influence of the maternal imagination upon the embryo was so great that its formation could be guided and interfered with by the sole means of this redoubtable power. In *The Secrets of Nature Revealed,* the author wrote that a child would look more like his father if the mother was thinking of him during copulation—and, at that, "the younger the woman the better." In the sixteenth century, Paracelsus warned us that "the imagination of women is stronger than that of men, and may, during their sleep, transport them in dreams."[60] And Paracelsus's French contemporary Ambroise Paré, dealing extensively with human malformations in his treatise *On Monsters and Marvels,* had explained very colorfully the causes of the problems that the doctor could have to deal with.

> How many persons does one see who on their face or other parts of their body have the form of a cherry, of a sorb-apple, of a fig, or a mulberry, the cause of which has always been referred to the very powerful imagination of the conceiving and pregnant woman moved by a vehement appetite or with the appearance of an unexpected touching of this fruit; so likewise because one sees some persons born from such imagination having in some part of the body the shape and substance of bacon rind, others of a mouse, others of a crawfish, others of a sole, and other such; which is not beyond reason, given the force of the imagination being joined with the conformational power, the softness of the embryo, ready like soft wax to receive any form, and given that when one might seek to examine all those who are thus marked, it will be found that their mothers have been moved during their pregnancy by some such appetite or happening. Whereupon we will note in passing how dangerous it is to disturb a pregnant woman, to show her or to remind her of some food which she cannot enjoy immediately, and indeed to show them animals, or even pictures of them, when they are deformed and monstrous.

Such cases were certainly taken very seriously in the sixteenth century, for, right after this explanation, the studious surgeon felt compelled to add: "For which I'm expecting someone to object to me that I therefore shouldn't have inserted anything like this into my book on reproduction." In his defense, he had a simple, stern sentence to offer: "I do not write for women at all."

Having issued this warning, Paré proceeded to write an entire chapter on "Monsters that are created through the imagination." The opening paragraph seeks to render this idea respectable through the opinions of "the ancients, who sought out the secrets of nature" (Aristotle, Hippocrates, and Empedocles), who offered as causes for the birth of monstrous children "the ardent and obstinate imagination that the mother might receive at the moment she conceived—through some object, or fantastic dream—or certain nocturnal visions that the man or woman have at the hour of conception." Paré's inventory included Heliodorus's description of how Persina, queen of Ethiopia, conceived by the Ethiopian king Hidustes a daughter who was white, "and this occurred because of the appearance of the beautiful Andromeda that she summoned up in her imagination, for she had a painting of her before her eyes during the embraces from which she became pregnant"; Damascene's report of a girl born furry as a bear due to the fact that her mother had looked too intensely at the image of Saint John the Baptist dressed in skins, along with his own body hair and beard, "which picture was attached to the foot of her bed while she was conceiving"; and Hippocrates's declaration that a princess who gave birth "to a child as black as a Moor, her husband and her both having white skin" was not guilty of adultery, but solely impressed by "the portrait of a Moor, similar to the child, which was customarily attached to her bed."

Consequently, Paré concluded, "it is necessary that women—at the hour of conception and when the child is not yet formed—not be forced to look at or to imagine monstrous things." Failing to observe this rule could lead to deformations such as "a monster . . . having the four feet of an ox; its eyes, mouth and nose similar to a calf"; or to the birth of a child with the face of a frog because, as the father explained to the doctors,

> his wife having a fever, one of her neighbor ladies advised her . . . to take a live frog in her hand and hold it until said frog should die. That night she went to bed with her husband, still having said frog in her hand; her husband and she embraced and she conceived; and by the power of her imagination, this monster had thus been produced.

Within the parameters of Western thought, which framed the debate on the role of semen in fertilization, the concept of telegony was as old as antiquity and still rather vigorous during the Scientific Revolution. In 1751, M. Vandermonde made it clear in his *Essai sur la manière de perfectionner l'espèce humaine* that the woman, as "the depositary of a precious fruit, who must soften its pains and support its life," should take added precautions during pregnancy:

> a pregnant woman should not make a movement that is not marked by reason. Her nourishment, her work, her sleep, all must contribute to the formation of a new creature . . . the very air she breathes may become the food or the poison of the fetus, and she must put a stop to her passions and stay away from the entice ments of her unleashed imagination.

Fig. 29. *Below, left:* Children affected by maternal imagination: Paré's drawings of a child born "black as a Moor" because the mother looked at a portrait of a Moor during conception, and of a girl born "furry as a bear" because the mother looked at a painting of Saint John the Baptist during conception. (From *On Monsters and Marvels*, 1573.)

Fig. 30. *Below, right:* Paré's drawing of a child with the head of a frog, born with this deformity because the mother was holding a live frog in her hand during conception. (From *On Monsters and Marvels*, 1573.)

The same worry over the conformational powers of the feminine mind was expressed by Nicolas Malebranche, the first theorist of preformation (see chapter 1); and this worry was discussed in his *Search after Truth,* the book that launched the Russian doll model of preformation. Among other examples, the French theologian described a man "born crazy and with all his limbs broken exactly in the place where the limbs of the criminals are broken," explaining that this had happened because his mother, while pregnant, had watched the public flagellation of one such criminal.

> The fibers of this woman's brain were shaken and perhaps broken by the violent spirits produced at the sight of such a terrible action, but they were strong enough to prevent the collapse of the entire brain. Contrarily, the fibers of the child's brain could not resist the flood of spirits and became entirely dissipated, and the damage was great enough to deprive him of reason forever . . . The view of the blows . . . went with great strength from [the mother's] brain to all the parts of her body corresponding to the criminal's body, and the same happened to the child. But, since the mother's bones were strong enough to resist the violence of the animal spirits, they were not damaged. Maybe she did not even feel the slightest pain . . . while they were breaking the criminal. But the rapid flow of the spirits was able to attain the soft and tender parts of the child's bones.

The leading spermists did not shy away from this concept either. The idea is clearly expressed in a fundamental work produced by one of them. When writing his treatise *Orthopaedia, or the Art of Correcting Deformities in Children,* first published in 1741, Nicolas Andry was still ascribing a crucial power to the feminine imagination in shaping the fetus. It is important to note that this book was not just another collection of obscure prescriptions and gothic antiques. Dr. Andry, a dynamic and highly controversial figure who defended his ideas with an ardor destined to make him unpopular in some quarters, was an active researcher whose writings include papers on bleeding, the chemistry of certain medicines, purging, and foods, together with the extensive treatise on animal parasites already mentioned in this chapter. With his *Orthopaedia,* he actually planted the seed of a great surgical specialty. The book contains many basic concepts resulting from his observations of pathological states of the musculoskeletal system. These have stood the test of time, and have come to form the core of the principles of orthopedic surgery. It was Andry who first coined the term "orthopedics." More importantly, he was also the first to understand the active participation of muscles in producing deformities of the skeletal system. The following considerations are, therefore, part of a enlightened and innovative body of work.

When a child is born with a nose excessively large . . . we may have hopes of curing it, provided in the mean time that the Mother, during her Pregnancy, has not been shocked with any Deformity of the same kind, whether in Masks, Pictures or Persons; for in this case the thing is quite incurable.

As we have seen, this concept of telegony had been extensively debated by the time Andry wrote his work on orthopedics. To back his statements, the French doctor quotes preferentially from the works of two of his learned countrymen, Scévole de Sainte-Marthe's *Paedotrophia* and Abbé Quillet's *Callipaedia*. The former wrote that "marks arise which appear upon the skin, in Children, in consequence, as is alleged, of certain Imaginations of the Mother, during her pregnancy"; and the latter argued that

the Imagination of a Woman with Child may imprint certain Marks upon the Child, without making any impression of the same kind upon the Mother. A Woman with Child ought carefully to shun looking at any thing, that may disturb the Order which Nature usually follows in the formation of the Foetus.

GOD THE PARENT

Could the debate on reproduction initiated in the seventeenth century ever have stayed away from God's intervention when Harvey himself stated that each child had three parents, mother, father, and God? And can we claim that this is nothing but an accidental comment, when I have heard it repeated at least once, at a baptism where I happened to be the godmother, and therefore had to go to a curious "preparation session" intended to make sure that we would bring up Rita as a spotless Catholic? We cannot even claim that this idea is an exclusively Christian fabrication, since the same concept appears in the Talmud, complete with a precise description of each party's contributions.

One of the first embryological debates registered in the Talmud concerns one of the most recurrent of worldwide perplexities, the initial site of the soul (see chapter 2): its writers were divided between the head and the navel, the latter being the best-supported hypothesis—but then, Adam had no navel, which made this option complicated. Later on, we encounter a sweet depiction of embryonic growth: the embryo goes through six different stages before its birth (the first one being the Golem, which we are going to examine at greater length in chapter 6), always keeping its mouth open to eat the mother's food, all the while being kind enough not to defecate, so as not to harm the mother. In any event, the mother alone could not construct the whole embryo. To be truly successful, this construction required a threefold effort. The father (whose contribution, as was plain for anyone to see, was white), contributed the bones, the tendons, the nails, the cephalic substance, and the white of the eyes. The mother (whose contribution was supposed to be the one that came in a red color), contributed the skin, the flesh, the blood, the hair, and the colored part of the eyes. Finally, endowing this agglomerate of substances with its ultimate purpose, God contributed the life, the soul, the facial expressions, and the function of the different organs of the body.

The importance of God's final intervention surfaces again, some centuries after the Talmudic period, in the teachings of the Cabala: by themselves, the mother and the father could achieve the formation of a human body; but this body would be stillborn were it not for a beam of light sent directly by God from the Heavens.

So we have to bow to the evidence that, in mingling spiritual factors with the carnal routines of reproduction, Western thought was never alone. Before and after Harvey, humans have insisted on perceiving their existence as an expression of some divine factor or another.

GOD AS LIGHT

Even assuming that the belief in spiritual insemination is universal and an intrinsic part of the human mind, it is still necessary to verify the solidity and origin of its roots in Christianity, the religious confession shared by all those who chose to believe in a seminal spirit they could not see, in preference to a sperm cell they could finally see. According to Drewermann, "the clearest, most extensive and many-faceted evidence of a belief in life after death is found in Egyptian rites. Their language found its way into Christian belief and remains the basis of many of our symbols and rituals. This is the first powerful expression of confidence that we human beings are really not composed of the dust of the earth but are essentially sun creatures, so that the golden light of the sun in the heavens breathes in us." If we borrowed immortality from the hieroglyphs of the Pharaohs, did we also borrow spiritual birth from them?

The belief that Jesus is God's son—true man and true God—cannot be derived from the Judaic texts. What these sources call the messianic king, the son of David, the servant of God, defines solely a "divine sonship," a reminder of the features bestowed upon the divine kings of the ancient Near East. The actual point of origin of the aforementioned myth is, at least according to the sources uncovered so far, located in ancient Egypt. Christianity was only lucky enough to expand initially through a Hellenistic milieu likewise familiar with the notion that all great personalities—Plato or Pythagoras, Alexander the Great or Augustus, Empedocles or Asclepius, even such founders of states as Romulus—were sons of gods. But even Alexander first had to travel to Egypt to be received as a god.[61]

The entire concept of the son of God, who is conceived by a virgin over-shadowed by spirit and light (Amun-Ra) and born into the world, was completely worked out in ancient Egypt *as an idea* thousands of years before Christ's birth. The pre-existence of such ideas had already been noted by Augustine, who mentions in his *Retractions* (I, 12, 31) that "what is now called the Christian religion existed even among the ancients and was not lacking from the beginning of the human race until Christ came in the flesh. From that time true religion, which had already existed, began to be called Christian."[62]

We first find evidence of this idea in a fairy tale from the seventeenth century B.C., which tells of the divine origin of the Fifth Dynasty. As A. Erman recounts, one day the sun god Ra is unhappy with King Cheops.

> And even if he is willing to grant him another grandson (the builders of the second and third pyramid of Gizeh), nevertheless after these a new race will care more for their gods than for their own giant tombs . . . Thus Ra will beget with the spouse of one of his priests, Red-dedet, a new race.

Ra proceeds to talk to Isis, Nephthus, Mesechenet, Heket, and to the ram-headed Cheops himself:

> Arise, go and deliver Red-dedet of the three children that are in her womb . . . They will build your temples, they will attend your altars and meals, they will make your drinking tables flourish, and they will make your sacrifices great.

Then these gods go in the form of musicians to the pregnant Re-dedet, and she bears three children, the first three rulers of the new dynasty.[63]

A POLITICAL REALITY

Why would the Gospels have to imagine the coming of the Redeemer of Israel in no other way than the primeval myth of the divine child, or the divine king, born of the blessed virgin?[64] And why would the new postulates of the Scientific Revolution take up the same leitmotif? What seems absurd to us now may have seemed obvious in different times. We, unlike Jan Swammerdam, live two centuries after the French Revolution, after the execution of the last absolutist king of the House of Bourbon. It may be hard for us to understand how these images of the virgin birth of a divine king could ever have been self-evident. But, as in the mythic picture that made its way into the narrative of Jesus' descent, the virgin birth is, in fact, inseparably bound up with the idea of the divine nature of the king, or any truly royal person. It is true that, in premodern European history, absolute monarchs were no longer more than occasional governmental episodes, reaching a zenith with Louis XIV and then quickly fading away—and even this smart French ruler understood the mentality of his era clearly enough never to claim any kind of divine status, satisfying himself with the more sober title of "Roi-Soleil," conveniently reinforced in the most subliminal manner by the several statues of Apollo dispersed through the gardens of Versailles.[65] But it is also true that, in the seventeenth century, when the *aura seminalis* emerged in scientific language, the memories of this political reality were still fresh, even if their mythological roots had all but vanished. Memories may never be more than unconscious frameworks, but, even at that level, they can certainly condition the way people think—and even more so the way they choose to express themselves in public.

Hopeful Monsters

The fear of the Lord is the beginning of knowledge
BOOK OF PROVERBS, 7

ERE IS THE TALE of how a little story became a piece of history. This story is not one of those well-known truisms that sometimes spring from history of science: some embryologists may never have even heard of the tale. But it is written in modern books, and is presented as a real fact of the past. The story typically occupies only one short paragraph. But in that paragraph, a vast number of complex issues of development hide between the lines, including the problem of explaining the generation of monsters. When reconsidering this problem, we are led through a maze of thoughts that include other related issues, such as the conformation of the earth, regeneration, hybridism, and heredity. And, when trying to understand why the story seemed so plausible as to go thoroughly unquestioned by contemporary writers who happened to come across it, we find ourselves once more face to face with the ultimate source of our joys and troubles. The survival of this paragraph was not made possible by humans. As we shall see, it was made possible by God.

WHAT DID THEY SAY?

Here is how Nicolas Andry explained the formation of monstrosities:

A spermatic worm seeks out the ovary, slips into the egg, closes the door behind him with his tail, and proceeds to develop. If several attempt to enter the egg at the same time they become enraged and strike each other, breaking and dislocating their limbs, and thus giving rise to monstrosities. Even at this stage the spermatozoa are endowed with the nature of the animals to which they will give rise, for those of the ram already live in flocks.[1]

If you are willing to play a little quiz game, find in one minute what is wrong with this picture.

If you noticed the use of the term "spermatozoa," you guessed right. The word did not even exist at the time of Andry's work, and entered the literature only two centuries later (see chapter 2). The paragraph reproduced above, though repeatedly quoted, is a fabrication of anonymous origin. The tale of the wounded sperm warrior never appeared in the spermist literature.

Here is what Andry really wrote on how the sperm enters the egg:[2]

> But how does this Worm get into the egg? How above all, amongst so many Worms that enter the Matrix, is there but one that ordinarily becomes a Foetus? . . . When the Egg comes down from the Ovarium, and falls into the Matrix, these Spermatic Worms, which are all of them in a continuous motion, go through all the Cavity of the Matrix; they meet with this egg, go round and over it; and the place by which the Egg breaks off from the Ovarium, resembles that by which the Fruit breaks off from the Stalk, that is to say, that place leaves a small opening; now it is easy to comprehend, that amongst so many Worms, it is not possible but that some of them should enter the Egg by this Opening. Then the Cavity of the Egg being little, and proportioned to the bulk of the Worm, which cannot bend to return back, 'tis obliged to continue shut up in the Egg, where in the mean time no other Worms can enter, because of the smallness of the place possessed.

The unfortunate excluded worms, "like Corn that does not fall upon good Ground," would "die for want of Nourishment."

As an alternative scenario, Andry then mentions a friend of his, "a Physician at the Faculty of Paris, a Man of extraordinary skill in Physic," in whose opinion

> at the opening of the Egg there is a Valvula, which suffers the Worm to enter the Egg, but hinders it to come out, because in the inside it shuts upon the outside. This Valvula is held fast by the Tail of the Worm which lies against it, so that it cannot open then either without or within.

But, in Andry's prose, this is not a certainty. He presents his friend's idea solely as "very probable."

The difference between the original writing and the apocryphal version becomes even more amazing when bearing in mind that Andry *did* write about fetal malformations—in a book called *Orthopaedia,* which launched modern orthopedic medicine (see chapter 3). In this book, children's deformities at birth were explained mainly through mechanical effects suffered by the mother during pregnancy that had pathological repercussions on the development of the fetus. This concept is totally accurate by modern standards, even if rendered in Andry's text with the characteristic ingenuity of pioneering works. In other words, the secondary accounts of the spermists' philosophy credit the first man to address the causes and cures of deformities in children in a serious and systematic manner with the invention of the most delirious tale of the production of such malformations. In the course of two centuries, a sober model for sperm entry became oddly transmuted into a fairy tale for the explanation of monstrosities.

This is not only amazing. It is also bitterly ironic. First of all, Andry's own real model has been confirmed by modern science. In some species—namely, among fishes and insects—the fertilizing spermatozoon fits tightly into a tiny pore at the egg surface (the micropyle) and ensures by this means alone the necessary blockage to the entrance of supernumerary sperm cells. But, most importantly, if Andry's apocryphal model were true, it would have allowed his camp to resolve something for which the other side had no logical explanation. For if the early spermists had actually gone so far in their proposals, they would have scored a point over the ovists. Through the model of the wounded warrior, they could have explained a crucial phenomenon that the ovists were at a total loss to address and that the epigeneticists had the exclusive privilege to cite as a strong supporting argument. If, as the ovists believed, all the future creatures were patiently sitting inside their own eggs waiting for the wake-up call of the *aura seminalis* (see chapter 3), or for the ring of some other kind of alarm clock set by God at the beginning of time, to unfold and start to exist, then it would be hard to explain how something could go wrong and lead to the emergence of deformities. By postulating that embryonic development occurred de novo in each generation, with a gradual formation of the different organs of the body, epigenesis seemed to be much better positioned to offer a clue to such aberrations. It has been said several times that the lack of an explanation for any kind of deviant generation was one of the weakest spots in the fabric of preformation. In *Investigations into Generation,* Gasking lists some of the efforts of the preformationists in this area:

> concentrating on hybrids, such as the mule [some authors] suggested that the young germ, being very delicate, was susceptible to alteration, and that the form was altered by the food. Since, in both the ovist and the animalculist views, the first food was provided by that parent which did not produce the germ, the intermediate character of the offspring was thought to result from its early nutrition. Some were prepared to credit some of the many folk myths, and relied on maternal impressions and the like. Bolder theorists cut short the whole search for explanation, and simply asserted that some germs were originally created in a different form. A dispute between Lemery, who held that monsters were the result of various accidents, and Winslow, who thought some germs must have been created as deviants, continued in the Académie des Sciences in Paris from 1724 to 1740.

The question of hybridism definitely posed a problem for epigeneticists as well as for preformationists, since it involved the delicate question of whether or not two different species could actually mate and produce viable and fertile offspring. But the same did not apply to the generation of monsters. In the midst of the tormented search for truth at this front, the model of the wounded warrior could have been a useful simplification. Although Gasking hints at this potential when noting that some preformationists also "held that variations were always due to accidents which befell the developing germ," meaning that "either two germs fused together to produce a monster with supernumerary parts or part of the germ would fail to develop," nothing is thereby explained as to how the addition or loss of parts might have occurred. Yet the idea of a war among sperm cells would have served this purpose perfectly, and in the same breath would have dispensed with "folk myths" and "the like," making the spermist theory sound more plausible. Had the spermists really considered the necessity of a microscopic battle for every birth, they could have provided a novel and dynamic explanation for congenital deformities. Such an explanation would have been made possible only by the discovery of the sperm cells, and that discovery had been made possible only by the use of the microscope. So, by the same token, theirs would have been a truly revolutionary breakthrough in revealing the secrets of an old and perplexing mystery, brought forth by a truly technological discourse, uncovering a system that ovism was unable to accommodate.

A very simple introspective deduction would have made this scenario even more credible. The early microscopists could not see with their simple lenses all that we can see with ours. But they certainly saw millions of creatures swimming around in a very small drop of liquid. These creatures seemed incredibly agitated. We all know that if you put lots of agitated people in an overcrowded, confined space, sooner or later a fight will break out. And, for all the spermists knew, the sperm cells were *animals.* By definition, they would be even more likely to engage in territorial fights than human beings. It made sense. Perfect sense. Yet the spermists never went that distance when the argument could have worked in their favor. And, by the time the concept of the wounded warrior was finally ascribed to them, the age of innocence was over and the tale no longer made sense. They did not say such a thing when they could have done so, and were quoted as having said it when the idea had become obsolete.

REDEMPTION

The 1990s brought some sort of poetic justice to this unfair outcome. Spermatologists are presently uncovering a number of "sperm wars" that actually occur before fertilization. Due to the attractiveness of the phenomenon (do we not instinctively like all those events that are easily relatable to human behavior?), the "sperm wars" theme did not remain restricted to the relative secrecy of laboratory benches for long: it is now a commonplace feature of pop culture, complete with cartoons, popular writings, and even guest appearances in television specials. In short, we derided the spermists for having conjured up all those absurdities (which, as we have now seen, they never did), only to find ourselves, as a society, buying into exactly the same kind of ideas—and greatly enjoying them. Here is what can be considered a good example of a bad example:

In 1991, Meredith F. Small addressed the issue in a long feature article in the July edition of *Discover* magazine. With the aim of popularizing modern scientific discoveries, the text sounds amazingly in tune with the tales attributed to the spermists. "Conception . . . is now seen as a do-or-die cellular brawl that determines which males will pass on their genes to the next generation," she wrote.

> Biologists find the new "sperm wars" view exciting because a lot of odd facts are
> falling into place. It means that many characteristics we think of as "male" . . .
> may have evolved as so many weapons to aid a male's tiny sperm. The concept
> of competitive sperm may thus explain the basic source of maleness itself and
> serve as a unifying theme for male evolution.

These wars, as the article explains, are fought mainly between sperm from different males, rather than among sperm from the same ejaculate. But they are vivid, vicious—and real. Some sperm swim faster than others. Some sperm bind to the egg's membranes better than others. Some sperm sacrifice themselves on kamikaze missions to further the success of their brothers, in a behavior reminiscent of the teamwork of colonies of bees or ants.

It is possible that, in most species, the ejaculates contain many more useless sperm than what the specialists now casually call "egg-getters." These ill-fated cells can be two-headed, two-tailed, tailless, motionless, able only to swim in circles, or afflicted by numerous other defects. Their presence in the semen was traditionally considered a by-product of the enormous numbers of spermatozoa generated in the testes, with the implication that the production of some tens of good "egg-getters" entailed the careless fabrication of hundreds of handicapped counterparts.

Within the more sober realm of basic science, numerous modern spermatologists deem these ideas highly conjectural. Yet, others insist just as seriously that it seems possible that the long-known "useless" sperm actually have a mission: to clear the way for the mighty swimmers and to block the passage of sperm from other males.

Moreover, several laboratories point out that some sperm may carry specific enzymes designed to exterminate or inactivate competitors from foreign ejaculates. And the idea that sperm approaching the egg also use this enzymatic artillery to numb their compatriots that are still in the race has not been totally ruled out. Evolutionary and chemical "sperm warfare" is still the subject of heated debate in developmental biology, and the data keep piling up. Just log in any day at the Internet address Spermail@cleo.murdoch.edu.au. You will be in the Australian-based Sperm-Net, where new discoveries in this field are reported firsthand (if not necessarily with absolute accuracy) at a steady pace.[3]

Still, this is not a posthumous homage to the clairvoyance of the spermists. If anything, it is a homage to the clairvoyance of the jokes made about them after their deaths.

HOW THE LEGEND CAME TO PRINT

In *Early Theories of Sexual Generation,* Cole suggests that the strange fate of Andry's convictions "probably originated as a satire on the credulity of the animalculistes." Probably. This ridicule would have made perfect sense, say, at the close of the eighteenth century, when spermism had definitely fallen from grace, but its memory was still fresh and vivid. It does not explain, however, why such putative mockeries persisted into our days in the guise of true history. Sheer inaccuracy may have played a part in it. But it is hard to conceive of a whole community of historians—and scientists, for that matter—as being so thoroughly inaccurate, with no second thoughts and no double checks. More likely, in my opinion, Andry's apocryphal tale survived the test of time because of the efficacious, albeit unconscious, convergence of two seductive qualities. It comes across as yet another amusing fabrication of agitated minds, and we like a good joke. Moreover, it tackles the problem of abnormalities— and we like monsters.

It is an old cliché of the art of popularization that when in doubt between fact and legend, one should always print the legend. Considering the particular appeal of the legend of the wounded warrior, we should then take it as normal that the tale has reached us as a fact. However, to fully evaluate what happened in this process, we should now retrace the main steps. What we find is an interesting puzzle, showing that the legend came to print through the unconscious aggregation of several disparate pieces.

In 1769, the idea that some imaginative spermist had claimed to find behavioral similarities between the spermatozoa of a species and the adults of that same species (stated, as we have seen, in the closing sentence of Andry's apocryphal tale) was already manifest in the secondary literature. Here is how Abbé Regley referred to it, in his introduction to Spallanzani's *Nouvelles recherches sur les découvertes microscopiques, et la génération des corps organisés:*

Our observers [the early microscopists] started by admitting that these beings who swim in the liquor [the sperm cells] are perfect animals; others, already bored with that idea, tried to distinguish in each the tendencies and moods of each species; in the dog, the vigor and force; in the hare, the weakness and fear; in the rooster, the fire, the vivacity, the audacity. Have we not seen Leeuwenhoek make this mistake, call his neighbors, and show them with enthusiasm the entrails of a ram where the young sheep already marched in herds, timidly following their leader, like the ones who graze the grass in the fields?

Sounds good. Sounds *cute*. But, if Leeuwenhoek ever called his neighbors to witness such a clever phenomenon, he did not leave us any written account of this little personal epiphany.

Also, Andry had other peers making room for future generations to develop their stories of little animals at war. One of his predecessors actually came closer to this model by introducing monstrosities into the picture. In his *Éssai de dioptrique*, Hartsoeker claimed to have reported the following to Malebranche, when the latter was working on his *Search After Truth*:

I think that each worm that we find in the seed of birds encloses a male or female bird from the species that produced the worm . . . When the male throws the seed over the ovary of the female, this seed surrounds the eggs found here: That each worm of this seed tries to enter one of the eggs, in order to be fed and so that it can grow; That each egg has but one single opening leading to what is called the germ . . . and, as soon as one single worm has entered, this opening closes itself and lets no other worm pass; That if by chance two worms enter the germ of an egg the two animals . . . may become united through some part of their bodies, thus becoming a sort a monster.

Hartsoeker applied this reasoning mainly to flies, butterflies, "all other flying insects," and birds. But he did not hesitate in speculating that "we can suppose that the same happens with the animals visible in the seed of men and other quadrupeds"—and, with this modest sentence, he did our imaginations a big favor. His brief passage on "flying insects" was most likely the source of inspiration for Andry's later account of how the spermatic worms had to force their way through the narrow passage to the feeding grounds inside the egg— although, and this is a significant difference, the French doctor's sole emphasis was on the mechanism of sperm entry, not on the accidents that could account for the formation of monsters. But the door to revisionism had been opened. In the modern literature, the trend toward the explanation of monstrosities, and away from the search for biological models of fertilization, is clearly visible in the condescending tone adopted by Needham in his influential *A History of Embryology*:

[The spermists'] usual view was that of Hartsoeker and Andry, who pictured each egg as being arranged like the Cavorite sphere in which H. G. Wells' explorers made their way to the moon, i.e. with one trap-door. The spermatozoa, like so many minute men, all tried to occupy an egg, but as the eggs were far fewer than spermatozoa, there were, when all was over, only a few happy animalcules which had been lucky enough to find empty eggs, climb in and lock the door behind them.

Notice that neither Hartsoeker nor Andry ever referred to trap-doors. But accounts such as Needham's were being written three centuries later, by people who had already decided that whatever spermism had proposed was naive at best and unworthy of closer scrutiny. Now we have Needham's colorful scenario as a source of inspiration. We live in an age when every writer dealing with this subject is familiar with the Darwinian concept of natural selection. These writers also now know for sure that all the ejaculated sperm cells that cannot fertilize eggs will generally die within less than an hour. The idea of the urge to go through the trap-door is thus ripe to become self-evident.

It must have followed, quite logically, that if only a "happy few" of those "minute men" could avoid death, a war of sorts was bound to take place among them. We all know that nobody wants to die, and this would be a stark example of the survival of the fittest at work. In the logic of modern days, then, Hartsoeker's and Andry's ideas of *sperm entry* became "the spermists'" ideas of fierce *sperm wars*. If there are wars, there are bound to be wounded warriors. If these wounded warriors are among the lucky ones, there are going to be abnormal newborns. The better they fight, the more likely they are to get to the egg, but also, the more likely they are to be wounded. Through the reasoning of the twentieth century's writers, it made perfect sense to attribute these ideas to the seventeenth century's naturalists. After all, our contemporaries were simply repeating what they so relentlessly accused the spermists of having done: they were letting their own imaginations run free, adding more and more little touches of color to the picture as they went, embellishing their subject without ever feeling compelled to verify its initial truth.

"Andry," wrote Celestino da Costa in the introduction to his 1933 medical textbook *Elementos de Embriologia,* used in Portugal, Spain, and France, "describes the competition of the spermatozoa trying to get inside the egg through an orifice, a kind of trap-door. But if by chance several spermatozoa penetrate the same egg, they fight among themselves until only one of them wins, but not without having lost in this battle an arm or an eye, or suffered some other kind of mutilation. We would therefore have the generation of monstrosities explained!"

This subject was a blessing for contemporary writers. Few aspects of the history of life sciences could be more appealing than the "old" tales of how "monsters" arose. But what were people's real views on this subject at the time Hartsoeker and Andry wrote their books?

An Enduring Presence

According to Judith Taylor Gold in her 1988 *Monsters and Madonnas: The Roots of Christian Anti-Semitism,* "monsters are very old. Perhaps as old as Man himself. Their pictures are found in prehistoric drawings in the walls of caves. They have come to us in the myths of Asia and Asia Minor; in the folktales of Europe; in the stories and legends of the New World. The idea of monsters exists in every culture; their prevalence is universal, their persistence nothing short of phenomenal."

The author's compilation to prove this point includes a representation of a horned creature, half-human and half-stag, found in the caves of the Ardennes and dating back over forty thousand years; the legends of the faceless people of the Upper Amazonian jungle, echoed in the myths of headless people recurrent through Nubia, Ethiopia, and upper Egypt; the stone sphinxes among the silent tombs of the Saharan sands; the giant, bodiless, man-eating craniums of Iroquois folklore; or the Oannes of ancient Babylon, half-man and half-fish, which must have been the prototype for the European mermaids and mermen. In this fabulous realm, giants abound in all cultures, from the Cyclops of Homer's *Odyssey*[4] to Maushop of Martha's Vineyard, who could eat a whale in one gulp.

As for the sky, it has always hosted myriads of prodigies, starting with the Phoenix, reborn from its own ashes, which made its way from old Egypt to biblical texts and has mythological counterparts among the Arabs, the Persians, and the Hindus. Many of these flying creatures are vampire-like, such as the Tornit of the Eskimos, a cannibal that spies on his victims from the greatest heights and them heaves stones at them; the bird-woman Harpies, with their lethal claws; the monkeylike Ahuizoth of Mexico, with four heads on its body and a fifth at the end of its tail, seeking the teeth, nails, and eyes of its victims; or the Japanese Gaki, with their swollen bellies and emaciated bodies, their huge mouths extending from ear to ear, cursed to remain always unsated. The Japanese folklore also gave birth to the Oni, hideous creatures that came from the infernal regions to drag down the sinners, generally portrayed with monstrous mouths, a third eye, three toes, and three fingers, and sometimes said to be giants able to devour the whole world. As for the Chinese, they knew better than to kill the Unicorn, whose ferocity and wildness could be tamed only by virgins; it never walked on green grass, never ate vegetation, was always present at the birth of Sovereigns (it also made a special appearance at the birth of Confucius) and would bring death upon those who would injure it.[5]

Some authors point out the distinction between monsters that exist as tribes, like the centaurs, and those that come as individual, solitary creatures. Among the latter, those that are mortal are of special interest, since they seem to exist mainly to allow for the rise of human heroes. "Thus Perseus," writes Peter H. von Blanckenhagen in "Easy Monsters," "kills Medusa . . . and before he delivers her head to Athena as her emblem, he uses his new weapon to kill another monster, a dragon which threatens princess Andromeda chained to the rocks."[6] It is the killing of the monster that transforms the anonymous person into the legendary hero.

Of course, most of the peoples who produced these creatures must have been thoroughly aware of their legendary status. We may include the Egyptians as a good illustration of the acknowledgement of this dichotomy, since, while they worshiped a vast menagerie of animal-headed gods, they apparently did not like to be confronted with monstrosity in the flesh. "Rabelais," recounts Henry G. Fischer in "The Ancient Egyptian Attitude Towards the Monstrous,"[7] "in the preface of the third book of 'Pantagruel,' recalls a tale from his classical reading, concerning the first of the Ptolomies, who attempted to enhance his popularity . . . by displaying, amidst the booty of his conquests, a double-humped Bactrian camel and a slave, one side of whom was black, the other white, divided in the middle. To his dismay, these curiosities were rejected as abominations. And he was forced to conclude that the Egyptians were more pleased by things beautiful, elegant and perfect, than those that were ridiculous or monstrous."

Still, the creation of mythological entities had to start somewhere in the human perception of the surrounding universe. "The origin of monsters is lost in antiquity," Gold states, "but it may be conjectured that it would be improbable for man to conceive of anything that does not or did not at one time exist, at least in some sense. Earliest monsterism may well represent the physical embodiment of psychological distortion, particularly distortion relating to the animal kingdom then extant." Obviously, there is a significant share of distortion connected with the ancient and medieval beliefs that only two regions on earth could be inhabited, ours and "theirs." This belief in some vaguely defined "otherness," a race of distant and threatening creatures beyond the reach of our eyes and the resolving power of our words, was made even more creative by the persistent assumption, presented by Claude Kappler as having survived until the onset of the Discoveries, that no connection could ever exist between us and "them": "None of us can go to where they live, and none of them can come to where we are."[8] Also, a substantial part of the same distortion could have had to do with the spiritual and religious feelings present at the beginning of human self-perception, which projected monsters one step past mere mistakes of evolution or embryogenesis to see them as deeper mysteries. The fact that most legendary monsters are half-human and half-animal is a good demonstration of this perception, which leads to a portrayal of a monster that is both superanimal and superhuman, and thus to its immediate psychological projection as an object of fear. These creatures were generally conceived of as being able to plot and scheme like humans, but not just any humans: necessarily cruel and cunning, they were endowed with evil powers and with the capacity of attaining the most sinister goals.

Another feature attached to monsters also testifies to this merging of animal and human characteristics: monsters, in several cultures, are often the result of some transformation, a sort of unnatural metamorphosis that, unlike caterpillars becoming butterflies or water becoming ice, breaks away from the laws of nature and even violates the concept of time. Such events, triggered by the utterance of a magic incantation by an external agent or by a simple snap of the fingers by the monster itself, include individuals who become animals, trees, rocks, and stars; creatures turned into salt or gold; heroes changed into serpents; or rocks turned into precious stones. Some transformations are not even visible, but still permit those whom experience them to acquire occult powers, always dangerous or threatening: a black cat that drives its guilty master to the gallows, a tiny lizard that suddenly is able to swallow a city, a snake that carries out the command of its master.

The transformation of the monster endured in our cultural development as much more than a folk tale. According to Paula Findlen in her 1990 article "Jokes of Nature and Jokes of Knowledge" (see below), the natural history of the sixteenth and seventeenth centuries still abounded with reports of apparently inexplicable metamorphoses, for the most part lifted directly from Aristotle, Pliny, and Ovid. Quoting his Jesuit master Athanasius Kircher, Gioseffo Petrucci wrote in 1677 that

> the works of Nature are prodigious. Whoever does not penetrate the reasons behind them, imagining it to be impossible, does not believe in them . . . Do not doubt that the most extravagant metamorphoses that we hear of occurring at all hours in lands distant from ours are jokes of ingenious Nature.

From this universal body of monstrous configurations, our own culture drew very early on a rich vein of iconography and philosophy. The tradition of writing on this subject started in classical antiquity, continued into early Christianity, and gathered momentum during the Middle Ages, bringing to the writers of the Renaissance and to the onset of the Scientific Revolution a vast array of possibilities to explore in new, more sophisticated lights. According to Kath-

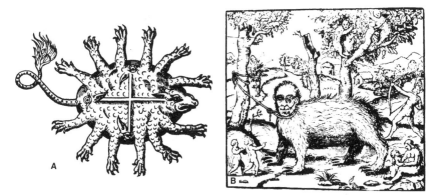

Fig. 31. Excellent candidates for jokes of Nature, presented by Paré in *On Monsters and Marvels* under the heading "Terrestrial Monsters": *(A)* "Figure of a very monstrous animal that is born in Africa" and *(B)* "Figure of a monstrous animal that lives only on air, called Haiit."

arine Park and Lorraine J. Daston in their 1981 paper "Unnatural Conceptions: The Study of Monsters in Sixteenth- and Seventeenth-Century France and England," three main threads can be identified in the earlier tradition:

> The first was the body of scientific writing on monsters which appears most characteristically in the biological work of Aristotle and his classical and medieval followers, notably Albertus Magnus. The second dealt specifically with monstrous births as portents or divine signs; the most influential pagan contributor to this tradition was Cicero, although later Christian writers relied overwhelmingly on the interpretations of Augustine and those he influenced . . . The third strain . . . was cosmographical and anthropological, and concerned the monstrous races of men widely believed to inhabit parts of Asia and Africa.

Within the frame of Christendom, the prodigious aspect of monstrous appearances was naturally linked to the will of God, since He alone could produce events contrary to Nature. Drawing directly from the classical tradition of divination as well as from Judaic sources, the Christian writers of the Middle Ages tended to associate these prodigies with large groups of spectacular events: comets, floods, earthquakes, rains of blood and stones, as well as monstrous births. Such packages, according to Park and Daston, were systematically seen as apocalyptic messages: "They presage world reformation, the overthrow of the wicked, and the vindication of God's elect."

The popular tracts on monsters sold for a penny during the Renaissance generally began with a catchy title, presented a woodcut of the person or animal being described, and then proceeded to render an account of God's message in that particular case, in prose or in poetry. Within this tradition, some authors left nothing to chance: the message conveyed by Boaistuau's (see below) monster of Cracow, covered with the heads of barking dogs, who died after four hours, saying "Watch the Lord cometh!" could not be more clear. Likewise, the anonymous author of a pseudo-Aristotelian book entitled *The Complete Master-Piece* described "a most strange, hideous and frightful Monster indeed . . . an Hairy child," born in Arles in 1587. And he added:

> it was all covered with Hair like a Beast. That which rendered it yet more frightful was, that its navel was in the place where his Nose should stand, and his Eyes placed where his mouth should have been, and its Mouth was in the Chin . . . He lived but a few days, affrighting all that beheld it. It was looked upon as a Forerunner of those Desolations which soon after happened in that kingdom, wherein Men were towards each other, more like Beasts than Humane Creatures.

Then he concluded, resorting to poetry,

> here Children thus are born with hairy Coats
> Heaven's Wrath unto the Kingdom it denotes.

THE WRATH OF GOD

The idea of the wrath of God seems to have always been present in Western writings—and to awaken a morbid fascination in readers of all times. Descriptions such as the one of female twins joined at the back, offered in a pseudo-Aristotelian publication from the eighteenth century, are always easy to find. Apparently, these twins lived several years; "but one outlived the other three years, carrying the dead one (for there was no parting them) till at last the other fainted with the burden, and more with the Stink of the dead Carcase." Every age has the tabloid press it deserves.

By the end of the sixteenth century, treatises on monsters had become a veritable genre, with a huge and avid following. The trend was started by some illustrious Medieval forerunners, such as *La nature et les prodiges,* by Jean Céard, and *Reveils et prodiges,* by Jugis Baltrusaitis. In 1560, Pierre de Boaistuau explained this fascination in the first volume of *Histoires prodigieuses*[9] by stating that monsters are creatures "wherein we see the works of nature turned artwise, misshapen and deformed, but (which is most) they do for the most part discover onto us the secret indignation and scourge of God, by the thing that they present . . . that we be constrained to enter into ourselves, to knock with the hammer of our conscience, to examine our offices . . . specially when we read that oftentimes the elements have been heralds, trumpeters, ministers and executioners of the Justice of God." Among other "sundry kings, bishops, and monarchs" whose "marvelous death" has been enacted by God as a vengeance for their wretched behavior, the author cites the notorious case of the Babylonian ruler Nebuchadnezzar, who conquered a vast part of the known world and submitted all of its peoples to his tyranny, only to die afflicted by a strange dementia that made him think himself a beast, walk on all fours, and eat grass.[10] "It is most certain," adds Boaistuau, "that monstrous creatures for the most part proceed of the *chatissement* and curse of God," so that their parents repent of having indulged freely in their bestial appetites, "regardless of age, place, time and other laws of nature."

To back this vision of monsters as ungodly beings, Boaistuau affirms that "you can read most at large" in the Bible that God told Moses "not to receive them to do sacrifice among his people." Here we find a hint of what is now termed "embryological theology," an approach that produced masterpieces such as F. G. Cangiamilla's 1700 *Embryologia Sacra,* which included instructions for the baptism of monsters, or the report issued by the Doctors of the Divinity at the Sorbonne in 1733, in which intrauterine baptism by means of a syringe was solemnly recommended. This legacy is still manifest in the internal rules of the Catholic Church: clause 241 of the current Canonic Law, by stating that "the Bishop shall only admit into the major seminary those who, through their qualities moral and human, spiritual and intellectual, psychical and *physical health* . . . can be considered apt to perpetually dedicate themselves to the sacred ministries," actually bans men bearing a vast inventory of deformities (from lacking one hand to being lame) from being active as priests. Further paragraphs of the same clause ban children of "adultery or sacrilege" from entering the ranks of the Church, and suggest that "hereditary conditions" must also be taken into consideration when selecting candidates for priesthood.

The spectrum of monsters considered by Boaistuau is certainly very broad, as becomes clear from these descriptions of how the simple sighting of such beings can change the course of historical events: "The emperor Adrian, chancing to see a Moor, assured himself to die immediately"; or "the soldiers of Brutus, being ready to engage in battle with the army of Octavus Caesar, having encountered an Ethiopian in their way prognosticated that they would lose the battle, which happened according to their imagination." But even in this loose context the author was already aware of counterfeits and falsifications. In fact, although the more learned texts insisted on the association of monsters with bad things to come, the curiosity of the masses seems to have easily sidestepped this issue and made monsters a popular source of entertainment, frequently displayed in local pubs or at country fairs. In "Unnatural Conceptions," Park and Daston tell the tale of the "Two Inseparable Brothers," immortalized as such in a popular ballad, in the following terms:

Fig. 32. Two inseparable brothers: Lazarus Coloredo and his parasitic twin John Baptista. (From Park and Daston, "Unnatural Conceptions," 1981; reprinted by permission of *Past and Present.*)

On 4 November 1637 . . . Sir Henry Herbert, Master of the Revels, granted a six-month license "to Lazaras, an Italian, to shew his brother Baptista, that grows out of his navel, and carries him at his side." Lazarus Coloredo and his parasitic twin John Baptista arrived in London at the age of twenty after appearances on the Continent. Lazarus' exhibitions were a great success. In 1639 he was still in London; he later appeared in Norwich, and, in 1642, in Scotland, on what appears to have been an extended tour of the provinces.

There is little wonder, then, that people with no other resources turned to fake monstrosity as a means of earning a comfortable living. As Boaistuau wrote,

These masked pilgrims, or rather absolute hypocrites, studying nothing but the philosophy of Satan, as soon as their children be born, while their skins and bones be tender and flexible, with small force will brake their arms, crush their legs and puff up their belly with some artificial powder, defacing their noses or other parts of their faces, and sometimes poking out their eyes, all to make them appear monstrous.

As knowledge progressed, the attention given to monsters slowly evolved from a "freak show" to a medical approach. The nineteenth volume of the collected works of the influential French royal surgeon Ambroise Paré, first published in 1573, was a book titled *On Monsters and Marvels,* copiously illustrated with reproductions of "monstrous" births and other related occurrences, from the "picture of a Marine monster having a human torso" to the "figure of a dead fetus carried in the mother's womb for twenty-eight years." In this sense, Paré's work emerged as an anticipation of Francis Bacon's recommendation to natural philosophers in 1620 that "a compilation, or particular natural history, must be made of all monsters and prodigious births of nature; of everything, in short, which is new, rare and unusual in nature. This should be done with a rigorous selection, so as to be worthy of credit." [11] Paré's images, presented in well-organized groups of "cases" and therefore conveying a solid sense of technical reliability, had a strong impact on the imagination of both the common reader and learned scholars. [12] For the first time, we glimpse a shift from monsters as prodigies to monsters as examples of medical pathology, as we find the primordia of a teratological taxonomy clearly explained:

Monsters are things that appear outside the course of Nature and are usually signs of some forthcoming misfortune, such as a child who is born with one arm, another who will have two heads, and additional members over and above the ordinary. Marvels are things which happen that are completely against Nature as when a woman will give birth to a serpent, or to a dog, or some other thing that is totally against Nature.

Paré also systematized the "several things that cause monsters:"

The first is the glory of God.

The second, His wrath.

The third, too great a quantity of seed.

The fourth, too little a quantity.

The fifth, the imagination.

The sixth, the narrowness or the smallness of the womb.

The seventh, the indecent posture of the mother, as when, being pregnant, she has sat too long with her legs crossed, or pressed against her womb.

The eighth, through a fall, or blows struck against the womb of the mother, being with child.

The ninth, through heredity or accidental illness.

The tenth, through rotten or corrupt seed.

The eleventh, through mixture or mingling of seed.
The twelfth, through the artifice of wicked beggars.
The thirteenth, through Demons and Devils.

He added that still other causes can explain "why persons are made with only one eye in the middle of the forehead or the navel, or a horn on the head, or with the liver upside down," as well as for "certain monsters which are engendered at the sea." These are left to a sort of miscellaneous final chapter, since "it would take too long to describe" all their peculiarities. This may all seem rather tenuous to us now, but it definitely paved the way for all the literature that followed. In the way Paré had systematized them, monsters still held some ground as prodigies, diffusing the boundaries between the natural and the supernatural; on the other hand, as part of the study of natural history, they brought the natural closer to the artificial. Reports on monsters and their causes would not change much during the following centuries, when preformation emerged in the generational debate.

NATURE'S GOALS

Much of Paré's legacy was still alive at the beginning of the eighteenth century, although some of its premises were now clearly open to debate. This continuity, submitted to questioning, is manifest in a book by the Flemish Jean Palfyn, *Description anatomique des parties de la femme qui servent à la Génération, avec un traité des Monstres,* published in 1703. Like Paré, Palfyn was a surgeon and an anatomist, but he was exerting his skills two centuries later. His section on monsters opens with a discussion on what should and should not be considered as such. The "figures of lions" or of "soldiers in battle" seen in the clouds should not really be considered monsters, but rather Signs and Prodigies. As for animals that are born from others of a different species, such as "a frog born out of a serpent" or a "white child born from an Ethiopian," they are not really monsters, but rather monster-like creatures. The same applies, in Palfyn's opinion, to "those men of Africa who are said to have only one eye, or only one foot, or not to have a head." From a clinical point of view, the author claims, a monster is "an animal (and particularly those that have something human about them) whose arrangement and disposition of the members is extraordinary, which are different from those that gave birth to them, and which are rarely born, such as a man with three hands, a pony with a human head, a child half-dog, a girl with two heads, or a centaur."

Showing that time has passed and thought has evolved since the days of Paré's publications, Palfyn stated that monsters were not signs of things to come. He explained that this reasoning, started by Aristotle and widely endorsed ever since, held that monsters were vices and defects of Nature, who had failed to fulfill her functions correctly; and, that if they were going to be signs of something, they had to be signs of something bad. "But," contradicted Palfyn, "I have known of very poor mothers and fathers who got out of their miserable lives and amassed vast quantities of money by showing a monster-child that had been born to them, and this happened without any damage occurring in their countries at that time." Although he believed that "the good Lord can produce by miracle all sorts of monsters," he did not think "that He does so to announce to men any kind of calamity to come." And here is why:

> In the deserts of Africa, where a great quantity of monsters is constantly being engendered, there are no men for them to announce future misfortunes to; and also, monsters are often born among peoples who do not know or revere the true God, and who therefore could not understand in these prodigies the sign of God's intentions; and yet that monsters generally result from malignant actions, so that it seems impious to me to assume that God would use them as messengers.

Nevertheless, Palfyn was still sure that human monsters exist and have always existed, citing numerous sources from antiquity. His very long inventory includes a huge variety of monster types, from the "several monsters who were left behind when the builders of the Babel Tower dispersed" to "a Roman servant who gave birth to a monster consisting of solely one hand and another who had her child from the canal used to pass out the excrements" and "a centaur that was born from a mare." An equally long list deals with similar abnormalities found among beasts. Then, the author proceeded to demonstrate that there are also "similar defaults of Nature" at the roots of "trees and plants" and even "in what they produce." These defects are more rare than those found among animals, but that is only because "the Nature of plants is more simple, and, since she does not require as many organs for the production of different species, she sins less frequently." These monsters include "several fruits of the same tree joined together," which has been observed in figs, peaches, plums, "and even lemons," and should be attributable to "an excess of seed, exciting several germs to grow simultaneously." Similar excesses, or the presence of all kinds of bizarre-looking "excrescences," were also detected in flowers and in cereals.

The anti-Aristotelian attitude of the Scientific Revolution notwithstanding, Palfyn stuck with Aristotle in assuming that "Nature does nothing in vain and has a goal in everything she does." Therefore, "monsters, being creations of Nature, have to have been simultaneously engendered by all the Causes, the Material, the Formal, the Efficient, and the Final." Nature aims, above all, to keep species alive and distinct from one another.

The goal she achieves in the production of monsters is solely a part of the goal she achieves in the production of perfect animals, because when she produces monsters she does not always attain the goal she aims to and desires, but only that part of that goal that is of most importance to her, the most essential, and that is the perpetuation of the soul.

MAKE THEM BEAUTIFUL

A more sophisticated approach to the theme of monstrosity appeared in France some decades later. In 1751, when M. Vandermonde wrote his *Éssai sur la manière de perfectionner l'éspece humaine,* heredity was becoming a subject of great interest to natural philosophers, and the author, then regent of the Faculté de Médecine de Paris, had a number of practical suggestions in mind. Arguing that "the beauty of men is but a reflection of the beauty that the Creator has dispersed all through the universe. The order, the arrangement, the proportions, the symmetry, are all in His works," he pointed out that "when we do not consult the portrait of Nature . . . we get lost and we become prey to very bad taste," offering as examples of this effect the small feet of Chinese women, the small waists of French women, the small ears of the Ethiopians, and the huge ears of the South American Indians, or the black teeth of some Indian women. Based on this principle, he advocated a sort of proto-eugenics: beautiful people should marry each other, so that their children are even more beautiful, and so forth. This program, in his opinion, was a strong argument for ruling out prearranged marriages.

Still concerned with ultimate beauty, Vandermonde then held that monsters are born out of the disrespect of their parents for the proper times for procreation, insisting that the ancients were already aware of this problem and wrote fables to illustrate it:

> They imagined that Jupiter, excited by the fumes of the nectar that had inebriated him, decided to give his spouse the marks of his love, and that Juno gave birth to Vulcan, that monster who was neither man nor god, and that was chased out of Olympus because he was the fruit of his father's imprudence.

This imprudence includes marriages between persons too far apart in age:

> If the repeated efforts of the older spouse give birth to a child, what should we not fear from this bizarre coupling?

Also to be avoided are marriages between pubescent children, aimed only at augmenting their parents' estates.

The unhappy victims of their parents' greed become the authors of their own destruction, and if one day they give birth to children it is only to impinge upon those children all kinds of infirmities that shall lead them to death.

Even the union of a big man with a small woman or vice versa will produce a deformed being. Vandermonde also noted that children born in big cities are more likely to have deformities because all kinds of vices are rampant in those places. The thing of beauty may be a joy forever, but it is a hard one to achieve; and the achievement is doomed to fail if generation is not kept under constant scrutiny.

THE BEAST WITHIN

As is evident from the range of works quoted above, explaining the generation of monstrosities had never been an easy matter. But the subject had certainly always been very appealing, and it commanded the attention of leading ovists such as Haller *(De monstris,* in the supplements of the *Elements de physiologie)* and Bonnet (in the eighth chapter of the *Considérations sur les corps organisés)*, together with other prominent scrutinizers of nature such as Buffon (in the fourth supplement to the *Histoire naturelle)*. Judging from their own words, the writers of the Enlightenment may have considered teratology the satisfaction of an "unhealthy curiosity," but they were exceedingly prolific on the subject. It appears that our "taste for the unnatural" tends to surpass by far any kind of ethical restraint, even as it expresses itself in different forms in different ages, according to the idiosyncrasies of the moment.

A particularly interesting explanatory model, rooted in antiquity and present ever since, reached its height with Paracelsus, who pointed out, in *Astronomia Magna,* that man has both a spiritual and an animal capacity, and that, therefore, when one calls a man a wolf or a dog, this is not a matter of simile, but a matter of true authenticity.[13]

This view was directly derived from Paracelsus's belief that the microcosm is affected by the macrocosm because both are made from the same substance. In turn, Paracelsus was deriving his own inspiration from a vast range of classical and medieval sources: before him, Arnauld de Villeneuve proposed a cycle of transformation of the elements and their qualities; then, the cabalists showed in man a reflex of the macrocosm, connecting us with the whole of Nature; and then, through alchemy, the hermetists recovered the notion of universal evolution and general transmutation, suggesting the possibility of an extended monadism. Paracelsus was also recovering from Hippocrates the correspondence between microcosm and macrocosm, since Hippocrates had already said, "Man is made of particles encompassed by the Whole." As rendered by Harris L. Coulter in his 1975 *Divided Legacy,* Paracelsus's theory, as presented in *Astronomia Magna,* could be summarized as follows:

> The Scriptures state that God took the *limus terrae,* the primordial stuff of the earth, and formed man out of this mass. Furthermore, they state that man is ashes and powder, dust and earth; and this proves sufficiently that he is made of this primordial substance . . . But *limus terrae* is also the Great World . . . *Limus terrae* is an extract of the firmament, of the universe of stars, and at the same time of all the elements.

This line of reasoning, implying that there is no real distinction between humans and animals, is further substantiated in Paracelsus's "De Homunculis,"[14] in which the author recalls his theory of man as microcosm to point out that, since man is made from the dust of the earth, he contains all the powers of Creation within himself; and thus, if he acts in a bestial fashion, he expresses the beast within, literally becoming the animal whose behavior he imitates.

However, the idea of the beast within was not entirely a Paracelsian creation. The same theme appears long before the sixteenth century, and is clearly embodied in the classical centaurs, a tribe of mortal man-beasts, tamed like horses but not civilized like men, who became notorious for misbehaving at festivities held by humans, particularly when they were drunk. Until the appearance of frescoes by the Greek painter Zeuxis, centaurs seemed to be exclusively male, thus leading to the suggestion that they could reproduce only through the rape of women. But Zeuxis brought centauresses and baby centaurs into the picture, giving the tribe full reproductive rights as a truly autonomous species. This, however, did not rob the centaurs of their role as living expressions of the animal inside the human. As von Blanckenhagen writes in "Easy Monsters,"

> The combination of the body of the horse with the torso and the head of a man results in something visually unified and acceptable, as if nature may indeed create such monsters . . . There emerges the convincing image of a powerful being transcending the power of horse and man separately, and thus it represents, as it were, what is beastly in man's nature. But what is beastly in man is not only wildness, fury, sensuality; it is also being part of nature itself, i.e. familiarity with all her powers, her secrets, her gifts.

This is why centaurs like Chiron can be the tutors of future heroes: Chiron teaches Achilles not only the secrets of attack and defense, but also the secrets of healing herbs and many other traits of natural instinct, the opposite of human intelligence.[15]

The concept of monstrosity through hybridism remained alive and well during the Scientific Revolution. This is clearly illustrated by the exponential growth of punishments related to any form of bestiality. Prescriptions for such punishments appear abundantly at the turn of the seventeenth century in the legal codes of many countries.

The approaches recorded at this time applied just as well to the birth of suspicious-looking children as to the genealogy of all the strange new animals revealed to Western culture during the Discoveries. As written in the logbook of a French ship arriving from Senegal in 1719,

> monsters . . . are produced with frequency in Africa, due to the accidental en-
> counters of a great number of different animal species at the side of the rivers and
> subjected to a climate which excites them to cross among themselves indiscrimi-
> nately and in disorganized manner.[16]

The "Camelopardalis" described in Gaspar Scott's 1667 *Physica curiosa, sive Mirabilia natura et artis,* presumably a giraffe[17] but considered at that time to result from intercourse between a camel and a leopard, is a good example of this hybrid monster variety. Interestingly, in the extensive zoological accounts from that period, such as Edward Topsell's *Historie of Foure-Footed Beastes,* these presumed hybrids are listed side by side with absolutely normal ani-mals: Topsell's inventory includes dogs, horses, and wild boars in the same pages featuring specimens like the Camelopardalis or the "Allocamellus"—most likely a llama, since the text mentions that it was discovered in Peru and that it came "from the land of the giants"—described as being the result of intercourse between a mule and a camel, because "it has the head, neck and ears of a Mule but the body of a Camel."

However, monstrosities related to hybridism were most often associated (at least in the secondary literature) with suspicious activities of a human mother. Javier Moscoso notes that this attitude has deep roots, as exemplified by the clarion of the French surgeon Jean Riolan, issued in 1605, that "a mon-ster is a perversion of the order which assures the continuity of natural causes, a rupture of the virtue; of the health of the people, of the authority of the king." Moscoso further points out that, although these matters were rarely clearly stated, the truth is that "whenever a monster was born with a certain similarity to a different animal species, the honesty of the mother was straight away taken into consideration."

It is true that Ambroise Paré seemed to regard both parents as equally responsible for such occurrences, since in his opinion "God permits that the parents produce abominations as a consequence of the disorders in a copula-tion that they practice, as brute beasts, conducted by their appetite and regard-less of the times and laws ordered by God and by Nature." But women, in

Chapter Four

Fig. 33. The "Chamoelopardalis:" Paré's drawing of the giraffe, considered by some naturalists to be a hybrid of camel and leopard. The original text reads, "in the kingdom of Camota . . . of Bengal, and other mountains . . . which are in interior India, beyond the Ganges river some five degrees beyond the Tropic of Cancer, is found an animal called, by the Western Germans, the Giraffe . . . It is spotted in several places, like the Leopard, which has given to some Greek Historiographers the argument that he should be given the name of Chamoelopardalis." (From *On Monsters and Marvels,* 1573.)

great part due to the sculpting powers ascribed to their indomitable imagination-tions (see chapter 3), were ultimately seen as the main culprits. According to Moscoso, as late as 1788, a certain Dr. Balbot was still writing that although a woman and a dog could not produce a monster directly through intercourse, considering the different times of gestation between the two species and the relative weakness of the animal seed, monsters with "canine appearance" were the result of "the imagination of a mother who, during her marital activities, had kept to herself the abomination of her crime." There was no way women could free themselves from generalized suspicion. When two girls were born connected by their heads (one died at the age of ten and the other followed her soon afterward, as a result of wounds suffered during surgical attempts to sever the dead body from the twin that was still living), Boaistuau considered

Fig. 34. Some of the cases of hybridism presented by Paré: *(A)* the "figure of a child, part dog," and *(B)* the "figure of a pig, having the head, hands and feet of a man and the rest of a pig" are both given as examples of "monsters created through the mixture or mingling of the seed." (From *On Monsters and Marvels,* 1573.)

this phenomenon a result of the habit that women have of joining their heads together when they are gossiping. In this case, the problem was simply that the mother of the conjoined twins was pregnant when she engaged in this well-known female pastime. According to such influential popularizers as Nicholas de Venette,[18] even a boy who resembled his father was not necessarily legitimate: "his mother could have produced that resemblance by the force of her imagination during her illicit sexual contacts." No solace for women here, make it popular or erudite.

CHEMICAL REASONS

These accounts, and their frequency in the secondary literature, provide an interesting testimony to the despised role occupied by women in the universal configurations of all natural philosophies (see chapter 7). For a number of alternative explanations concerning the birth of monsters have emerged through the ages, and yet we seldom hear of these other proposals.

First of all, we must consider an explanation for monstrosities that the ovists could have used to their own advantage—just as the spermists could have used the model of the wounded warrior—but, like the spermists, never did. In the thirteenth century, Albertus Magnus of Cologne and Bolstadt, one of the masters of Thomas Aquinas, attributed the birth of deformed creatures to problems concerning the egg alone, through a sort of primitive chemistry. His reasoning starts with knowledge of the artificial incubation of eggs:

> For the alterative and maturative heat of the egg is in the egg itself and the warmth which the bird provides is altogether external since in certain hot countries the eggs of fowls are put under the surface of the earth and come to completion of their own accord, as in Egypt, for the Egyptians hatch them out by placing them under dung in the sunlight.

It follows that, if the eggs can develop by themselves under the influence of external stimuli, it is understandable that some may develop in a wrong manner. Thus monsters, according to Albertus Magnus, result from corruption of the eggs, which can occur in four different ways: by decomposition of the white, by decomposition of the yolk, by the bursting of the yolk membrane, ^r by the aging of the whole egg. "And from the second cause," he concludes,

> it sometimes happens that in the corruption of the humors certain igneous parts are carried blazing out of the shell of the egg and distribute themselves over so that it shines in the dark like rotten wood.[19]

Three centuries later, Albertus Magnus's ideas still intrigued Ulysse Aldrovandi, whose *Ornithologia,* published in Bologna in 1599, includes a passage suggesting that monsters come from eggs with chemically abnormal yolks.[20] Albertus Magnus's legacy certainly influenced Réaumur to pursue the efforts on artificial incubation reported in 1749 in his *De l'art de faire éclore les Poulets,*[21] the first attempt to introduce to France the Egyptian technique of artificial incubation of eggs. In this work, presented in two volumes with a total of 478 pages and fifteen plates, Réaumur showed how to preserve eggs by smearing them with fat, and how to hinder the evaporation of "spirituous liquors" through the use of mercury. The influence of these ideas and methods is still traceable as late as the nineteenth century, as is apparent from this brief passage, written in 1809:

> During the period that I was at College, the late Sir Busick Harwood, the ingenious Professor of Anatomy in the University of Cambridge, frequently attempted to develop eggs by the heat of his hotbed, but he only raised monsters, a result which he attributed to the unsteady application of the heat.[22]

Following a different line, the Bohemian Marcus Marci of Kronland published in 1635 a book called *Idearum Operatricium Idea,* in which scientific contributions to the understanding of optics were combined with speculative theories of embryology. His rationales combine the old Aristotelian ideas of seed and blood; the new rationalistic mathematical approach to generation that Descartes was trying to use; the progress of experiments in optics; the cabalistic mysticism of light as the source and origin of things; and, finally, the postulate of the existence of "centers of radiant energy" inside the embryo. From this melting pot emerged an explanation of the complexity of the generational seed by analogy with lenses, which also produce complicated beams from a simple light source. Similarly, according to Marci, the formative force radiates from the geometric center of the body, creating complexity but retaining its power. Within this frame, monsters appear as the consequence of accidental doubling of the radiating center, or of abnormal reflections or refractions at the periphery.[23]

A BLESSED INTERVAL

From all the evidence presented above, a reality emerges that adds one more ironic touch to the tale of the wounded warrior: the spermists entered the reproductive debate right at the moment of change, when the whole perception of nature, and hence of the meaning of monsters, was undergoing a profound remodeling. Had they seized the occasion to offer an explanatory model for this particular phenomenon, the modernity of their claims would have been even more striking, cutting straight through one of the greatest perplexities of their time. As we shall see in this section, the use of the microscope contributed a brand new seriousness to the old theme of the jokes of Nature. The leading spermists were, fundamentally, fine microscopists. If they had used that device of modernity par excellence to study monstrosity, they could have—at least temporarily—provided a concrete explanation for the existence of monsters amid the turmoil that surrounded the issue, and by those means scored points in their favor. But, at what would have been for them a blessed interval, none really took that path.

The perplexities involving the classification and handling of monsters were many, and they appear in virtually every form of discussion during the sixteenth and seventeenth centuries. Civil and canon lawyers debated the marriageability of hermaphrodites, and whether both heads of Siamese twins deserved baptism. Meanwhile, theologians were busy debating whether human monsters possessed rational souls and, if so, in what form they would be resurrected.

These new worries mirror a historical moment when the religious implications of monstrosity had suddenly become larger than life. During the troubled days of the Reformation, the rebels inside the church seized upon the potential of monster lore to serve their cause—just by adding to the old tales of bad presages the inspired twist of their presentation as signs of how urgently God wanted His own religion to change.

On this front, the most influential push came straight from Luther's pen. In 1523, with the help of Melanchthon, he published the polemic short pamphlet "Baptstesels zu Rom und Munchkalbs zu Freijberg in Meijsszen funden." According to Park and Daston,

> the pamphlet was a pointed attack on the church. It began with two woodcuts, one of the "monk-calf," an actual calf born several months earlier in Freiburg with what looked like a cowl around its neck, and the other . . . of the "pope-ass," a composite and clearly fictitious monster reputedly fished out of the Tiber in 1496. The "pope-ass" . . . represented the "Romish Antichrist," its various bestial parts corresponding accurately to the bestial vices and errors of the church. The monk-calf . . . symbolized the typical monk—spiritual in externals, but, within, brutal, idolatrous, and resistant to the light of Scripture. Both monsters were prodigies prophesying the imminent ruin of the Roman church.

Luther and Melanchthon had no scruples against putting these new polemical devices to work in more than one religious war; in their hands, monsters

Fig. 35. God produces monsters mocking the institutions of the Catholic Church to announce that the time has come for Reformation: Luther's "pope-ass." (From Park and Daston, 1981; reprinted by permission of *Past and Present*.)

Das Mvnchkalb zu freyberg

Fig. 36. Luther's "monk-calf." (From Park and Daston, 1981; reprinted by permission of *Past and Present*.)

were used against Calvinism during the French religious battles, against Rome in late-sixteenth-century England, and against separatism during the English Civil War.

Then came the use of the microscope, and, at the same time, another development took place on the monster scene. In the mid-seventeenth century, largely under Bacon's impulse, the trend in the study of monsters moved clearly toward an emphasis on natural over supernatural causes. Although he still believed in monsters as God-sent presages, Paré himself had already hinted at this new attitude, saying, for instance, about a strange creature reportedly found in Africa, that he could offer no explanation for its occurrence other than that "Nature was playing to make us admire the greatness of her works." Here Nature is no longer a mere mirror of God's will, but a creative entity in her own right. Monsters became increasingly treated as jokes of a playful Nature rather than divine prodigies. "They signified [Nature's] fertility and invention," write Park and Daston, "and through her God's own fertility and creativity, rather than his wrath."

The idea of Nature displaying her sense of humor in numerous creations had been launched by several writings of antiquity, most notably those of Pliny, who held very clearly in his *Natural History* that "Nature, in her ingenuity, has

Fig. 37. Nature enjoying herself with the game of similarities, as depicted by Giambattista Della Porta in *De humana physiognomia,* 1610. (From Findlen, "Jokes of Nature and Jokes of Knowledge," 1990; reprinted by permission of *Renaissance Quarterly.*)

created all these marvels . . . as so many amusements to herself," singling out shellfish and flowers, among others, as examples of Nature's "sportive mood," expressed in their variety of shapes and colors. This classical heritage had long been a central point of Renaissance natural philosophy, and the concept endured until the middle of the seventeenth century. Everything that was hard to explain by current standards, and yet was obviously not an aberration, could be classified in the taxonomic realm of scientific jokes. This classification even applied, as Findlen reports,[24] to cases such as the giant cabbage detected in a hospital garden in Nuremberg in 1697. This cabbage, noted an entry in the *Galleria di Minerva* of the same year, "is privileged in the natural order, because it is not according to the usual laws of nature."

This sentence denotes a slight detour in the philosophical rationale of the new times: at the close of the seventeenth century, Nature was seen not only as playing jokes to amuse herself, but also as an artisan fulfilling the duty of adding colors and shapes to the world outlined by God. To illustrate this point, Ulisse Aldrovandi's *Musaeum Mettalicum* in Bologna was filled with examples of Nature's final touches: stones resembling human appendages, and stones containing crosses, shields, swords, or torches. "Nature has joked uncommonly in all the outward appearances of natural things," wrote the Danish physician Olaf Worm in the 1651 catalogue of his museum.[25]

As Findlen points out, the generalized use of the microscope in the mid-seventeenth century only enhanced the seriousness of Nature's enterprise, shifting attention away from Pliny's concept of jokes as a source of variety and toward the seriousness of Nature's hidden structures. In the Jesuit Athanasius Kircher's 1661 *Diatribe de prodigiosis crucibus,*

> the microscope became an instrument of religious as well as scientific revelation, for its magnifying powers allowed the viewer to discover more easily crosses formed naturally in fruit, rather than using the naked eye. In fact it seemed that nature had left her brushstrokes everywhere, if only one looked closely enough.

From the idea of Nature playing tricks to enliven the work of God to the trend of learned men playing tricks to imitate the jokes of Nature, there was but a very small distance to cover—and it certainly was joyfully covered, without the slightest trace of concern about possible unethical behavior by the naturalists. In her article, Findlen tells the story of the English virtuoso Philip Skippon on a trip to Modena in 1664, where he visited the ducal palace and was given a tour of the Este collection. He noted the presence of

> A *Hydra* with seven heads, the middlemost of which was biggest, and had two canine teeth, and six little ones between, and a long tail, two feet, with four claws on each, and five rows of tubercles on his back . . . Very probably this *Hydra* was fictitious, the head being like that of a fichet . . . , the body and feet were of rabbit or hare, and the tail was made of common snake's skin, the back and the neck covered with the same.

There is no whistle-blowing in Skippon's account. These forgeries were deeply embedded in the spirit of their time, and either way you looked at them, you won: you were invited either to participate in the joke, understanding the subtle transition from natural to artificial, or to be deceived by it, becoming in a certain sense part of the joke, and thus part of the allied efforts of God, Nature, and man to embellish the created world.

Bacon pushed Paré's early hints one step further by holding that monsters both illustrated the regularities of nature, for "he who has learnt her deviations will be able more accurately to describe her paths," and instigated the inventions of art, since "the passage from the miracles of nature to those of art is easy." [26] The idea that unraveling the innermost secrets of nature required the close study of nature's anomalies dominated several seventeenth-century studies on singular phenomena, such as a double refraction in Iceland spar—and these phenomena were often described as "marvels" or "monsters" of nature by Bacon and his followers. This kind of approach, as Bacon had warned from the start, required extreme care in the verification of the authenticity of the marvelous things recounted. As put by Park and Daston,

> reporters of monsters were to identify the authority or witness from whom the description originally derived, to assess the reliability of the source, to state how the source had come by the information (eyewitness, oral or written), and to judge whether additional corroboration was required.

As the English Royal Society proceeded to follow Bacon's instructions, in France the Académie des Sciences took an even more sober approach to the problem, leaving the participation of God further and further behind. Here, much emphasis was placed on the dissection of cadavers. Selected, carefully chosen anatomists dissected not only monsters, but also a wide variety of normal animals, following the reasoning that some structures key to the understanding of teratology could be hidden or hard to detect in one species, but clear and self-evident in another. The resulting conclusions verged much more on the physiological than the theological: the surgeon Jean Méey, for example, based a whole theory of fetal nutrition on his studies of a monstrous fetus without a mouth.[27]

In any event, by the late eighteenth century, jokes of nature had disappeared almost entirely from the discourse of naturalists. The Baconian idea of using deviance to understand the global taxonomy of the world was also sliding into oblivion. Mountains could no longer be considered examples of monstrous confusion in Nature, but were simply part of a tightly defined geologic plan. The outcome of the Galilean, Cartesian, and Newtonian postulates had produced a new brand of natural vision that was bound to take over and become known as science. Within this new scientific frame, Nature submitted herself to a structure that was increasingly rational and selective, therefore less and less inventive and freely poetic.[28] The model of the wounded warrior could no longer be of any help within this new context. The spermists had not used it at the right moment, and now even spermism itself was vanishing from sight.

HAUNTED BY THE CRAYFISH

We must admit at this point the possibility of having wasted tens of pages on the analysis of a paper tiger. If we assume that the spermists could have been better positioned than the ovists to conjure up biological models to explain the birth of monsters, we are implicitly making the assumption that the birth of monsters presented a complex problem for all preformationists. This assumption has been stated explicitly in numerous secondary sources. But, in fact, the preformationists had to confront problems of much broader proportions, and they still managed to pull through—although this is rarely stated often enough. The case of the regeneration of the polyp, for example, represented a much more challenging puzzle for those who believed in God's original encasement of all creatures. But preformation dealt with this issue rather nicely.

Although it had long been known, through simple empirical observation, that lizards and salamanders could regenerate their tails if they were cut off, no one really anticipated what Abraham Trembley would discover when he started his observations on the freshwater hydra, a microscopic creature then generally called the "polyp," a term chosen by Réaumur because of the animal's similarity to the macroscopic octopus. Unbeknownst to Trembley, this creature had already been described by Leeuwenhoek, who considered it an animal and noted its reproduction through budding. But the report of the Dutch microscopist, published in a 1703 issue of the *Philosophical Transactions of the Royal Society,* had gone practically unnoticed. And, most importantly, Leeuwenhoek had not noticed the dramatic reproductive behavior that the Swiss naturalist was about to report.

After much debate about whether the polyp was animal or plant,[29] Trembley established that this organism was indeed an animal after he noticed that it caught food with the tentacles and delivered this food to an interior stomach. After watching one of his creatures ensnare a tiny eel with its arms and stuff the prey into a central cavity, he happily wrote to Réaumur: "They are carnivorous, and certainly very avid!" He also noted that the polyp reacted to touch, and even to the agitation of the water; and that it was able to walk, using a primitive foot in a type of head-over-heels locomotion, similar to that of inchworms. Moreover, the polyp reacted to light, consistently moving toward the warmest and brightest areas of the jars in which the author kept his specimens—and this reminded Trembley of the ability of birds to migrate.

On the other hand, Trembley noticed that not all of his polyps had the same number of tentacles, which was reminiscent of the uneven number of branches and roots found in plants. Divided in his mind between the animal and vegetal options, he decided to cut the presumed animal in two, assuming that only a plant could regenerate after this kind of treatment. Assuming that the cut polyps would die, he hoped to establish, by these means, their true animal nature. Much to his surprise, however, each one of the halves regenerated a full polyp. The results were the same whether the polyp was cut crosswise, lengthwise, or even in several pieces. Imperturbably and infallibly, new entire polyps always regrew from those pieces. In a last test of these astonishing capacities, Trembley turned the animal inside out, inserting a bristle into its gut and pulling the body back. The polyp did not seem to care, and just grew a new outside over what had been the former inside.[30]

Trembley announced his perplexing discoveries for the first time in De-

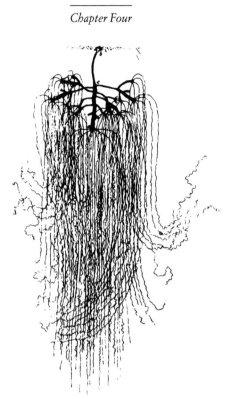

Fig. 38. Abraham Trembley's drawing of a freshwater polyp reproducing by budding. (From *Mémoires pour servir à l'histoire d'un genre de polypes d'eau douce à bras en forme de cornes*, 1744; as reprinted by Roe in *Matter, Life, and Generation*, 1981.)

cember 1740. He formally published his results in 1744,[31] stirring the passions of a community of naturalists still a century away from the development of the cell theory and knowing nothing about the organization of colonies of unicellular animals.[32] The preliminary report published in 1741 by the *Histoire de l'Académie des Sciences* could not have been more dramatic:

> The story of the Phoenix who is reborn from its ashes, fabulous as it is, offers nothing more marvelous than the discovery of which we are going to speak. The chimerical ideas of Palingenesis or regeneration of Plants & Animals, which some Alchemists believed possible by the bringing together and the reunion of their essential parts, only leads to restoring a Plant or an Animal after its destruction; the serpent cut in half, & which is said to be rejoined, gives but one & the same serpent; but here is Nature which goes farther than our chimeras. From one piece of the same animal cut in 2, 3, 4, 10, 20, 30, 40 pieces, & so to speak, chopped up, there are reborn as many complete animals similar to the first.[33]

The subsequent uproar was impressive. "Still I confess," wrote Réaumur in his *Histoire des insectes,* describing the effect of the discovery on him, "that when I saw for the first time two polyps form little by little from the one I had cut in two, I could hardly believe my eyes; and it is a fact which I am not at all accustomed to seeing, after having seen it over again hundreds and hundreds of times." The trend of cutting in pieces any animal suspected of being able to undergo regeneration became so popular that Réaumur pleasantly commented that "during the summer of 1741, lizards, frogs, worms, snakes, crabs, butterflies, and lobsters were threatened by dangers to which they had never been exposed before."[34] How little the preformationists feared that these discoveries would destroy their argument is clearly indicated by the fact that most of the naturalists who followed Trembley's lead and looked for other regenerative organisms were among the leading defenders of the theory: Réaumur and Bonnet soon turned their attention to freshwater worms as likely candidates for regeneration, in what seems to have been a conscious decision to rule out the possibility of the polyp being nothing but an isolated joke of Nature.

Bonnet did most of the work on this front, and, in August 1741, first reported to Réaumur that worms cut in two halves were able to regenerate the missing half. Réaumur read this letter to the entire Académie des Sciences, and the tone of the correspondence exchanged by the two men reveals an enthusiasm that is far from suggesting any kind of conflict with their preformationist beliefs. In November of the same year, Bonnet announced that he had divided worms in three, four, eight, ten, and fourteen portions, nearly all of which grew back a head and a tail. He also noted that not all the segments grew at the same rate, but that apparently the ones closest to the original tail grew the most slowly.

This phenomenon was much more complex than the ability of reptiles or crustaceans to regenerate missing parts, since in those cases the severed parts would die,[35] implying that they lacked the fundamental organizing principle, the "soul" that kept life coherent. The case of the polyp now seemed to suggest that the soul was not located in one precise fulcral part of the body, but was rather disseminated throughout. Or, bolder yet, and much to the satisfaction of the materialistic tastes of Diderot and de La Mettrie (1709–1751; see below), these findings might imply that no soul existed at all—that the properties of life were merely disseminated through all matter. To think along these new lines was hard for any naturalist of the period. And it would certainly appear at first to be impossible for those who defended preformation, since it would mean that the calm continuity of the original encasement was being perturbed by occasional responses to accidents.

The spell cast by regeneration over preformation seems to promise a wholesale devastation when one considers the fate of Hartsoeker's beliefs. He was one of the first spermists, one of the leading philosophers of spermism, and certainly one of the most outspoken representatives of the idea. Yet, in 1722, he pronounced this theory a failure and withdrew from the ranks. Hartsoeker had long agonized over the structure of spermism, and of preformation altogether. But his last straw was the crayfish. "The intelligence which can reproduce the lost claw of a crayfish can reproduce the entire animal,"[36] he sadly wrote, admitting with this statement that regeneration was making any simple preformation theory untenable.

Preformation seemed at this point to be in real trouble. But the situation did not last long. Soon Bonnet gave the old theory a much more malleable and abstract logical mechanism (see chapter 1). His new concept of the "germ" could incorporate all the regenerations of the world. He actually used the process of regeneration as an excellent example of how the development of the germ worked, commenting that "this multiplication that seems to us so extraordinary, would seem to us the most natural, since it is the most simple, if we had never witnessed any coupling in Animals." In parthenogenesis (see chapter 3) or in regeneration, in all those species that are not "submitted to the Law of Coupling," the individual had inside itself "the Principle of Fecundation": "He is furnished with organs that separate from the mass of its Blood the subtle Liquor which must enact the development of the Germs. These Germs are nourished, grow and perfect themselves like the other Parts of the Animal." Therefore, even the polyp could be considered as regenerating from a germ: if the germ, as the author urged, was understood in "its widest sense," then the polyp was just regrowing out of a "secret preorganization."[37]

The absence of any real threat to the credibility of preformation by the regeneration of the polyp is also well illustrated by the fact that those who defended epigenesis did not seize the occasion to deride the postulates of their rivals. A good example lies in the approach taken by Maupertuis (1698–1759; see below) when discussing this subject in his book *Vénus physique*. The French savant had conjured a model for reproduction that was entirely his own and encompassed much unorthodoxy compared with the more classical ideas of Harvey or Wolff—but he supported development de novo at each generation nonetheless, and was clearly not in favor of the general preformationist view. Yet, when considering the regeneration of the polyp, the sole worry he expressed had to do with the source of pleasure in this operation. Maupertuis believed that pleasure was the main driving force behind all reproductive acts. "Nature has the same interest in perpetuating all species," he wrote. "She has inspired in each the same theme, and that theme is pleasure. It is pleasure that, in the human species, drives everything before it—that, despite a thousand obstacles opposed to the union of two hearts, a thousand torments that must follow, conducts lovers towards the purpose that nature has ordained." So how did this universal driving principle apply to the hydra? "In other animals," reasoned Maupertuis, "nature has attached pleasure to the act that multiplies them; could it be that nature has endowed this creature with some sort of voluptuous feeling when it is cut into pieces?"[38]

Besides, epigenesis did not have a sound explanation for regeneration either. During his brief epigenetic period (see chapter 1), Albrecht von Haller wrote that "pre-existence is coming to an end. From the observations done in the polyp, one has to admit that certain animals re-form themselves from their heads, their arms, any of their organs . . . where it is not possible that a miniature being exists."[39] The truth, however, is that epigenesis, as it then existed, could not explain the regular patterns followed by the regenerative process. Nothing in the initial forms of this theory could account for the fact that an animal with a severed tail regenerated a new tail in its place, instead of growing, say, a leg or a head. And the philosophical grounds of early epigenesis were equally unequipped to deal with the fact that, once regeneration was completed, the entire process of growth came to a stop—as if nature was able to grasp that its regenerative task was no longer required.

In fact, epigenesis could have gotten deeper into hot water than preformation by trying to address this problem. Since lost parts or entire polyps were being regenerated de novo, how could the defenders of epigenesis not resort to occult organizing forces to explain the phenomenon? Such forces, unexplainable and invisible as they were, had to be accepted on philosophical grounds. But the mid-eighteenth century, with its strong preference for the mechanical ideas set forth by Descartes and ultimately accepted even by those who recognized the insufficiency of mechanism to explain reproduction, was not the ideal time to propose intangible driving impulses. The problem is clearly illustrated by the suspicion that surrounded the publication, in 1687, of Isaac Newton's (1642–1727) *Principia Mathematica*. Although Newton's genius was generally recognized and admired, many scholars (particularly Leibniz and Huygens) considered the idea of universal attraction "a manifest stupidity," responsible for "turning all the operations of Nature into a perpetual miracle." On the Continent, there was frank irritation with the combination of the "rational" (mathematics and physics) with the "irrational" (the notion of force not reducible to mechanical impact). This irritation is well expressed in Leibniz's complaint to Huygens about Newton's "obscure ideas which involve continuous divine intervention" for explaining how parts of matter could act on other parts without any material impact:

It is men's misfortune to grow, at last, out of conceit with reason itself, and to be weary of light . . . What has happened in poetry happens also in the philosophical world. People are grown weary of rational romances . . . and they become fond again of tales of fairies." [40]

For the longest time, Western philosophers had nurtured all kinds of explanations for how inert matter could animate itself under the influence of various stimuli. In the realm of reproduction, an interesting idea included in the book *The secrets of Nature Revealed* offered what might be viewed as a precursor of urine-based pregnancy tests, and was repeated in several other popular books by anonymous authors:

> Keep the urine of a woman close in a glass three days, then strain it through a fine cloth. If you find small creatures living in it, she is most assuredly conceived with child; for the urine, which is part of her own substance, will be generated as well as its mistress.

The first period of the Scientific Revolution had dispensed with all these suggestions as old-fashioned absurdities. The case of the polyp brought a new respectability to the old ideas of vitalism that the Cartesian model of nature had tried so painstakingly to undermine. It clearly indicated that animals were not simple machines designed by God according to preordained patterns. If a creature could regenerate itself in such a spectacular way, it was harder to view the world as evidence of God's rational plan.

After 1741, zoological thought underwent a pronounced shift, culminating with such gems as Julien Offray de la Mettrie's *L'homme machine.* But this was an a posteriori repercussion. In the heat of the debate first raised by the discovery of regeneration in the polyp, vitalism was still considered an obscure thing of the past. At that point, in the minds of most of the contenders, the universe was held together by God-given natural laws, eternal and unchangeable. The book of Nature was the manifest and visible expression of the book of Revelation, and only on the rarest occasions would God interfere with the admirable clockwork set in motion at Creation. Invisible forces were no longer in fashion. Although many authors assert that the hydra was a bonus to epigenesis, this bonus came into being only after the fact. It does not imply that matters were all that transparent when Trembley first communicated his discovery to the world.

A TROUBLED JESUIT

If preformation could survive challenges as extreme as the regeneration of the polyp, does it make sense to assume that the whole theory could have been mortally wounded by the challenge of a problem as inexplicable, for all contenders alike, as the birth of monsters? Since all phenomena can be used by factions that are fighting for the establishment of their own truth as universal, it is possible that the contemporary adversaries of preformation tried to build cases around monstrosities. But such cases, when we consult the original writings, never seem to have been put forward with much fervor.

In 1744, the French physicist Pierre-Louis Moreau de Maupertuis, one of the major crusaders for Newtonian ideas on a mainly Cartesian Continent,[41] became interested in heredity when an albino Negro was exhibited in Paris in 1743. Maupertuis had long been interested in human races, and actually believed that a "missing link" between man and apes lived in the "Sudland," adding that he should prefer making the acquaintance of these beings to making the acquaintance of the most distinguished persons in Paris. The case presented at this particular exhibit evoked profound Western anxieties. Albinism among peoples of dark races, as well as among animals, was already well known at the time, but its occurrence in humans perplexed those who believed that the white race was a direct result of living in the most blessed regions of the earth, whereas blacks had been relegated to the Tropics, while the Arctic regions were inhabited by giants and dwarfs. The source of this perplexity was social as well as scientific: if lesser humans from lesser earthly regions could emerge even whiter than Caucasians,[42] then maybe the racial categories were not as comfortably fixed as the studious Europeans would like to believe, and the abominable danger of Caucasians eventually giving birth to "Ethiopians" could not be totally excluded.

As we have seen earlier in this chapter, heredity was at this point an intriguing teaser for the philosopher's minds. Maupertuis's countryman Vandermonde would take it into account in his pursuit of beauty in *Essai sur la manière de perfectionner l'espèce humaine,* arguing, among other things, that "People with problems such as gout should not father children at all"; "the germ of the gout is transmitted to the fetus, and the child is born with the body shrunken and the limbs malformed"; and, even worse, that when a blind child is born to parents with good eyesight, "to find the first germ of these vices we must search back several generations," since such "vices" are prone to become hereditary although not always expressed in each generation. Vandermonde was also concerned about the possibility of hybridism being disguised in one generation and reappearing in the next, and quoted Réaumur's second volume of *L'art de faire éclore les oeufs des poulets* to state that polydactylism (which Réaumur had studied in a family from Malta) could disappear in one generation only to reappear again afterward. He also stressed the importance of mixed breeding, pointing out once more the role of geography in sorting out the things of beauty: in his opinion, maritime cities always held the most beautiful and talented peoples because they attracted many foreigners, allowing ugly races to improve their traits. He remarked that Jews were forbidden by their laws to marry foreigners, "and that is maybe one of the physical reasons that caused the degeneration of this people." A change to a better climate could always improve things, because "ugly parents can bear beautiful children" there.

Under these combined stimuli, Maupertuis tried his hand at the mysteries of the life sciences, undertaking selective breeding of his own pets and analyzing genealogical data from four successive generations of families suffering from polydactylism. As a result, he came up with a revised version of the old double semen theory,[43] combining his firm belief in epigenesis with his staunch defense of gravitation, as fully detailed in his book *Vénus Physique,* published in 1745. According to Maupertuis, the key to reproduction involved electrical attraction between seminal particles with different charges, finally producing a complete body. These particles were carried by both sperm and eggs, and represented the scattered and desegregated parts of the future embryo, although they came in numbers much larger than what the embryo would actually need. The selective attraction between them would command their selection, their sorting out, their precise joining, until a new creature was fully formed. Maupertuis held that this aggregation would proceed in the fashion of the *Arbor Dianae.* This tree-shaped inorganic structure, which fascinated the chemists of the eighteenth century, results from chemical attractions. As Maupertuis wrote, "when one mixes silver and spirits of nitre with mercury and water, the particles of these substances come together themselves to form a vegetation so like a tree that it is impossible to refuse it the name." If nitre and mercury could do it, why not seminal particles, too?

Maupertuis certainly knew how to fight for an idea—in order to promote his Newtonian beliefs, he directed an expedition to Lapland that confirmed Newton's suspicion that the earth was not spherical, but rather flattened at the poles—and he was more than a little suspicious of preformation, which he regarded as unable to explain either the birth of monsters or the phenomena of inheritance. Yet, when he voiced this suspicion, he never even mentioned the monstrosity factor, and for the rest adopted an almost amiable joking tone:

> Can one imagine that the spermatic worm, because it has been nourished by the mother, will come to have her characteristics? Wouldn't it be more ridiculous to believe this, than to believe that animals have to resemble the foods that nourish them or the places they inhabit?[44]

When dismissing the claims of the ovists, he remained equally placid, questioning mainly the divisibility of matter:

Eggs destined to produce males each contain only a single male. But an egg with a female contains not only the female, but also her ovaries, in which other females, already fully formed, are enclosed—the source of infinite generation. Can matter be divisible to infinitude; can the form of a foetus that will be born in a thousand years be as distinct as the one that will be born in nine months?[45]

The same moderation echoes in the bitter controversy between Voltaire and the English priest John Turberville Needham, one of Buffon's associates in exploring the model of reproduction through attraction of organic molecules (see chapter 5)—a concept much closer to epigenesis than to preformation. The polemic between the English naturalist and the French popularizer was strikingly long-lived. Voltaire constantly scorned Needham's ideas, and in the end managed to turn him briefly into a private joke among learned Europeans. Since Voltaire was also fond of satirizing preformation,[46] Needham could have done the same to gain some ground. But he never went to any radical extreme on this issue, even when deriding preformation could have helped his cause.

The first clash came at about the time of Trembley's discovery of the polyp's amazing capacities. In 1744, while teaching philosophy at Lisbon's Roman Catholic English College, Needham became well known to all the Portuguese fishermen in the vicinity of the city. He constantly scrutinized their nets, looking for a specimen that could be considered the macroscopic counterpart of the hydra. Such a creature, when discovered, would allow for a detailed dissection, shedding a brighter light on the mystery of regeneration. The driving force behind Needham's pursuit was the belief, common among several microscopists of the time, that the newly discovered microscopic world must constitute a miniature version of all visible things. When populating the microscopic realm, the idea went, the Creator had used the same blueprints applied to the macroscopic creation, changing only the scale. Thus the minuscule entities that the microscope was now revealing should be invisible replicas of some visible analogue. Needham was very serious about applying this concept to the study of the polyp's regeneration. Voltaire was not. In his celebrated *Conte philosophique* he mocked the whole concept, calling it the "Micromégas," or the little-big, the name that ended up being used to refer to this theory in our times.

The relationship between the two men never improved from this starting point. In 1756, Needham published several small pamphlets in Geneva, aimed at rebuking Voltaire's *Lettres sur les miracles*. Voltaire, to put it mildly, overreacted—even by his own incendiary standards. He proceeded to publish a long series of pamphlets starring a "fictional" persona incidentally called

Needham. This Needham was a hysterical Irish lunatic, a charlatan who pretended to create eels by sprinkling holy water on spoiled flour.[47] This character traveled around Europe, "disguised as a man," as an underground agent for sinister Jesuit conspiracies; wherever he went, this pathetic "eel-priest" ended up being submitted to all kinds of outrageous ridicule by the local populations.[48]

Finally, after enduring the laughter of the entire learned society of Europe behind his back, Needham, in his *Idée sommaire* of 1776, delivered his self-defense against Voltaire. As noted above, he could have used one of the few notions they shared in common—the disbelief in preformation—as a building block to regain his credibility. But he chose to remain rather mild, stating solely among his reflections:

> The numerous absurdities which exist in the opinion of pre-existing germs, together with the impossibility of explaining on that ground the birth of monsters and hybrids, made me embrace the ancient system of epigenesis, which is that of Aristotle, Hippocrates, and all the ancient philosophers, as well as Bacon and a great number of savants . . . My observations also led me directly to the same result.[49]

Another example can be found in the writings of Caspar Wolff on this subject. Again, he could have used the birth of monsters much to his benefit, since he was the sole strong defender of epigenesis in the academic community, standing alone before the enormous power of the ovist Albrecht von Haller.[50] Wolff carefully studied the collection on monsters available in the museums of Saint Petersburg, and intended to write an extended examination of the subject. When he announced his intentions in a letter to Haller, the announcement could have been read almost as a threat. Yet he just said politely that "monsters are not the immediate work of God, but of nature," and, referring to a human monster possessing one eye without an optical nerve, he commented:

> that the Creator should have delineated on purpose . . . such a structure in the germ cannot be imagined . . . Monsters are produced by the forces of nature, not evolved from created germs. And if indeed these particular examples of their origin are sufficient, they in turn are at the same time arguments for epigenesis.[51]

Considering the caustic sarcasm that all these authors so often flung at each other, there does not seem to be much ground for arguing that the birth of monsters was a poison dart in the heart of preformation. Besides, Haller himself never showed any kind of restraint in publishing descriptions of monsters, and never pretended that those findings represented anything other than

Fig. 39. Caspar Friedrich Wolff. (From Roe, *Matter, Life, and Generation,* 1981; reprinted by permission of Cambridge University Press.)

Fig. 40. Wolff's drawing of a chicken with four feet. (From "De pullo monstroso," 1780; as reprinted by Roe in *Matter, Life, and Generation,* 1981.)

monstrosities. His book *Pathological Observations, Chiefly from Dissections of Morbid Bodies,* published in England in 1756, includes a good number of examples. Here we find a chapter concerning "A Foetus with a deformed head"[52] (Haller did not hesitate in referring to the "monstrous appearance" of this fetus, enhanced by a bizarre proboscis "projecting an inch beyond the lips").

Other chapters described "A lamb with one eye"[53] (again, Haller refers several times to this lamb as a "monstrous animal"); "A chicken with three legs" (whose description starts with the words "in the class of monsters, I may reckon a chicken about six weeks old, no bigger than one just hatched, which had a supernumerary leg hanging down from its rump"); and, finally, "A deformed lamb without a mouth" (once again, referred to as "a monstrous lamb"). If Haller had any problem with monsters, it was because of the authorities, rather than because of epigenesis: during a stay in Paris, working with the famous anatomist J. B. Winslow, he got into serious difficulty with the police and barely escaped the gallows by paying to have the dead bodies that he had secretly bought quickly hidden away.

Other phenomena, such as the shared traits of both parents manifest in the offspring, were easier to explain by epigenesis than monsters. Therefore, heredity was easier to use as a lethal weapon against preformation. Yet, even in this case, the preformationists managed to confront the challenge with some success. As is clear from the debates described earlier in this chapter, definitive and fully satisfying explanations for monstrosities were still beyond anyone's reasoning abilities at the close of the eighteenth century, regardless of the convictions of the naturalists trying to address the problem. Yet, somehow, the idea that the birth of deformed creatures could not be explained by preformation, and therefore dramatically exposed the inadequacy of the theory as a model for reproduction, is always stated prominently in our modern thoughts on this subject.

GOD SHALL NOT PITY THEM

The widespread idea that monsters posed a tremendous problem for preformation provides an interesting case study, since the blame for this overstatement cannot be ascribed to historians of science or even to the scientists who have seriously studied the history of embryology. This fallacy resides in an error of the laity, carried mainly through oral tradition. Our science professors told it to us when we were in college. They pretended to know it for certain, although, since our teachers were not historians and certainly had not wasted much time researching the issue, their main source of information must have been all those vague introductory chapters in the embryology textbooks. We thought it made perfect sense, and casually repeated the story to our friends as an interesting historical anecdote. We too pretended that we knew it for certain. The basic core of old institutions does not change half as fast as we would like to imagine, and the *magister dixit* (otherwise known as "teacher says") still rules our judgment when we sit in our embryology classes. Unwittingly, we passed along an error that had nothing to do with the debates of the Scientific

Revolution described above. In the oral account, the problem of preformation went straight to the source—God himself.

The argument seems at first perfectly solid. In Roe's *Matter, Life, and Generation,* the issue is clearly summarized as follows: "monsters presented a problem for preformation, because, if monsters are formed by accident, then it must be admitted that chance can interfere with God's preordained program for development. And on the other hand, if monsters are preordained, this challenges the wisdom of God, for why would he create malformed organisms, most of which live only a very short while?" In the folk version of the problem, the grounds were even simpler: if God had created everybody Himself, and if God was good, how could one justify His creation of severely deformed people? We are repeatedly told that the preformationists had a very hard time with this question. As a matter of fact, before I took up this study, I myself said the same to my embryology students in medical school. And they took notes. However, I have not found any reliable proof that this question played a central part in the disputes on generation that occurred during the seventeenth and eighteenth centuries. Mea culpa. And everyone else's.

Besides, nobody ever said God was fair. A gentle Creator, meaning only well and concerned above all things with the happiness of His chosen offspring, would never have indulged in the pagan delights of all those unspeakable cruelties of Hell that illustrate visions of the afterlife in Christian art. Hell may have been crucially important in stabilizing and legitimizing the Christian faith, but all those fumes of sulphur never spelled out an unquestionable kindness on the part of the Almighty. Throughout two thousand years of Christianity, millions of Christians have been inventing gothic versions of Hell's horrors, all the while assuming and restating, ever since Thomas Aquinas, that only "very few" would be saved from it.[54] This is a clear demonstration that such unrestrained kindness never entered the fabric of most Christian paradigms.

According to Paul Johnson in *A History of Christianity,* "the three most influential medieval teachers, Augustine, Peter Lombard and Aquinas, all insisted that the pains of Hell were physical as well as mental and spiritual, and that real fire played a part in them. The general theory held that Hell included any horrible pain that the human imagination could conceive of, plus an infinite variety of others . . . Jerome said that Hell was like a huge winepress. Augustine said it was peopled by ferocious flesh-eating animals, which tore humans to bits slowly and painfully . . . Adam Scotus said that those who practiced usury would be boiled in molten gold. Many writers refer to a continuous beating with red-hot brazen hammers. Richard Role . . . argued that the damned tear and eat their own flesh, drink the gall of dragons and the venom of asps, and suck the heads of adders."

Even if one tries to argue that such unspeakable torment[55] was supposed to be the doing of the Devil and not of God, the truth remains that hundreds of Christians wrote and spoke of Hell as if they had been there—and the argument that God has nothing to do with Hell vanishes before the paucity of similar accounts of the pleasures and rewards of Heaven. Actually, this silence flies in the face of yet another cherished truism of Christendom, the enjoyment experienced by those comfortably installed in Heaven while witnessing the sufferings of the damned.[56] Paul Johnson quotes a number of interesting examples. First we have the Scottish preacher Thomas Boston thundering, "God shall not pity them but laugh at their calamity. The righteous company in heaven . . . shall sing while the smoke riseth up for ever." Then comes another Scottish congregation being assured that the Son bore infinite pain "from the vindictive anger of God . . . pure wrath, nothing but wrath: the Father loved to see him die." Then we read of some preachers arguing that the damned may have been created mainly for the purpose of completing the heavenly bliss. And finally we encounter William King speculating, in his 1702 "On the Origin of Evil," that "the goodness as well as the happiness of the blessed will be confirmed and advanced by reflections naturally arising from this view of the misery that some shall undergo."

Obviously, many other voices tried to temper these vindictive enthusiasms. But their widespread existence provides sufficient proof that, regardless of how we look at it, heavenly fairness has never been at the heart of Christian beliefs. God, as various Christian sources remind us, plays His own games and abides by no rules. Every now and then, He enjoys sending us a monster just to remind us of His absolute power.

Paré had no doubt about this when he wrote, in his chapter dedicated to examples of "the wrath of God": "such marvels often come from the pure will of God, to warn us of the misfortunes with which we are threatened." These "misfortunes" could be wars, regicides, or all kinds of natural and social calamities. But great historical events were not necessarily required for God to unleash His signs. According to Paré, the causes of these divine manifestations can be much more pedestrian: because it is "a brutish and filthy thing to have dealings with a woman while she is purging herself," it is well known that, through the Judgment of God, "women sullied by menstrual blood will conceive monsters" (see chapter 7). As for his sole example of "the glory of God" as a teratological agent, Paré quotes the Gospel of John, citing the man who, as Jesus explained to His disciples after performing the miracle of giving him sight, was born blind not because of any sin committed by his parents, but simply "in order that the works of God might be magnified in him."

In this light, preformation would not have faced any problem in postulating that we had all been engineered simultaneously during the first six days of Creation, and that God himself had predetermined our appearances right then. It is undeniable that the willingness of God to allow catastrophic events to happen amid His creations has always been a complex moral problem, so much so that even Darwin had to address it when defending the theory of natural selection.[57] But, on the other hand, the most respectable Christian sources never said that God would not consider including a fair share of monsters in the Russian dolls of humanity, so that we would have to confront His unlimited power throughout the generations and therefore remain modest in our self-perception. Consciously or not, the argument that preformation could not explain monstrosities if we had all been arranged by God himself thus becomes the fallacious expression of a double standard, under which the authors contend that God, in all His goodness, would never had created deformed people—but only when this argument serves their purposes.

THE CREATOR'S MECHANICS

Moreover, preformation came into existence at the onset of the Scientific Revolution. If we bear in mind the modern refrain that this was the golden age of mechanism, the period in which life was systematically understood in terms of organization so that "living beings" became synonymous with "organized bodies," then a body would be nothing but the sum of its parts. Therefore, the bodies of monsters could easily be considered exciting sums of interesting parts. The end result could have been made even more interesting by the evident fact that these anatomically aberrant sums of parts were physiologically viable, since they were alive. Following this reasoning to the extreme, monsters would be even more remarkable than normal creatures—and, with that, they would in fact constitute an added proof of the power of God. As the French anatomist du Verney wrote in 1706, "the inspection of monsters reveals the richness of the Creator's Mechanics . . . the regularity of the apparent irregularity." Or, as his countryman Jean Palfyn wrote below a plate showing conjoined twins, "The miracles of God are present in all of His productions."[58] In a sense, as de la Fosse would write in his section on hermaphrodites for Diderot and D'Alembert's *Encyclopédie,* these beings, far from being imperfect, must necessarily be more perfect than any "normal" creature. In 1739, during his first preformationist incarnation as a spermist, Haller repeated that monsters are preformed, and that they constitute, in their own special way, further evidence of God's design.[59] And Haller went on to become an extremely influential man.

It is hard to find, within this background, where preformation could have gotten into any kind of serious explanatory trouble. For all of their excellent rhetoric, what the preformationists had to say in defense of their model may not have been truly satisfactory to the public; but this still does not imply that their rhetorical weapons were less sharp than those of their opponents. It is undeniable that monstrosities posed a problem for the preformationists—but only as much as they posed a problem for all the undeveloped physiology of the time. It is clear that the spermists had a real problem with the waste of sperm cells (see chapter 2) because all of their important representatives attempted over and over to produce models to explain such an apparent contradiction between God's designs and teachings. This suggests that they were under pressure from their peers, feeling compelled to justify something that really seemed unjustifiable. But nothing of that sort is visible in the case of monsters, indicating that the pressure for viable explanations must have been much less. Which leads to the likely possibility that, in our reevaluation of studies of generation, we saw a problem where only a question mark existed. Based on our own perceptions of this past time, we decided that a great dilemma concerning monsters had tortured the health of preformation. And, in my humble opinion, we were wrong.

Frogs with Boxer Shorts

Serene and immobile
The frog
Stares at the mountains

JAPANESE HAIKU BY MATSUÔ BASHÔ

ERE IS SOMETHING that I clearly remember learning in high school, just before I turned sixteen. We were going through a list of scientific heroes of the past, the men who had shaped modern knowledge with an observation or an experiment that established some sort of truth destined to remain with us forever—beacons of reason, dispersed lighthouses emerging one by one to illuminate our path. Spallanzani's name was on that list. I remember learning that he had been a priest, and then memorizing something much more appealing to my juvenile sensibility: he had done some fancy experiments involving frogs dressed up in boxer shorts. This, obviously, was the part we all liked. The part we were supposed to retain as Spallanzani's grand contribution, however, was much less pedestrian. He had been an influential player in the overturn of spontaneous generation. In that process, he had ended up establishing some sort of fundamental truth that made possible all our subsequent understanding of the biology of fertilization. By means of the boxer shorts, we were told, Spallanzani had demonstrated beyond any reasonable doubt that semen was indeed indispensable to reproduction. In other words, I first heard of Spallanzani as the bearer of the greatest support that the spermists could ever have dreamed of.

But, as we now know—and this might seem paradoxical—Spallanzani was an ovist.

Are we going to have to claim again that our teachers erred in hasty interpretations of debates too old and too obscure to deserve better scrutiny? Interestingly, not in this case. Ovist that he was, Spallanzani really did perform crucial observations on the function of the semen. These observations really did involve procedures as delicate and tedious as fitting dozens of frogs into tight taffeta pants. All this was meant, initially, to support the case of the ovists. But the true meaning of the results has been retained to our day as proof of what is now undeniable: without semen, there is no fertilization; and the semen's contribution is material, rather than spiritual, as Swammerdam would have liked to believe.

We have already seen in chapter 1 how Spallanzani's approach to the natural sciences paved the way for modern research. One final touch now has to be added—all to Spallanzani's honor: another mark of his modernity was his reminder to us that, in performing real scientific experiments, we cannot expect to force biology into a docile accommodation of our preconceived notions. Results will often tell us that nature's mechanisms are just the opposite of what we had assumed. Spallanzani was fighting for the egg, but his final legacy was an enhanced awareness of the importance of the sperm.

THE INTERIOR MOLD

We have already made the acquaintance of the charismatic Comte de Buffon in chapter 3 of this book. We have discussed some of his prolific contributions to the natural history of the Enlightenment, but we have not yet analyzed the details of his theory of reproduction. Buffon proposed a new concept that, together with those of Descartes (see Prologue) and Maupertuis (see chapter 4), appeared as an alternative to the dominant ideas of epigenesis and preformation. Unlike those of Descartes and Maupertuis, however, Buffon's system of reproduction came a step too close to resurrecting the old idea of spontaneous generation, which many leading naturalists, and most notably the preformationists (since a belief in spontaneous generation would necessarily invalidate their system), were trying very hard to undermine.[1] This was where Buffon had to measure his own kaleidoscopic theoretical powers against the formidable experimental skills of the preformationist Lazzaro Spallanzani.

It is always interesting to see how prominent popularizers try to disseminate an idea that strikes their fancy. In his *Essai sur la manière de perfectionner l'espèce humaine,* published in 1751, Dr. Vandermonde, then regent of the Faculté de Médecine de Paris, claimed that most matters presented in his book were "directly inspired in the Natural History of M. de Buffon, whose experience and enlightenment assure my principles."

And here is how he described Buffon's observation of male semen under the microscope:

> He saw something that less enlightened eyes would have considered worms, but he understood that he was looking at living particles, which he named *organic particles,* because they seemed to him to be rudiments of the organization of the body.[2]

Having reached this conclusion, Vandermonde explained, Buffon guessed that females should have some "prolific liquor" of this sort, too.

Fig. 41. Georges-Louis Leclerc, comte de Buffon.

> Vallisnieri . . . had studied the anatomy of the ovary in viviparous females, and
> looked in vain for eggs that Nature had not put there. This was enough to excite
> the curiosity of M. de Buffon and make him suspicious of the system of the egg.

Guided by this suspicion, the Count then examined "those small viscera that
they call the ovaries of women" and discovered that "they were nothing but
glandular bodies containing vesicles with a particular liquid inside, in brief,
true testicles." Microscopic examination of this liquid from dogs, cows, and
rabbits in heat, and of several different fishes, soon revealed the presence of
"organic molecules in perpetual movement, in all resembling those of the male
juice."

Since Buffon did not like those theories that would "lead necessarily to
the idea of the actual existence of geometric and arithmetic infinity," and thus
was not pleased by even the most sophisticated forms of pre-existence, he
proceeded to put his "organic particles" to use in a new theory of reproduction that involved their constant cycling, being released by an organism at
death and promptly reformed into new organisms. As he wrote in his *Histoire
naturelle,*

> There exists in nature an infinity of little organized beings, and . . . these little
> organized beings are composed of living organic parts which are common to ani-
> mals and vegetables . . . Organized beings are formed by the grouping of these
> parts, and . . . therefore reproduction and generation are nothing but a change
> of form which comes about simply by the addition of these similar parts.

According to Buffon's system, this endless mass of organic particles gained form under the stimulus of a force the author called the *moule intérieure,* or "interior mold," and which he likened to Newton's gravitational force—a notion that Buffon had strongly endorsed, and whose minute details he had painstakingly come to master when translating Newton into French. Generation, development, and growth were all controlled by this interior mold. As John Farley explains in *The Spontaneous Generation Controversy,*

> in early life the organic particles were absorbed and employed in augmenting the different parts, the *moule intérieure* assuring that each part of the body would receive only those molecules it could use. At the higher levels, therefore, there were different groups of particles representing the different organs. After puberty, when growth stops, the excess particles were passed to reservoirs—the testes and the ovaries . . . When decomposition occurred, the organic molecules, no longer constrained by the mold, found liberty and later became incorporated into other organisms.

A particular kind of spontaneous generation could thus be considered in lower organisms, in which free organic molecules from a decomposing body would become agitated and mingle with "brute," passive matter to form such creatures as worms, mushrooms, and all kinds of microscopic organisms. Taking the idea one step further, Buffon also proposed that the same phenomenon could occur in living organisms, finally providing a sound explanation for the ever-problematic existence of internal parasites:

> When there are several malfunctions in the organization of the body, which prevent the absorption and assimilation of all organic molecules in the food by the interior mold, these excess molecules, unable to penetrate the interior mold of the animal, reunite with several particles of brute matter in the food and form, as in putrefaction, organized bodies. This is the origin of tapeworms, ascarides, flukes, and all the other worms that are born in the liver, stomach and intestines.

Buffon's contribution to the renewed respectability of spontaneous generation cannot be viewed solely as a brilliant outcome of his personal inspired vision. Old truisms die hard, and even more so when they come across as soothing simplifications of matters too complex to be easily grasped. It was fashionable to discredit Aristotle during Buffon's time, but this generalized attitude did not really apply, even among the most learned men, when reproduction was at stake. Harvey modernized reproduction by revamping the Aristotelian idea of epigenesis, but, like the philosophers of antiquity, he still believed that "inferior organisms" could have no parents but the wind. We have seen in chapter 3 how Francesco Redi's experiments dismissed the idea that flies were children of decomposed matter; but, even after Redi had published his simple and illuminating conclusions, spontaneous generation still had numerous supporters. We have only to bear in mind that it was still defended after Louis Pasteur's work on microbial contamination, and retained sympathizers who wrote entire books on the subject until as late as the early decades of the twentieth century,[3] to realize the enormous appeal of this age-old concept.

When Buffon proposed his theory of organic molecules to the world, learned men such as the English naturalist Ross were claiming that "to question that beetles and wasps are generated in cow dung is to question reason, sense and experience."[4] Among the commoners, almanacs and manuals of all sorts circulated with recipes such as this one, destined to produce a good swarm of bees:

> Take a young bullock, kill him with a knock on the head, and bury the carcass under the ground with the horns sticking out. Leave him there for a month, then saw off his horns—and out will fly your desired swarm of bees.[5]

THE VEGETATIVE FORCE

John Turberville Needham has also already appeared in these pages, but it is now time to explore his partnership with Buffon and how the joint efforts of both men prompted Spallanzani's legendary experiments on semen. Needham had an explanation for reproduction that rested on "vegetative powers, which reside in all substance, animal or vegetable, and in every part of these substances as far as the smallest microscopic points,"[6] rather than in the interior mold proposed by Buffon. However, Needham and Buffon held a strong conviction in common: nature had a real productive force, and this force was enough to account for form and growth in the living world. This idea seemed so coherent, and was brought forth with such persuasive power, that it produced numerous echoes even during the nineteenth century, when authors defending new models of spontaneous generation did not hesitate to consider themselves beneficiaries of Buffon and Needham.

That both the laity and the selective academies were ready to embrace spontaneous generation is clearly exemplified by the enthusiastic reaction to Needham's famous experiments with flasks containing boiled meat broth, described in chapter 1. More than one learned society considered inviting Needham to join its ranks. He was elected a Fellow of the Royal Society and an Associate of the Académie des Sciences. Eventually, this widespread fame took the priest to Paris, and fate did the rest. Before long, Needham was under Buffon's wing.

The cherubic Catholic priest with the large nose and the handsome Count dressed in spectacular outfits must have made an entertaining odd couple; but, as in the case of the alliance between Haller and Bonnet, the combination of their disparate personalities proved to be extremely efficient. The Count was rich, a gifted writer, and largely at ease with mathematics. He did not have much patience for the microscope; but here the humble Father Needham could fill in the blanks. As De Kruif puts it in his popular rendition of these events,

> Buffon wore purple clothes and lace cuffs and he didn't like to muss up on dirty laboratory tables, with their dust and cluttered glassware and pools of soup spilled from accidentally broken flasks. So he did the thinking and the writing, while Needham messed with the experiments. These two men set about to invent a great theory of how life arises, a fine philosophy that everyone could understand, that would suit devout Christians as well as devout atheists . . . [They] deluged the scientific world with words . . . [And] in a little while Vegetative Force was on everybody's tongue. It accounted for everything. The wits made it take the place of God, and the churchmen said it was God's most powerful weapon. It was the Force, prattled Needham, that had made Eve grow out of Adam's rib. It was popular like a street song or an off color story—or like present day talk about relativity.[7]

SCIENCE IN THE MAKING

While the whole world marveled at the Force, back in Italy, in the semi-obscurity of dusty laboratories, Spallanzani was raising his eyebrows in disbelief. Malebranche had suddenly awakened to the splendors of geometry by reading Descartes's *Treatise of Man*. Bonnet had found his mission in life by reading Pluche's *Le spéctacle de la nature*. It was yet another book, devoured in its entirety during a single lonely and fateful night, that made Spallanzani see the light—and this time, quite appropriately for the fight that followed, the book was Francesco Redi's *Experiments on the Generation of Insects*. It is said that Redi's elegant experiments demonstrating that flies did not arise spontaneously from decomposing meat (see chapter 3) were the inspiration that led Spallanzani to the bench. Flies were no longer Spallanzani's concern; but, having deduced from his reading that spontaneous generation was nothing but a myth—or, at best, a misconstrued outcome of careless experiments—he was determined to establish once and for all that microscopic animals also had to have parents. In the process, he had to learn everything from scratch. In the beginning, he did not even know how to use a microscope. He cut his hands several times trying to grind lenses. He broke expensive flasks. He forgot to clean the glasses properly and had to endure the frustration of seeing nothing in the end but foggy, undecipherable images. But he never gave up.

We have followed in chapter 1 Spallanzani's step-by-step deconstruction of Needham's discoveries concerning the spontaneous birth of microorganisms in boiled meat broth inside sealed flasks. Spallanzani first assumed that perhaps Needham had not boiled his infusions long enough. To address this problem, he divided his flasks into two groups. He boiled one group for ten minutes, as Needham had done, and boiled the second group for one hour. This division of the original sample gave the world an added understanding of microscopic life's secrets: the flasks boiled for ten minutes grew new organisms after a few days of culture, whereas the group boiled for one hour remained sterile. Thus, we came to realize that not all microorganisms are equally resistant to heat: some can endure extremely harsh conditions and still survive. Time, together with heat, became a crucial element in sterilization. Spallanzani has not been considered Pasteur's mentor for nothing.

One of Spallanzani's trademark precautions during his laboratory research, however, is even more remarkable in what it brought to the dynamics of modern science. He wrote that "while we are in quest of one truth, it generally happens that others offer themselves as it were spontaneously," and he kept his eyes open for those other truths. There is a constant pattern in Spallanzani's approaches that is seldom mentioned, but that pattern brought about a new type of investigation. It is worth retracing the emergence of this method from the struggle against spontaneous generation.

Spallanzani had suspected from the beginning that the main factor behind Needham's erroneous conclusions was the defective sealing of his rival's flasks. Therefore, instead of using only a cork, as Needham had done, he made sure that his own infusions were totally isolated from the external air by melting the flask's glass around the cork. He had reasoned that any exposure to air would allow all sorts of invisible airborne creatures to dive into the "sterilized" soup, giving the observer the false impression that they had grown there by themselves. Spallanzani's assumption proved to be right on target, for nothing regrew inside his tightly sealed flasks. But he also thought it important to take an added cautionary step, and this step rendered the experiment even more meaningful. He set aside another group of samples, also boiled for ten minutes and one hour, but sealed only with corks. As now seems obvious, these samples all regrew animals, regardless of their heating times.

Moreover, when Needham argued that the excessive heat had killed the Vegetative Force, Spallanzani did not rest on his laurels and call the claim a bluff. Instead, he dared to question his own discoveries. He went back to his flasks, filled them with different kinds of soups, and then, instead of just boiling those soups, he baked them in a coffee roaster until they were reduced to ashes. If there was indeed a Vegetative Force, and if this Force was indeed heat-sensitive, now one could assume beyond all doubt that the Force had been thoroughly exterminated. However, even in this extreme case, all the flasks that were left exposed to the air soon presented new legions of swimming microscopic creatures—while the liquid inside the sealed flasks remained lifeless and quiet.

Needham's argumentative power still had a reply for this. Faced with Spallanzani's new twist, he claimed that the Vegetative Force needed, most of all, a very elastic air to help it accomplish its goal; and that, with all his dramatic sealings of the flasks and heatings and roastings of the soups, Spallanzani had killed that elusive elasticity. Spallanzani could have just rebuffed such arguments with a couple of clever remarks and let the issue be solved only on paper. After all, that was what Needham had done all along from the onset of their controversy. But he did not do so. He was determined to leave nothing to chance. Once more, he took up Needham's challenge and set out to prove *experimentally* that his opponent was wrong.

He placed some more soups inside flasks, sealed them with the flame, heated them for one hour, and then cracked off the neck of one of the bottles. He repeated this procedure twice. In both cases, he could hear a distinct whistle coming from the orifice, meaning that air was either coming in or going out—in other words, meaning that, after all, Needham was right in his assumption that the air inside the sealed flasks differed from the open air outside. He lit a candle and held it close to the newly cracked orifice of a third bottle. The flame was sucked inward toward the opening. That meant that air was going in—an observation that could certainly support Needham's concept of a less "elastic" air inside the sealed container. Could Spallanzani have been fooled by his own experimental conditions? Was there at least some veracity to this strange notion of lost elasticity in the air? In light of the knowledge available at the time, there was only one possible counter-theory. Spallanzani's flasks had very wide necks, which meant that sealing them with the flame took a lot of heat to melt a lot of glass; maybe this heat was driving most of the air out of the bottle before that bottle had been totally sealed, thus creating a vacuum inside. If this were true, then it would make perfect sense for the outside air to rush into the bottle as soon as a little opening was produced.

This theory made sense, but it was not easy to prove experimentally. Only an extremely stubborn researcher would go the distance, but Spallanzani did. He filled another flask with soup, and then rolled the flask's neck around the flame until the neck had melted down to a very tiny opening—very tiny, but still allowing contact with the air outside. He let the newly shaped container cool down, until it could be assumed that there was no difference between the air inside and the air outside: they were both at the same temperature, and were still in contact with each other. At this point, he finally sealed the flask— a procedure that now required but a couple of seconds, and involved the use of a very tiny flame. With his new flask thus sealed, he boiled it again for one hour. Afterward he cracked a small opening in it, again holding a lit candle against the opening. He could hear the whistle once more. But, this time, the candle's flame blew *away* from the flask, instead of inward. If there was any difference in elasticity, now it was the air inside that was *even more elastic* than the air outside. However, several hours later, the soup contained in the bottle had remained totally sterile. Needham's Vegetative Force was doubtlessly in dire need of something to do its work. But the possibility of that something being "elastic air" had just been ruled out.

This succession of experiments was certainly a decisive blow to the credibility of Needham's doings.[8] But, as their final legacy, these particular studies gave us much more than that. They represented the introduction of the concept of *control groups* in science. Any undergraduate student now knows that, when investigating a biological problem, one cannot simply perform an experiment designed to verify one's convictions. One always has to play the devil's advocate, performing at the same time a parallel experiment that could prove the contrary. Today, when we design our laboratory protocols, we immediately ask ourselves, "What can we use as a *control* for this experiment?" If we try to avoid this issue, no matter how beautiful and convincing our data appear to be, any journal will return our paper with the sacrosanct question, "What *controls* have you done for this?" But this is the end of the twentieth century, and the golden rules for doing science are established and laid down in writing. Back in the days of the Enlightenment, when modern science was still in the making, no one had thought of elevating this practice to the status of a golden rule. So here's to Spallanzani—even when our controls prove us wrong and we go around hating life for days on end.

PASSIVE VERSUS ACTIVE

At this point, basking in the glow of his triumph, Spallanzani could have been rather cruel. But he retained his gracious demeanor. Commenting on the outcome of his long battle with the Vegetative Force in *Nouvelles recherches sur les découvertes microscopiques,* he cordially remarked that "Needham, being English, has an astonishingly good capacity of expressing himself." With this elegant bow, Spallanzani's quest could have come to a peaceful end. Instead, it led him to the endless investigations that supposedly reached a climax with that famous experiment concerning frogs with boxer shorts.

The crucial problem was that Buffon, as we have seen in the beginning of this chapter, had revitalized spontaneous generation through a claim that Spallanzani was not willing to leave unquestioned. Buffon's theory of organic molecules implied that the male testes and the female ovaries were nothing but storage sites for the surplus production of such entities. Among other things, this meant that the semen, a fluid so obviously related to fertilization, was just another soup of undifferentiated particles. Not to doubt such a claim was an implicit agreement with the random paths of spontaneous generation.

When Jean Sénébier translated Spallanzani's *Opuscules de physique, animale et végétale* from Italian to French, he dedicated this translation to Bonnet and Horace Benedicte de Saussure—according to him, men in whom "your compatriots in Geneva see their benefactors, because you enlighten them with your lights and you make them illustrious with your successes, whereby all those who study natural history and philosophy have taken you as their models and their guides." The choice of the dedication was not without reason, for Bonnet and de Saussure were Spallanzani's allies in the effort toward a description of the semen that would dethrone Buffon's assumptions. Moreover, as we shall see later in this chapter, de Saussure's observations were crucial in giving a push to the experiments leading to Spallanzani's final conclusions on the part played by the semen in fertilization—but, until the Italian priest came to de Saussure's rescue and certified the authenticity of his findings, Buffon's supporters had been eager to mock and deride the Swiss naturalist.

Next, Senebier clearly explained, in very simple terms, the pivotal dilemma in the clash between Spallanzani and the joint forces of Buffon and Needham:

> M. de Buffon, in his *Histoire naturelle,* and Mr. Needham, in his *Recherches sur les animalcules et la reproduction dans les êtres organisés,* try to establish the passive nature of those beings [the "spermatic animals"] but Mr. Spallanzani shows in this work the animality of the Spermatic Beings and leaves nothing uncovered concerning their history.

This clash was made even more acute by the fact that Buffon and Needham had been the first scholars to undertake a systematic study of spermatozoan functions since the death of Leeuwenhoek. Extrapolating from his findings in the squid sexual apparatus,[9] Needham suggested that the "alleged animalcules" found by others in the semen of different animals[10] could be nothing but little machines, a view that would suit much better his concept of the Vegetative Force and Buffon's concept of the organic molecules shaped by the interior mold:

If I had ever seen the supposed Animalcules in the Semen of any living Creature, I could perhaps be able to determine with some Certainty, whether they were really living creatures, or might possibly be nothing more than immensely less Machines to those Milt-Vessels [in the squid], which may be only in Large what those are in miniature.[11]

To this Buffon promptly agreed, and added that

the supposed animalcules . . . could be only very slightly organized . . . they could be at the most the rough outline of an organized being; or, to put it more clearly, these supposed animals are only the living organic parts of which we have spoken . . . or, at most, they are only the first reunion of these organic parts.[12]

After having examined, with Needham's collaboration at the microscope, the seminal fluids of man and some other animals, Buffon introduced a new theory in his *Histoire naturelle,* proposing that the semen consisted of a large number of filaments, undergoing longitudinal separation during the course of a few hours. These filaments were accompanied by a string of rounded globules, which at some point would start detaching themselves from the parent filaments, to which they remained attached by only a small thread—what his opponents, in their silly insistence on the animal nature of the particles, insisted on calling "the tail." When the globules finally managed to set themselves free from the filaments, their next task was to get rid of the thread, and that was the reason why they displayed those bizarre oscillatory movements in all directions. The fact that they became motionless after about an hour of microscopic observations meant simply that they had finally shed the thread. The matrix of filaments, meanwhile, was constantly producing new globules with new attaching threads, so that in living animals this process went on endlessly.

There is undeniably some truth in this report, for spermatozoa undergo a complex and continuous process of maturation in the testis, starting out as large, oval, or rounded cells (called initially *spermatocytes,* and in subsequent stages of maturation, *spermatids*) and only later losing most of their cytoplasm and acquiring a tail. Unbeknownst to them, Buffon and Needham may have observed just that,[13] and then filtered their observations through their a priori convictions. But Buffon was definitely entering the realm of imagination when, further coloring his report, he declared that he had seen threadless globules "marching," seven or eight abreast, "following one another without interruption, like troops in line"; and that these structures had moved far more briskly than those that still had the little thread attached, gradually decreasing in size as they marched on.[14] This was a perfect scenario for his conclusion that "semen is composed of particles in search of organization," which in turn fit perfectly his grand theory of the organization of life.

Thus Spallanzani had a philosophical problem on his hands, a problem of passivity versus activity in the generational organs, that could be addressed only by detailed scrutiny of the function of sperm and eggs. As is clear from Senebier's dedication, he was not the only naturalist to sense the urgency of this question: from across the border, Bonnet had been urging him to dedicate some much-needed attention to the mysterious "seminal animalcules." Still, Spallanzani's respect for Buffon was strong enough to make him hesitate in making his own early discoveries public. Were it not for Bonnet's insistence, he may never have focused his attention on sperm cells. But when he finally did so, he soon gained enough self-assurance to attack Buffon and Needham quite forcefully. Regarding the understanding of spermatozoa, Spallanzani complained, "we have descended from the observations of Leeuwenhoek to those of Buffon." [15] Later, in his *Nouvelles recherches sur les découvertes microscopiques,* he got even bolder:

> the multitude of admirable writings which all those famous men have left us . . . show us that all living substances are born from an egg; that after having been tightly locked inside that shell, they want only to set themselves free and to develop in succession; finally, that it is through this mechanism that each substance maintains and propagates its own species. This is the System adopted by the Sages, to explain to us in a satisfying manner the Great Work of Generation, with no need to resort to *plastic forces;* they reject those forces as totally useless and inefficient, not to mention the darkness in which they are wrapped. As for myself, I am strongly persuaded that today we would reject them with even more repugnance and distance . . . if the two famous scholars, MM *de Needham and de Buffon,* had not joined their efforts . . . to give a second life to those almost forgotten *plastic forces* . . . it is true that nowhere in their writings we find the use of this term . . . but, reading what they say about that force, it is evident that they are referring to what the Ancients called the *plastic force,* and that our two modern Physicians did nothing but to give a new name to an opinion very old and very well known.

In his own studies, Spallanzani was extremely careful. He now had microscopes much better than those used by Leeuwenhoek, and made the most of their possibilities. He introduced the clock and the thermometer to the list of mandatory laboratory equipment, and kept logbooks constantly updated. He extended his observations to numerous animals species, such as horses, dogs, rams, carps, bulls, and, most memorably, frogs and newts. He also did numerous tests to determine the nature of the semen, exposing it to vacuum, desiccation, violent agitation, and treatments with fumes of candles, paper, and tobacco, as well as mixtures of chemicals such as common salt, vinegar, oil, and spirits of wine. He noted the difference in the sperm's response to ice and to high temperatures. He carefully compared preparations of spermatozoa with preparations containing all kinds of other microorganisms, and listed their differences in detail, refuting Buffon's and Needham's claims that the creatures in the semen were nothing but another variety of organic molecules.

"Why, one may ask, why has Mr. de Buffon, who has so much genius and discernment, put the spermatic animals and the animals from infusions in the same class, when there is a real difference between the former and the latter?" Spallanzani asked in *Nouvelles recherches sur les découvertes microscopiques.* And he ventured his own answer: "if I had to answer that question, I would say that this great man has listened in those matters to nothing but his great love for the theory he produced."

Dwelling deeper and deeper in the difference between spermatozoa and other microorganisms, Spallanzani also explored what allowed them to live and what caused them to die, further substantiating the concept of their "animality." In all cases, regardless of the type of treatment, he observed that, dead or alive, the "worms" never lost their tails, fully contradicting Buffon's theory of the mature globules shedding their "threads." He investigated the nature of the sperm in different sections of the testis, and noted that these "animals" could swim as promptly as those obtained from ejaculates, thus debunking Buffon's idea of the need of the "molecules" to "form" from the filaments before they could have any kind of movement. At the very beginning, he had promised to "follow this race of little animals to the end." Some people keep their promises.

BLOOD PARASITES

Even with all his remarkable care and his newly introduced precaution of double checks, Spallanzani still turned out a considerable number of erroneous conclusions. He accurately remarked that, although true spermatozoa were already present in the semen before it left the testes, once the semen was ejaculated, those spermatozoa were not able to live long when exposed to the air. But this led him to the assumption that sperm cells were genuine parasites of the generative organs, passed from one generation to the next during the copulation of the hosts.[16] Moreover, since parasites had to come from somewhere inside the body, this view implied that the seminal worms could already be present in the organism even before the formation of the testes. Perhaps under the guiding perspective of the old Galenic assumption, according to which semen results from the ebullition of the blood (see chapter 2), Spallanzani's own suggestion was that the precursors of the "worms" were initially circulated in the blood, and were attracted to the testes right before fertilization—a phenomenon that could be linked to the agitated behavior most animals display at the onset of the mating season.

He presumed that these parasites had a system of reproduction. This belief was made easy by the fact that he had observed numerous microorganisms undergoing several cycles of vegetative propagation after their careful isolation in small drops of distilled water.[17] He had been prompted to undertake these observations after de Saussure published a paper reporting the astonishing events of bacterial division. This paper had been strongly criticized, and Spallanzani wanted to settle the issue. Moreover, if de Saussure's report was true, he wanted to see it for himself. He did, and certainly got his fill. In his own descriptions of vegetative propagation, and of the methods he invented to isolate microscopic creatures, Spallanzani's enthusiasm almost jumps off the pages. He seemed to like his observations so much, however, that when applying them to the sperm, the fact that he had never actually seen any of those "parasites" divide did not appear to encumber him with much doubt. Spallanzani's belief in the irrelevance of spermatozoa to fertilization is clear in this passage from "A Dissertation Concerning the Generation of Certain Animals":[18]

> having often observed the seminal liquid of the toad, I found it very full of spermatic worms . . . Upon two occasions I had been greatly surprised at finding this fluid totally destitute of such inhabitants. I was induced to try, whether it was also destitute of fecundating virtue, but I found that it was just as effectual in this respect, as that which most abounds with the diminutive animals.

Whether this fertilization without spermatozoa was just wishful thinking or the result of some unverified factor affecting Spallanzani's generally carefully organized experiments, we will never know—but either case serves to prove the author's profound conviction about the parasitic nature of sperm.

To err is human. To achieve error through experiments as powerful and imaginative as Spallanzani's is divine. When he was done with the sperm, nothing in developmental biology would ever be the same again.

THE REAL SEMEN

Spallanzani's observations of frog eggs, described in the opening section of "A Dissertation Concerning the Generation of Certain Animals," were clearly made from the viewpoint of a scrupulous naturalist, totally leaving aside any simplistic waving of the ovist flag. Actually, he went as far as stating, right in the introduction, that the artificial fecundation of various animals "is accomplished by means of the seminal liquor of the animals themselves; and I have

succeeded as well as if the male himself had performed his proper function"—
thus implicitly assuming that "the proper function" of "the male himself" was
necessary for reproduction.

> The *amours* of this species[19] begin in April, and end in May; they are, however,
> influenced by the temperature of the atmosphere[20] . . . In autumn and winter the
> immature eggs lie all in the ovary . . . The ovary of this, as well as many other
> species, appears to be externally covered with black points, which have been the
> source of a memorable mistake; for Vallisnieri, and other celebrated naturalists,
> have supposed them to be rudiments of the tadpole. The error arose from their
> being contented with first appearances, and searching no further; for the spots
> lying close to the eggs, might easily induce the observer to believe that they form
> part of them, and are therefore so many fetuses. But . . . if the membrane of the
> ovary be examined by the microscope, it will then be seen that these points are
> black spots of irregular shape. They are not peculiar to this membrane, for they
> are found in the area of the mesentery, and upon the heart.

In making these statements, Spallanzani was not only discrediting the
claims of his fellow ovists. He was explicitly including Vallisnieri, the mentor
of his youth and the man who had convinced his father to let him become a
naturalist rather than a lawyer (see chapter 1), in the group of those who had
been guilty of hasty conclusions. This makes it even clearer that he was willing
to abandon his strong ovist opinions for the sake of reporting the truth. The
same attitude surfaces in the ensuing paragraphs, in which the role of the male
in reproduction appears to gain more and more importance:

> If the eggs be again examined in spring, we shall still find them in the ovary, but
> considerably enlarged . . . and they will be found to be mature when the male is
> coupled with the female . . . And, although Vallisnieri asserts, upon the authority
> of a single experiment, that frogs do not discharge their eggs in close vessels, it
> is certain that numberless facts prove the contrary . . . It is true, as he asserts in
> another place, that this does not happen when the female is kept constantly sepa-
> rate from the male. I say *constantly;* for if they be pulled asunder when the eggs
> are descended into the cavity of the uterus, they will be discharged, though the
> female is kept separate, but they will be barren.

His following experiments, designed to determine whether eggs collected
from the ovary, the oviduct, or the uterus could develop by themselves,[21]
seemed to drift further and further away from any kind of supporting proof
of ovism.

> I repeated my experiments to satiety; and in my journal I find, that I have opened
> one hundred and fifty-six females while they were not embraced by the male,

of not one of which did the eggs ever bring forth young, though I immediately placed them in water; whereas those that were excluded spontaneously by the female, were all prolific. I have even taken further pains. The discharge of eggs lasts about one hour; during this process I killed a female, and put the eggs that remained in the body in the water containing those that had been discharged by the animal; but the latter became tadpoles, while the former became an offensive putrid mass. From these facts I concluded, that the fecundation of the eggs does not take place within, but without the body.

This conclusion, as Spallanzani claimed after he had repeated the same observations in green tree-frogs and in the "ugly and disgusting" toad and had the same results, proved that many a great naturalist had been wrong. This number included Linnaeus, with his thesis that "Nullam in rerum natura, in ullo vivente corpore fieri fecundationem vel ovi impregnationem extra corpus matris" ("In Nature, in no case, in any living body, does fecundation or impregnation of the egg take place outside the body of the mother"). It also included some more obscure observers, bearers of much more bizarre ideas, such as a certain Professor Menzius, "who supposed that, while the male embraces the female so closely, the seed is emitted from the fleshy prominence of the toe, and passing through many windings unknown to us, penetrates into the thorax, and there impregnates the eggs."

Spallanzani's attention next focused on this process of external fertilization that had gone unnoticed for such a long time.

> As soon as the eggs begin to be discharged, the agitation of the female is extreme; she darts backward and forward, rises toward the surface of the water, and then sinks, keeping the hind legs constantly stretched out, and croaking in a low voice. The male keeping his hind legs close to his body, throws himself into strange contortions, and accompanies the croaking of the female with a kind of interrupted noise, which I cannot express by words. I moreover observed, that an obtuse tumid point, which I suspected to be the penis, was elongated, and now and then brought toward the eggs nearest the vent.

Although he initially could not see any semen coming out of the presumed penis, Spallanzani soon settled this question by placing seven pairs of frogs inside empty and dry vessels.

> The male is so much attached to the female, that notwithstanding his being taken out of his natural element, he persists in doing his office. I now saw that there darted a small jet of limpid liquor from the small point in the vicinity of the anus, upon the eggs hanging out at the vent of the female . . . The eggs being afterward put into water, and bringing forth young, I hesitated not to suppose, that the liquor emitted by the male was real semen.

In these lines, Spallanzani had done more than finally identify the elusive frog semen:[22] by basing the proof that this was really semen on the fact that the eggs that came into contact with it had "brought forth young," he was once again equating semen with fertilization, and seemingly making the case for ovism more difficult at each step. It was at this point, in his effort to further explore the true nature and function of this semen, that the famous boxer shorts appeared.

Interestingly enough, the idea that we currently ascribe to Spallanzani was not really his. Before his description of those experiments, the author mentions his correspondence with Réaumur and the Abbé Nollet concerning fertilization in frogs, and recalls a letter from Nollet, received by him some nine years before. In that letter, Nollet referred to experiments that he and Réaumur had pursued in vain thirty years before, trying to isolate frog semen—and, to illustrate how frustrating that pursuit had been, he confided to Spallanzani that "I remember putting breeches of waxed taffeta on the male, and watching a long time, without perceiving any appearance that denoted an act of fecundation."

In his "Dissertation Concerning the Generation of Certain Animals," Spallanzani did not waste a single line explaining how he fashioned the breeches and how he managed to put them on the males. True to his nature, he went straight to the main points:

> The idea of the breeches, however whimsical and ridiculous it may appear, did not displease me, and I resolved to put it into practice. The males, notwithstanding this encumbrance, seek the females with equal eagerness, and perform, as well as they can, the act of generation; but the event is such as may be expected: the eggs are never prolific, for want of having been bedewed with semen, which sometimes can be seen in the breeches in the form of drops. That these drops are real seed, appeared clearly from the artificial fecundation that was obtained by means of them.

The same observations, he added, were true for semen of green tree-frogs and different kinds of toads. In the case of aquatic salamanders, the briefs were not even necessary; since the males ejected their semen into the water, Spallanzani had only to separate them from the females when the latter were laying their eggs. "As expected," these eggs were barren. But, if the males were allowed to share the same vessel with the same females during the next round of egg extrusion, the whitish mass of semen would promptly cloud the water—and all the eggs would be fertile.

THE DEATH OF THE SPIRIT

How could an intelligent and open-minded man go through so many observations supporting the need for semen to achieve fertilization and still think of the spermatozoa as nothing but parasites? We could recite once more all the possible clichés concerning the tortuous functioning of human nature and engage in speculation. Or, alternatively, we could assume that Spallanzani was at ease with Swammerdam's concept of the *aura seminalis* (see chapter 3) and blindly believed that an invisible spirit coming from the semen was the actual driving force presiding over development. But even the definitive dismissal of this idea was Spallanzani's doing.

As a matter of fact, Spallanzani's enormous pool of observations had provided enough evidence to convince him that real contact between semen and eggs was necessary to trigger development. His experiment to prove this point seems at first clear and simple, but it is actually extremely difficult to perform: modern scientists who want to demonstrate it to their students prefer to resort to computer-generated simulations rather than to go through all the aches and pains of the real thing. But Spallanzani was not afraid of experimental aches and pains—even when, in the final balance, they were likely to work against his theoretical standpoints.

In his first approach, he tied about twenty eggs to a fine string, and suspended the string over a watch glass containing freshly ejaculated frog semen. If a spirit came out of the semen to awaken the eggs from their long sleep, sooner or later these eggs would show signs of development. But, no matter how close to the watch glass he placed the string, his results were all negative. This, however, was not yet a definitive refutation of the *aura seminalis:* all things considered, there was no evidence whatsoever that the spirit would ascend vertically. It could have become dispersed in the surrounding air. Another possibility was that this exposure to the air had simply killed the semen, the eggs, the spirit, or all of them. Also, maybe heat was necessary—physiological spirits tended to be perceived as vapors, and vapor, before anything else, was a trademark of boiling liquids.

Spallanzani's next round of experiments considered all these adverse variables and managed to exclude them all from the picture. He kept the eggs inside their gelatinous coats, and let these coats adhere to the sides of a vessel. Inside this vessel he placed a smaller one, containing fresh semen. This time, the whole was carefully sealed inside a large container, and then kept under gentle heat, so that the eggs could now beyond all doubt bathe in the vapor from the evaporating semen—if there *was* a vapor, and a vitalistic vapor at that. Once again, no matter how many times the experiment was repeated, and regardless of the distance between the eggs and the semen, development did not occur.

So, the *aura seminalis* was a myth. Apparently, semen was needed not only as an inspiring spiritual presence, but as an actual physical agent. Contact between semen and eggs had been proved indispensable for fertilization. To further strengthen this point, in the case of the boxer shorts and the eggs on a string alike, Spallanzani took some of the isolated semen and touched some eggs with it. These eggs developed. The others did not. In a final set of fine experiments, he also tested the effect of separating the components of the semen, filtering it through cotton, chiffon, paper, and other materials. In all cases, the liquid that passed through the filter had no power over the eggs. The viscous white mass retained inside the filter, however, was as powerful a fertilizer as ever (and a single microscopic observation would reveal that this was the component containing the "seminal worms").

THE "SUPPOSED EGGS"

Was ovism dead after these observations?

Amazingly enough, Spallanzani did not seem at all impressed with the direct relationship between lack of contact of egg and sperm and failure of the egg to develop. He just stated his findings, verified them in several different species, and then, "having discovered everything relative to the situation and manner of impregnation"—and apparently not the least bit shaken by it—he turned his attention to the egg proper, determined to observe it "till the young animal should appear."

> The egg is round, and has a smooth surface, of which one hemisphere is black, and the other white . . . When the hot season is far advanced, the observer soon perceives the lineaments of the tadpole. The egg grows for some hours without losing its round shape, it is next elongated; the white hemisphere becomes darker, and the black changes into a longitudinal furrow, terminated by two perpendicular processes . . . By tracing thus the progress of evolution, we come to perceive that these bodies are not eggs, as Naturalists suppose, but real tadpoles. The furrow and the processes become longer; the supposed egg assumes a pointed figure; the whitish hemisphere dilates, and the black is curved inward. The pointed part appears to be the tail of the tadpole, and the other the body. Further, the opposite end takes on the appearance of the head, in the fore part of which the form of the eyes is visible, though they are yet close . . . as likewise [become evident] the vestige of the aperture of the mouth, and the rudiments of the gills.

Certainly, Spallanzani was giving here a very accurate description of the initial stages of larval development in amphibians. But his real point lay elsewhere: he was subtly guiding us toward a notion of the egg as a preformed tadpole, thus bringing the basic core of ovism back into the picture.

The reader will probably be surprised at this description, since it appears, that the tadpole does not come out of the egg, but that the egg is transmuted into a tadpole; or, to speak more philosophically, that the egg is nothing but the tadpole wrapped up and concentrated, [reaching its final form] in consequence of fecundation . . . [therefore] it is fit to call these globules tadpoles or fetuses instead of eggs; for it is improper to name any body an egg, which, however closely it may resemble one, takes the shape of an animal without leaving any shell, as is the case of all animals that come from an egg.

To further secure his point, Spallanzani next undertook a careful comparison of the anatomy of these "globules" before and after contact with semen. He boiled them, peeled them, preserved them in vinegar, cut them in halves, smeared their membranes under his microscope's lenses. His final conclusions could not have been more clearly stated.

Hence the identity between the impregnated and unimpregnated globules is manifest. But the former are nothing but fetuses of the frog, therefore the latter must be so too, and consequently the fetus exists in this species before the male performs the office of fecundation. Hence we are led to other consequences of no less importance:

1—As these supposed eggs existed in the ovary before their descent through the oviducts into the uterus, and long before fecundation, the fetus existed in the mother's body long before fecundation.

2—Although the evolution of these fetuses is never so considerable and quick, as after fecundation, it is, however, perceptible before; for let it be considered, that the fetuses in the uterus are above sixty times larger than they were a year before, when they were in the ovary, as I have found by comparing them.

3—Besides the fetus, the amnion and umbilical cord exist before fecundation.

Furthermore, when observing the "amours" of two different species of aquatic salamanders, Spallanzani noted that, although a complicated courtship seemed to take place, during which the male emitted a "jet of semen" that became mixed with the surrounding water and then possibly "entered the female body through her vent," these animals never came into direct contact in the course of the entire mating. "Upon the whole," concluded Spallanzani, "it appears then, that copulation is not necessary to the fecundation of these animals." Also, once again, his observations seemed to indicate that the "supposed eggs" were in reality preformed baby newts.

As for why these processes required semen at all, Spallanzani decided that its function in animals was to promote growth through stimulation of the fetal heart. This proposal was made even more credible when a vast number of other substances failed to substitute for the semen in this task. According to Gasking's *Investigations into Generation,* "Spallanzani tried to fertilize frog's eggs with a great number of different things, starting with blood from the adult heart and juices extracted from the heart itself, and passing on to try the effect of electricity (which was said to speed up the growth of fertilized eggs, but to have no effect on non-fertilized specimens) and numerous chemical stimulants. He tried vinegar and spirits of wine diluted in water and urine, which merely made the eggs decay more quickly instead of fertilizing them. The juices of lemons and limes, and the oil from their skins, was equally powerless to bring about fertilization. Only the juices from the testicles themselves had the same effect as the semen."

This outcome could have been disappointing to Spallanzani's insatiable curiosity, but it was certainly very reassuring in what concerned nature's mechanisms of self-control. To all of those who had strongly opposed, or at least mildly doubted, Buffon's and Needham's notions of spontaneous generation, one of the most distressing aspects of both the organic molecules and the Vegetative Force was that, in such a scenario, nothing could stop nature from constantly developing new and unexpected life forms. A living world without precise regulation entailed the possibility of endless surprises in generation, without any guarantee that all those surprises would be pleasant. Now, everything seemed neatly organized in little boxes. The egg alone had the power to bring about a new individual. And the semen alone had the power to stimulate the heartbeat of that individual, allowing it to grow to its familiar adult form. As long as one endorsed the principle that the semen's role was solely to stimulate the heartbeat, any ovist could hold that semen was crucial and still remain an ovist.

And so the school of ovism breathed a long sigh of relief.

THE FATHER OF INSEMINATION

We have already noted in this chapter that Spallanzani was a pioneer of artificial insemination, the man who first brushed frog eggs with frog semen and watched those eggs develop. Again, he did not restrain himself to a single species: he was also successful with toads and silkworm moths, and even with a spaniel bitch. Quoting his statement about this last experiment, "the success . . . gave me more pleasure than I have ever felt in any of my other scientific researches," is almost mandatory in all secondary literature on this subject.

In the opening statements of "A Dissertation Concerning the Generation of Certain Animals," Spallanzani acknowledged that "the first attempt to effect artificial fecundation was made by my immortal countryman, Malpighi." However, Malpighi tried the experiment only on silkworm moths, and failed. But Spallanzani was convinced that the failure in moths did not make the project "seem unlikely to succeed in those animals in which fertilization takes place outside the body of the female, as in frogs and toads"—and his friend Bonnet was constantly urging him to try this avenue, in order to "throw new light upon the natural history of animals, and more especially upon generation."

So Spallanzani rolled up his sleeves.

I began with the terrestrial toad with red eyes and dorsal tubercles . . . The female, with the male on her back, discharges slowly at the anus two shining viscid cords, full of black globules. These globules are minute tadpoles, which the male fecundates at the time of expulsion, by besprinkling them with semen. As therefore the tadpoles are at this period best disposed for fecundation, I tried to effect it in the following manner. Just before parturition, of which I was apprised by the excessive swelling of the belly, I parted the male from the female, and set the latter by herself, in a vessel full of water. In a few hours the two cords began to appear; as soon as about the length of a foot was excluded, I cut them off, and left one in the vessel, while I took out the other, in order to wet it with semen, which I procured from the male that had been just separated from the female.

To obtain this semen, in Spallanzani's opinion, was easy enough "for any one who has the slightest skill in comparative anatomy."

In this animal, [the seminal vesicles] lie below the testicles, and cover part of the kidneys. At the time of coupling, they are always full. I laid open the vesicles, and receiving the liquor, which had the transparency of water, I spread it on the piece of cord with a pencil . . . after the operation, I placed this piece in a vessel of the same water as that in which the unimpregnated portion lay.

Five frustrating uneventful days passed. "It was not till the sixth day that I began to conceive hopes, that the application of the seminal liquid had not been ineffectual."

Many of the tadpoles . . . over which the pencil had passed, began to assume an elongated figure, while [those of the piece of cord left in the other vessel] preserved their round form. The seventh day was still more favorable to my hopes; for together with a manifest elongation, an increase in bulk became visible; these appearances grew more evident every day, insomuch that there no longer remained any doubt of a considerable evolution [preformation] of the tadpole. On the eleventh day, I perceived them moving within the amnion, and on the thirteenth they quitted the membranes and swam about the water. On the other hand, the unimpregnated tadpoles began to corrupt, and in time they were quite decomposed, and turned putrid.

Although excited by this new achievement, Spallanzani still paused to consider why some of the eggs in the cord exposed to semen had not developed. He concluded that not all had been touched by the pencil, and tested his assumption by cutting a smaller cord from another female and fully submerging it in semen from two males. This time the success rate was much increased. In the course of these experiments, Spallanzani also noted that impregnation was equally successful when carried out in a dry environment, as long as the impregnated eggs were returned to water after contacting the semen. His next *morceau de bravoure* was to determine whether artificial methods of contact led to a slower development.

> Having two toads coupled, I waited till the female had discharged part of the cords, and the male had besprinkled it with semen. Then I removed the male, and, cutting off the cords . . . left them in water. As soon as the female had discharged another portion of the cord . . . I cut it off, and impregnated it with the semen remaining in the vesicles of the male; and this piece was put in the same vessel . . . Both evolution and animation kept an equal pace in both.

These results were all news to the scientific world, and quite impressive indeed. Spallanzani could have stopped his quest here, but his mind was teeming with other questions. For instance, could eggs still contained inside the uterus display the same generative power? The cords of eggs inside the uterus, just passed from the ovaries through the oviducts, were "very much entangled," but "it is not, however, difficult to draw them . . . out . . . with a forceps, if they are managed gently."

> I opened the abdomens of several coupled females . . . and . . . I bathed [a cord] with seed, and set it in a vessel of water. The rest of the cord was put at the same time into another vessel of water; but not one tadpole contained in the latter portion was evolved, whereas all those that had been impregnated became complete tadpoles.

This allowed for his momentous conclusion: "It is therefore to be inferred that these germs, by the time they get into the uterus, have arrived at such a state of maturity, as renders them capable of being fecundated."

The same did not apply, he noted, to those eggs contained in the oviducts: only the largest, situated nearest the uterus, responded to the seminal activation. Spallanzani concluded that those eggs located closer to the ovary, and the ones contained inside the ovary proper, did not yet have the required surrounding layers of jelly needed for the early nourishment of the tadpole; therefore, even if such "tadpoles" could be activated by the semen, they would soon die of starvation. This belief seemed confirmed by his fortuitous finding of ectopic eggs within the abdominal cavity of the female: "their color was black, as it is when they are arrived at a state of maturity, but they were without gluten." Therefore, even after they were moistened with semen, they never developed—"as the reader will naturally suppose." Another supporting finding was that ovarian eggs carefully brushed with semen without being removed from their location remained unable to develop even after they were normally discharged, although covered at this point with the usual layers of jelly:

> I was obliged to infer, that the gluten is so indispensably necessary for the nutrition of these delicate beings, that should it not be present at the time of aspersion with seed, they will perish, even though it should be afterward supplied.

Having reached all these conclusions, the author repeated the same kinds of experiments with the water newt and the "fetid terrestrial toad," always with the same results. He proudly proclaimed, in his concluding remarks: "The same law that presides over the birth and evolution of the species mentioned in the preceding dissertation, extends likewise to this, that the fetus belongs to the female, and that the aspersion of the seed of the male is a condition necessary to the animation and evolution of the fetus."

A Tale of Three Puppy Dogs

Malpighi had tried to impregnate silkworms artificially, and he had failed. After his repeated successes with amphibians, Spallanzani decided to take up the challenge. His goal was far less mundane than trying to outdo Malpighi: with his experiments on the silkworm, Spallanzani intended to "try to fecundate some of those animals, in which it is certain, that impregnation takes place internally."

At first, he did not succeed either. He kept females away from males and collected their eggs when they were discharged, and in other cases collected the eggs directly from the uterus. Having done this,

> I moistened some of them with a large quantity of seed, and others very sparingly, but to no purpose. The fecundated eggs of this insect assume a violet color; the others remain yellow. My eggs, before they were moistened with the seed, were of this color, and retained it afterward; they moreover became soft, and an incavation appeared upon the surface, signs that never failed to attend sterility.

Then he remembered that he was working during the month of July, whereas the eggs of this species hatched only once a year, in the spring. He turned to another species, "which is most cultivated in Lombardy," with three hatching cycles per year.

My labor was not unprofitably bestowed upon the eggs of this species . . . I put the females under the receiver of an air-pump; the males were in the outside; and either the sight or the smell of the females lured them to the vessel, for they were constantly fluttering about it. By these means, I supposed that the male organs of generation would be filled with seed. As soon as the eggs were discharged, I bathed them with seminal fluid. Many of them, which were at first yellow, began in a few days to turn brown, and at length assumed a brownish violet hue; in about a week they produced little worms; those which had not been moistened with seed remained yellow, grew flaccid, and spoiled. I procured fifty-seven at two trials.

Spallanzani had long been nurturing the idea of trying the same kind of experiment in a viviparous animal "of some size," such as "the cat, the dog, or the sheep." Even before he artificially fertilized frog eggs, he had been toying with the idea of artificially fertilizing mammalian eggs. In his own words, "the event of my experiments on silk-worms, in which impregnation is internal, rendered my expectations even more sanguine, and I immediately set about to bring them to an issue."

He chose for this attempt "a bitch spaniel of moderate size," suspecting "from certain appearances" that she might soon be in heat.

I confined her to an apartment . . . For greater security . . . I fed her myself, and kept the key the whole time. On the thirteenth day, she began to show evident signs of being in heat; the external parts of generation were tumid, and a thin stream of blood flowed from them. On the twenty-third day she seemed fit for the admission of the male, and I attempted to fecundate her in the following manner. A young dog of the same breed furnished me, by spontaneous emission, with nineteen grains of seed, which were immediately injected into the matrix, by means of a small syringe introduced into the vagina. As the natural heat of the seed in animals of warm blood may be a condition necessary to render fecundation efficacious, I had taken care to give the syringe the degree of heat which man and dogs are found to possess, which is about 30°C. Two days after the injection, the bitch went off her heat, and in twenty days her belly appeared swollen . . . the swelling of the belly increased . . . and sixty-two days after the injection of the seed, the bitch brought forth three lively whelps, two male and one female, resembling in color and shape not the bitch only, but the dog also from which the seed had been taken.

THE DEBATE CONTINUES

Some questions about Spallanzani's work remain unanswered—another tribute to his modernity. He certainly became a master of the art of artificially inseminating frog eggs; and we know now, two centuries and millions of experiments later, that these eggs are indeed quite easy to fertilize. However, because his descriptions of amphibian artificial fertilization are immediately followed, in the same dissertation, by reports concerning silkworms and mammals, it is not rare to find modern accounts of Spallanzani's work presented as if he actually had managed to carry out in vitro fertilization in mammals (the case of the famous spaniel bitch). In his original renditions, Spallanzani called all of those acts "impregnations," but, by our century's standards, it is more prudent to draw a clear line between the different kinds of techniques. In frogs and newts, in which fertilization is external, Spallanzani's "impregnations" were truly what we would now call "in vitro fertilization" (or IVF), meaning that sperm and eggs were brought into contact and produced a new embryo outside the animal's body. In the case of the dog, in which fertilization is internal, Spallanzani's "impregnation" was rather what we now call "artificial insemination," meaning that sperm was artificially introduced inside the female's body at the appropriate moment during her ovarian cycle.

The confusion between these two techniques is further enhanced in the secondary literature by the fact that Spallanzani fertilized silkworm eggs outside the female body, an experiment undeniably hard to perform, and which bears clear testimony to the skills of its perpetrator. But to assume that Spallanzani could have done mammalian IVFs is totally out of the question. He was extremely talented, but, although he was a priest, he was not God. Mammalian gametes are difficult to handle, even by modern standards. Experimental procedures for in vitro fertilization did not become generalized until the mid-1970s. In vitro fertilization in some species, such as mice and cows, has proved relatively easy, when we use adequate incubation media at proper levels of temperature and humidity. Rabbits, however, remain just about impossible to use for such protocols, and no one knows exactly why this is so. To make matters worse, the eggs of canids, the taxonomic family of dogs, are fertilized at a point in the cell cycle that differs from that of all other mammals; although some in vitro fertilizations in foxes have been reported, mainly by northern European laboratories, the success rate remains low. With this in mind, to assume that Spallanzani hit it right, on the first try and in the face of so many adverse factors, is a delirious leap of faith. To start from scratch and obtain artificial insemination in the dog is wondrous enough. It means that Spallanzani knew his material well enough to inseminate the bitch at the exact moment of ovulation, and to be aware of the need to keep his instruments heated according to the temperature of the mammalian body (although this temperature is really 37°C, not the 30°C he estimated necessary). Perhaps we should thank God that he did not try to repeat this experiment. Sheer luck seems to have played such a crucial role in getting all the variables right that he could just as easily have failed in the second round. And our understanding of fertilization's mechanisms would have been delayed by several decades.

Even more problematic is the question of artificial activation. Numerous modern writings on Spallanzani's experiments, albeit complimentary on the whole, suggest the possibility that he did not always manage to fertilize the frog eggs involved in all of his experiments. Since he used needles to touch the eggs with semen, the argument goes, it is quite possible that he merely *activated* them with this treatment,[23] triggering their parthenogenetic development. In other words, Spallanzani could have induced frog eggs to develop without any participation of the sperm; and this could have misled him in some of his conclusions, and promoted confusion about the real meaning of his observations.

This claim must have been initially made by someone with strong persuasive powers, since the question of artificial activation is so frequently raised in the secondary literature. But does it make sense to suppose that an observer as keen as Spallanzani would have disregarded the possibility of a needle-induced effect if he had any supportive evidence for this possibility? After all, Spallanzani was an ovist. He was familiar with parthenogenesis in aphids, as described by his friend Bonnet (see chapter 3). If he could claim that frog eggs were able to resume development without semen, why would he not do so? And, bearing in mind his care with controls, if he had any reason to suspect that the needles were doing something by themselves, would he not consider testing possible differences between the use of a needle and the use of other instruments, such as a brush (which he also employed to bring eggs into contact with semen)?

Moreover, what we now know about frog eggs stands strongly against this possibility. These eggs lack centrosomes, the fundamental organizers of the cytoplasmic structure, including the apparatus that presides over cell division. The fertilizing sperm contributes the centrosome, and thus makes the first division of the embryo possible.[24] If there is one thing that developmental biologists know well, it is that frog eggs *cannot* be activated by needles (or by heat treatments, or by electrical pulses, or by any other general means of inducing parthenogenesis, for that matter). If you use nothing but a needle, the egg will show the first signs of parthenogenetic response. Most noticeably, the cytoplasmic components will change place, causing a different color pattern to appear at the surface. But the response will stop there: the egg will not be able to divide. Spallanzani, however, wrote that he consistently saw his eggs "bring forth young," not just "show a variation of the surface pattern." Also, he wrote that he "touched" his eggs with a needle, not that he "pricked" them—and mere touch will not induce any effect. Was he lying? Was he being careless with words? Was he assuming that, once the first signs of response had appeared, fertilization could be taken for granted and no more checks were needed? It is plausible, but highly unlikely, considering Spallanzani's character.

The modern speculation about needle-induced activation leaves out yet another important factor. As we have seen, frog eggs can respond to stimuli of activation, but they cannot resume development because they do not have centrosomes. So, if they are pricked by a sterile needle, they will not go beyond early responses. But if the needle carries any centrosome-containing cell at its tip (and practically *all* cells have centrosomes, eggs being the most notorious exception to the rule), the egg will incorporate this centrosome, and development will proceed. For instance, it is possible to trigger cell division in frog eggs by touching them with needles previously immersed in blood, as the French biologist Jean Eugene Bataillon demonstrated in the experiments he carried out from 1910 onward. And Spallanzani's needles were carrying *sperm* cells with them, by far the ones with centrosomes best fit to sustain embryonic development. Yes, he could have claimed parthenogenesis and scored a point for ovism. He did not say it because he did not see it. It is much more likely that he simply saw the truth.

Is it not remarkable that we are still debating the experiments made by a priest two hundred years ago? Is it not moving that we have to recruit the language of modern developmental biology to fully explore the stakes in this debate? And, while these matters remain unsettled, Spallanzani's name remains tenaciously attached to our ideas about sperm. Was it not a fabulous gift to all of us that Spallanzani believed so strongly in the egg?

The H Word

6

We must be silent about the things we cannot say
LUDWIG WITTGENSTEIN

E TEND TO ASSUME that spermism is somehow a part of our general knowledge. We may not know many details about it. In fact, we may know absolutely nothing. But we all have a certain picture clearly lingering in the backs of our minds. Somewhere, sometime, we have seen a comical representation of a sperm cell, dating back from a vague decade in the seventeenth century. There it is, the long tail and the bulky head. And, inside the head, a little man is tightly curled up, representing a person in some generation to come, waiting for his time to stretch and push himself into existence. Those of us who studied biology—myself included—could never escape the reproductions of that drawing in our textbooks. The subtitle always read something like "Hartsoeker's 'observation' of a spermatozoon, with the *homunculus* in the sperm head."

How could we forget this association? Even the entry on "ovism/animalculism" in the 1984 edition of the *Dictionary of the History of Science* tells us that "The spermatozoa, or animalcules, were thought by Animalculists to contain in miniature future generations. Nicolaas Hartsoeker . . . propounded the idea of the homunculus, a tiny man supposed to be embodied in the sperm."

Poor Hartsoeker. Pick the introductory chapter of any current book in embryology. Here are some typical examples. Oppenheimer and Lefèvre's 1984 edition of *Introduction to Embryonic Development* states that "In 1664 Niklaas Hartsoeker drew a figure of a miniature human *(homunculus)* inside a sperm, presumably representing what he saw under the microscope," and then seizes the occasion to remind us that "like the 'canals' on Mars, observations such as this demonstrate that we see what we look for, not at we look at." Bruce Carlson's 1981 edition of *Patten's Foundations of Embryology* includes the famous drawing, with the legend "Reproduction of Hartsoeker's drawing of a spermatozoon showing a preformed individual (homunculus) in the sperm head," while stating that the term "homunculus" is from Hartsoeker's "Essay

Fig. 42. Hartsoeker's drawing of a sperm cell with a presumed little person curled up inside the sperm head. (From *Éssai de dioptrique,* 1694.)

de Dioptrique." And John Farley's 1982 *Gametes and Spores,* once again sporting the ubiquitous drawing, starts with the recognition that "it is to Nicolas Hartsoeker that we owe the most explicit statement in support of sperm pre-existence." We also owe to Hartsoeker "that most extraordinary claim that the sperm actually contained a fully formed miniature adult coiled up within the famous homunculus." This association also appears in modern versions of landmark embryological treatises, such as Oscar Hertwig's *The Biological Problem of To-day.* In his preface to the 1977 edition, Joseph Mazzeo states that "those microscopists who saw a tiny homunculus in the head of a spermatozoan were deceived both by their imagination and their instruments." Moreover, Hartsoeker's drawing was also the very first figure presented in the first edition of Scott Gilbert's *Developmental Biology,* which has been the standard textbook in the field since 1985. The passage "Nicolas Hartsoeker . . . drew what he also hoped to find: a preformed human ('homunculus') within the human sperm" has remained unchanged through the four editions of the book.

It is no wonder, then, that we grew up believing that the little man inside the sperm had been described by the spermists as a "homunculus."

But we may all have been wrong.

And this misunderstanding may have played a substantial part, albeit a subtle one, in our patronizing dismissal of preformation. For we may never have paused to think about it, but—as I will try to demonstrate in this chapter—the term *homunculus* has never been, by any means, an innocent one. And, bearing this in mind, would it not have been a very strange move for the spermists to saddle their little man-to-be with the burden of a name so easily subject to ridicule?

Just think of the debate going on presently in the ranks of artificial intelligence and cognitive psychology. The former field started from an apparently clear-cut premise: if our ability to think is nothing but the output of neural networks, then it must be possible to mimic those networks artificially to create intelligent machines. One of the models for this operation, developed during the 1950s and 1960s, resorted to abstract entities, which were presumed to be the symbolic representatives of the different modes of brain activity, and therefore had to encompass a will, or soul, or whatever you choose to call it. This position opposes an older one, which reduces mental processes to purely symbolic manipulations akin to digital computation. In the fight that followed and is still unfolding, the old neural network "determinists" tried to discredit the newcomer AI theorists by calling them—what else?—"homuncularists." As in "the dummies who believe that we have a little man inside our head telling us what to do."[1]

Why is a problem of the seventeenth century being abruptly intercepted by a modern debate unfolding outside the realm of reproduction? Basically, it is an attempt to demonstrate that, even today, the term "homunculus" is still used mainly when one faction is trying to ridicule the other. Those who believe in a place for "free will" in neuronal networks, like Fodor and Minsky, do not call themselves "homuncularists": it was the opposite "determinist" faction that pinned that label on them. They did so to make their opponents sound silly. The point is that, right up to our own days, using the term "homunculus" is a joke on our enemies. It is a term of derision, because it has remained largely permeated by derogatory connotations. If you have a theory and you want the world to take you seriously, you definitely should not associate your theory with this word. Some words can kill.

THE EGYPTIAN TRINITY

It is just about impossible to calculate how many embodiments we might find for the idea of the "homunculus." We can almost see it in the Ka of the Egyptians, the double miniature image contained inside each living body, which did not need to die with the breath of its bearer. Since the Ka would survive all the more completely if the flesh were preserved against hunger, violence, and decay, complex mummification processes, and even more complex constructions of pyramids and other elaborate tombs, ensued. And, since this body had to be fed, clothed, and served after the death of its frame, images of servants, foods, fertile fields of grain, and busy artisans of all sorts were painted on the walls to keep the Ka as happy as possible.[2] In some cases, even lavatories were provided for the Ka's convenience. The presence of the Ka inside the tomb of its former, larger body seems to be of such unquestioned importance that a funerary fragment clearly expresses anxiety over the need of this creature, for want of food, to feed upon its own excrements.[3]

To better grasp the connection between such an entity and the homunculus, it should be noted that this creature is not a metaphor for the soul, which was conceived as a spirit wandering within the body, as a bird flits among trees. It is rather a third entity in its own right, a fulcral part of the Egyptian Trinity expressed in each human being.

Insofar as thinking of the Ka as a homuncular entity—and, for that matter, one that could be perceived as some sort of material soul—an interesting connection to present-day perceptions could be established here. What if we assume that the "homunculus" is the soul? According to Nelkin and Lindee in their 1995 *DNA Mystique,* modern genetics has been using analogous rhetoric to say that the genome is soul-equivalent. It forms your character, it is the essence of your being, it cannot be changed after fertilization, and it can be used after your death to resurrect your body. Maybe the Egyptians perceived something that eluded us for thousands of years afterward.

DREAMS AND NIGHTMARES

The connections discussed above could well be the only redeeming factors associated with the term "homunculus." But these are subtle, almost imperceptible tools of redemption. By and large, the word that cursed the spermists was a nest of malignant connotations. To better understand this problem, let us try to summarize briefly what the term "homunculus" had stood for until the days of early microscopy.

Basically, it stood for the artificial creation of life—human life, to make matters worse. And our culture, in its literary and folk forms alike, is full of warnings against this type of enterprise. Homunculi, regardless of whether or not they go by that name, are basically very scary creatures. Goethe never told us why he decided to make a homunculus sit on Dr. Faust's shoulder during the second part of the book, but the presence of this mute entity is definitely not reassuring. Likewise, when the two characters of Beckett's *Waiting for Godot* start reminiscing about the homunculus in the root of the mandrake, they certainly are not trying to cheer up the audience. And, needless to say, the creature of Dr. Frankenstein was by no means the first example of the unspeakable dangers that we face when we decide to cross the line and play God. Mary Shelley drew her monster from a long legacy of similar stories.

What is that legacy? Basically, it is the dream of man-made acts of Creation, and the perpetual moral conclusion that such dreams will always turn into nightmares in the end. Such is the spell of the homunculus; and that is the reason why the spermists of the seventeenth century should have known better than to call their "little man inside the sperm" a "homunculus." If you have a theory and you want your peers to accept it, you generally do not center your model on a word evoking fear, black magic, evil doings, and the like.

JÂBIR'S VISIONS

One of the earliest approaches to the homuncular concept appears in the texts of Jâbir ibn Hayân, whose ghost dominates alchemy as one of its most prominent inspirational figures. The reality surrounding his existence is elusive to the extreme. Most likely, a person by that name did live during the eighth century, in Arabia or in Iraq. He launched a school that continued his research and produced, during the following two centuries, a corpus of work that in the end included about 3,000 titles, of which some 215 survive to the present.[4] Jâbir was, at least in legend, supposed to have been the first Hermetic explorer to conduct proto-homuncular experiments, discoursing on how to animate statues based on a vision of all natural phenomena as the effect of a subtle energy that penetrated and animated inert matter—a concept traceable to Greek antiquity, when Daedalus gave movement to a wooden Venus by placing mercury in the wood, and doubtlessly tributary to the primitive belief, held as a true theological principle in old Egypt, that gods or demons, through a proper magic ritual, would come to inhabit their own statues.[5] According to Paul Kraus,

> the Middle Ages and the Renaissance dreamed of the automated man, the homunculus. But this problem has seldom received attention as 'scientific' and detailed as that of Jâbir . . . It is characteristic of his science not to admit any limit to human thoughts . . . Assuming the ancient belief that 'art imitates nature', Jâbir applies this definition to natural sciences: the human artisan imitates the Demiurge Creator of the Universe, exerting, like Him, a creative power.[6]

A fervent believer in spontaneous generation, Jâbir gave several examples of marvelous creatures encountered by himself or described by others, and used this belief to back his exhaustive description of an apparatus "of glass, crystal, or some species of stone, as thick as a finger," built specifically to produce human beings and versatile enough to allow for "the combination of the body of a young girl with the face of a man, or the intelligence of a man with the body of an adolescent, or other variations in form,"[7] in a coincidental anticipation of the computer-generated morphism that is so fashionable in present-day video clips and the graphic arts of slick magazines. If correctly constructed, during the right number of days, this apparatus "will function until the end of time."

In 1317, Pope John XXII condemned alchemy and forbade all men of faith to engage in such undertakings. His act provided a clear signal of the dangerous connotations acquired by this science, viewed as an attempt to transgress the natural laws established by Divine order. However, the call to arms came too late, and was all but useless. By this time Jâbir's torch had been carried in all directions by followers of all sorts. And the quest for human life without natural generation appeared everywhere.[8]

FEAR THE WRATH OF THE GOLEM

On the occult side of Jewish mysticism, the rich interval between 1150 and 1250 witnessed, in Germany, the unfolding of the second period of the Hasidim. A first brief Hasidic movement had occurred in Talmudic times among some individuals of exceptional piety. A third movement would start in the early eighteenth century, launched by Rabbi Israel ben Eliezer of Mezbizh, known as the Baal Shem Tov, Master of the Good Name. It expressed the revolt of the poor and destitute against the tyranny of learning as the only avenue to God; and in this form it expanded and reached our times, constituting, by the second decade of the twentieth century, a large percentage of the population of Eastern European Jewry. But it was the second cycle, during the Middle Ages, that gave birth to the strange myth of the Golem.[9] It is certainly not a coincidence that the legend of this "magical homunculus," as it is called by experts such as Gershom Scholem,[10] was developed by "the only branch of Judaism that surrounded human beings with the power for magic creation."[11]

It can certainly be argued that a giant made of clay would hardly be considered a "homunculus," since we tend to connect this term mainly with miniature creatures. Nevertheless, the Golems are an interesting case study in our atavistic fear of artificially created life: contrary to the current belief, they were *not* invented by the eighteenth-century rabbis of Prague to protect the Jewish people, and did not grow to gigantic proportions, eventually turning against their own creators and threatening the entire city. That is the legend. The reality (as Scholem clearly points out) is that they were conceived by the rabbis of the twelfth-century Hasidic period as a means of getting closer to God by repeating His most perfect creation through mystical trances. The real Golems (if we can call them that) were expressions of piety; legend turned them into monsters. Is this not an excellent parable for the spell of homuncular danger?

For those less familiar with the tale of the Golem, let us examine the most salient features of this interesting story.

The three men who started the trend were all members of the Kalonymides family, beginning with Samuel the Hasid. Drawing on his heritage, his son, Rabbi Jehuad the Hasid, rose to what Baer described in 1938 as a "historical position akin to that of his Christian contemporary St. Francis of Assisi." [12] Like the Baal Shem Tov, he appears to have been "a man both learned and charismatic, a folk healer, one of those who went about curing the sick by invoking the various mystical names of God." [13] Jehuad died in Regensburg in 1217, and the task of preserving and spreading his doctrines fell mainly on the shoulders of his faithful disciple, Eleazar of Worms. At this point the term Hasid was often used to mean "devout but otherwise not remarkable men," since the value of the Hasidim was not measurable by any intellectual standards, but only within the framework of *Hasiduth* (piety) itself. [14] It was part of this piety to believe that words and names carry a magical potency. [15]

The legacy of Eleazar of Worms discloses the oldest recipes to create the Golem, blends of magical letters and practices obviously intended to produce ecstatic states of consciousness. The starting point was, necessarily, the "Sefer Yetzirah," or "Book of Creation," [16] which ended up functioning for the Jewish mystics as a manual to be used for the act of creation itself. According to Sherwin in *The Golem Legend,* the ritual requires the presence of two or three adepts. A magical circle is drawn to circumscribe the space in which the Golem is created. Virgin soil, taken from a mountain, is kneaded in running water. From this, the form of the Golem is made. Over this form, various combinations of letters of the Hebrew alphabet are invoked. The formulae for these combinations are sometimes recited while walking around the circle, animating the Golem limb by limb. Separate letter combinations and permutations are required to create either a male or a female Golem. Reciting the permutations in reverse order deactivates the Golem and transforms it back into inert matter.

That the Golem is created by language implies that language also has the power to destroy the creature. It is also language, or rather the lack thereof, that distinguishes this living man from real men: the absence of speech indicates the absence of a human soul. In its initial conception, the Golem could have life only as long as the ecstasy of its creator lasted, expressing a particularly sublime experience felt by the mystic who became absorbed in the mysteries of the alphabetic combinations from the "Sefer Yetzirah." [17]

But it did not take much for folk tales to set the Golems free from their mystical prison. Actually, the spreading of such stories started while Jehuad was still alive. During the following centuries, legends kept piling up about how mean the Golems could be.[18] In the eighteenth century, when the motifs found in earlier versions of the legend coalesced around Judah Loew, the Maharal of Prague, we encounter the story of a Golem running amok through the streets of the city, wreaking destruction wherever it passed.[19] These tales overcame reality to such an extent that, in our own century, Collins and Pinch adopted the Golem as a metaphor for science:

> What, then, is science? Science is a Golem.[20]

And it is so because, in their opinion, there is a clear analogy between the scientist and the sorcerer's apprentice:

> A Golem . . . is a humanoid made by man from clay and water, with incantations and spells. It is powerful. It grows a little more powerful every day. It will follow orders, do your work, and protect you from the ever threatening enemy. But it is clumsy and dangerous. Without control, a Golem may destroy its masters with its flailing vigor.

Furthermore,

> in the medieval tradition, the creature of clay was animated by having the Hebrew "EMETH . . ." inscribed in its forehead—it is truth that drives it on.[21] But this does not mean it understands the truth—far from it.

These authors are not alone in examining the metaphoric applications of the concept. There have been novels, poems, ballets, movies, and science-fiction tales inspired by the Golem. Even comic books introduce characters such as the "Galactic Golem." Children's games like "Dungeons and Dragons" include a Golem. *God and Golem, Inc.,* by Norbert Wiener, is an essay on the relationship between humans and machines, the latter being "the modern counterpart of the Golem." And, when Louise Brown, the first "test-tube baby," was born in England in 1978, *Time* magazine compared this event to the creation of the Golem.

It is, once again, only another one of those ironies that so often occur in history. The original concept of the helpless, selfless, and indifferent Hasid, who craves no form of power, ended up capturing the popular imagination as an enormously powerful being who can command the forces of all elements. And Rabbi Jehuad, the introspective saint devoted to his ideal and to the loving care of his community, ended up being perceived as the bearer and dispenser of all sorts of magical powers, able to obtain everything exactly because he did not want anything for himself. In the process, even the concept of the Golem lost a large part of its original religious implications. According to early Talmudic legend, Adam was first created as a Golem, meaning a body without a soul, during the first twelve hours of his existence (Sahn. 38b). However, even in this state, he was accorded the vision of all generations to come (Gen. R 24 : 2), as if there was in this Golem a hidden power to grasp or see, bound up with the element of earth from which he was taken. Nothing could be more poetic. Yet, in the Yiddish brought from the Eastern European ghetto, the term "Golem" is now used for "any fool who knows neither his own strength nor the extent of his clumsiness and ignorance."

Wherever the homunculus goes, so does the cloud of confusion that surrounds him.

THE BOOK OF COWS

There is a constant theme here, from Jâbir to Jehuad, driving the Hermetic pursuit further and further: the notion that by enabling Art to imitate Nature we get one step closer to the Almighty who created the latter. This theme has been repeated over and over, in many different forms and latitudes, against the background of different cultural heritages. And, more often than not, the resulting legacy strikes us now as perturbing and weird. The intriguing *Liber vaccae,* wrongly ascribed to Plato,[22] is a striking example.

Organized as a collection of magical and necromantic experiments, the book opens with a chapter dealing at impressive length with methods for creating "a rational Being" inside the uterus of a cow (an "ape" or "other beast" would also do), literally by shoving mixtures of ingredients inside (including the perpetrator's "own water while warm" and "the stone which is called the stone of the sun . . . that shines at night like a lamp") and letting the invaded womb function as an incubator. This incubation should be carried out in a dark house, and every week the cow should be given the blood of another animal to eat. If his work succeeded, the creator of such a Being could then take it out of the cow, cut it up, eat it, and have its intelligence transferred to the diner.[23]

Here we are unapologetically in the realm of pure magic, and the incorporation of recipes for fabricating "rational beings" bathes the concept of artificial intelligence in a scary glow. In the following recipes, the idea of manipulating forces of nature stretches as far as recommending the use of a crow "submerged in water until it dies," in order to feed it, on the third day after the sacrifice, with the water in which it was drowned, to a very black dog imprisoned in a dark house. By the eleventh day, when only the whites of the dog's eyes show and it is unable to bark, the animal should be fed with some plant until it barks again, only to be then tied "hand and feet" and boiled in a big pan. The resulting broth should have the power to bring about rain.

If there are any doubts left about the Hermetic propositions of this book and the intent of putting them to use in concocting pseudo-humans, consider the magic lamps. One of them makes a man appear in any form desired. Another makes a house seem to be full of snakes. All of this is very exotic, but none of it is good company if we want the artificial creation of life to be viewed with respect.

PRETTY LITTLE DOGS

On the more playful side of the connotations associated with the homunculus, fascination with smallness could overcome interest in producing life without generation. In 1558, Giambattista della Porta proposed in the Second Book of his *Natural Magik* to show "how living Creatures of divers kinds, may be mingled and coupled together, and that from them, new, and yet profitable kinds of living Creatures may be Generated." Fulfilling this promise, he presented us with the ultimate recipe for "how to generate pretty little dogs to play with."

In these pages, we first learn that "frogs are wonderfully generated of rotten dust and rain"; "serpents may be generated of man's marrow, of the hairs of a menstruous woman, and of horse-tail, or mane"; "how a scorpion may be generated of basil"; "that new kinds of living Creatures may be generated of divers beasts by carnal copulation." Then comes chapter 6, and with it, the author brings in the dogs. Under the title "How there may be Dogs of Great courage, and divers rare properties, generated of divers kinds of beasts," Porta includes such items as "a strong indian-dog may be generated of a Tygre" or "a strong and swift dog, generated of a kind of Wolf called Thos." By now we are ready for chapter 7, dealing with the "pretty little dogs," who "are in such the kind to dogs as Dwarfs are around men" and therefore "are much made of, and daintily kept, rather for pleasure than for any use."

The trick is in the conjugation of the "straightnesse of the place wherein they are kept" with the "scarceness of their nourishment." The "parts" will still be "prettily well knit together"; for "nature performs her work, notwithstanding of the place." The resulting creatures will be "no bigger at their best growth than a mouse" and should be coupled "with the least you can find, that so lesse may be generated."

Now, if you care for "a dog that will do tricks and feats"—well, it's simple: "you must first let them converse and company with an Ape, of whom they will learn many sportful tricks; then let them line the Ape; and the young one which is born of them two will be exceeding practiced to do feats."

We instinctively associate smallness with the homunculus. But the means prescribed to achieve small sizes were not always as frivolous as Porta's tricks. Certainly not in the plant kingdom. Do you want to graft a lemon tree onto an olive tree in order to obtain lemons the size of olives? Here is Ibn Washya's method, to be performed at a certain conjunction of the sun and the moon: "The branch to be grafted must be held in the hands of a very beautiful maiden, while a man is having shameful and unnatural intercourse with her; during coitus the girl grafts the branch on to the tree."[24]

So, even when we think small, we end up thinking dubious and dangerous. Don't try this at home.

THE TRUE HOMUNCULUS

So far, we have been retracing the most prominent entities that can be associated with the concept of the homunculus. All of these propositions certainly cast a strong spell over human minds. But, alluring as it might be, the idea of artificial life had also always been extremely controversial, raising all kinds of religious perplexities, and responses from wonder to fury, from philosophy to hysteria. In the process, somehow, the name "homunculus" seems to have crystallized in the famous recipe produced by Aureolus Philippus Theophrastus Bombast von Hohenheim, known to most people simply as Paracelsus.

Paracelsus's body of work, filled with all kinds of reflections about non-human human look-alikes, emerges from the long line of reveries analyzed up to this point. In his "Treatise of Nymphes, Sylphes, Salamanders and other beings," Paracelsus explains that "these beings, although they have a human appearance, are not descendants from Adam; they have an origin completely different from both that of man and that of animals. However, they can mate with man, and from that union human beings are born." And then there are all those "beings as light as spirits, and that mate like man, they have its appearance, its habits."[25] From this fertile ground rises the most prevalent of all images of little people in little bottles.

Let us start with the recipe proper, the core of the uproar that echoed through the centuries and was so often reprinted and recounted that it has reached us as something of a pop icon. Most likely, the procedure was first published in *De Natura Rerum* in 1572.[26] The text is introduced with the statement that we should not "by any means forget the generation of homunculi," followed by a short and incisive sketch of the philosophical grounds for the enterprise:

> For there is some truth in this thing, although for a long time it was held in a
> most occult manner and with secrecy, while there was no little doubt and ques-
> tion among some of the old Philosophers, whether it was possible to Nature and
> Art, that a Man should be begotten without the female body and the natural

womb. I answer hereto, that this is in no way opposed to Spagyric Art and to Nature, nay, that it is perfectly possible.

And here is how to do it:

Let the semen of a man putrefy by itself in a sealed cucurbite with the highest putrefaction of the horse stomach for forty days, or until it begins at last to live, move, and be agitated, which can easily be seen. At this time it will be in some degree like a human being, but, nevertheless, transparent and without a body. If now, after this, it be every day nourished and fed cautiously with the arcanum of human blood, and kept for forty weeks in the perpetual and equal heat of venter equinus, it becomes thencefold a true living infant, having all the members of a child that is born from a woman, but much smaller. This we call a homunculus; and it should be afterwards educated with the greatest care and zeal, until it grows up and starts to display intelligence.

For those casually acquainted with the work of Paracelsus—in other words, for the overwhelming majority of us—the little freakish footnote in the bottom of our minds stops here. Yet there was more to it. Paracelsus did not view the homunculus as a mere pretty curiosity to play with. His is a much more complex and grandiose perspective:

Now, this is one of the greatest secrets which God has revealed to mortal and fallible man. It is a miracle and a marvel of God, an arcanum above all arcana, and deserves to be kept secret until the last of times, when there shall be nothing hidden, but all things shall be manifest. And although up to this time it has not been known to men, it was, nevertheless, known to the wood-spirits and nymphs and giants long ago, because they themselves were sprung from this source; since from such homunculi when they come to manhood are produced giants, pygmies, and other marvelous people, who get great victories over their enemies, and know all secrets and hidden matters.

THE FASCINATION

This recipe soon went well beyond the boundaries of scholarly knowledge, as illustrated by the folk tale developed after Paracelsus's death: having grown old, Paracelsus had himself cut into small pieces and buried in horse manure, intending to resuscitate himself as a handsome young man. Unhappily, a servant opened the grave two days too soon and thus put an end to his master's dream.[27] Simultaneously, the homunculus had also captured the scholarly imagination, as illustrated by the number of pieces written on the subject.

In William Maxwell's 1679 *De Medicina Magnetica*,[28] the Scottish physician claims to prove the possibility of creating a homunculus in the resurrec-

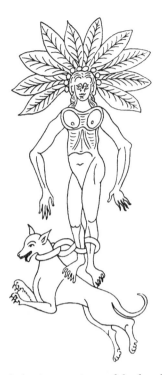

Fig. 43. The root of the mandrake, in a specimen of the female sex. The dog was necessary to pull the root from the ground because, during this operation, the plant was said to yell in such a monstrous fashion that anyone hearing the noise would immediately die. The dog, chained to the root and called from a distance by its owner, who meanwhile had covered his ears and hidden behind a tree, was also supposed to die while performing this duty. (From a thirteenth-century manuscript in the British Museum; as reprinted in C. J. S. Thompson's *The Mystic Mandrake,* 1934.)

tion of a plant from its ashes. He states that the salt of blood, if properly pre-pared, is the supreme remedy; and just as salts of herbs can reproduce the likeness of the herb in the test tube, so the salt of human blood can show the image of a man—"the true homunculus of Paracelsus." In his 1638 "Rare et Curieux Discours de la Plante Appelée Mandragore," Laurens de Castelan discussed at great length the embodiment of yet another homuncular concept, the well-known story of the mandrake: according to the legend, this plant grows from soil fertilized by the urine emitted in the last agony of an innocent man hanged for the crime of theft, developing roots with the shape of a man, including prominent sexual organs. Castelan was careful to admit that these ideas had been "denied by many," and that Paracelsus's homunculus could be "a bit of diabolical magic."²⁹ But caution did not stop Christian Friedrich Garmann, author of the book on the miracles of the dead *De Miraculis Mortuorum,* published posthumously in 1709, from writing a shorter piece in 1672, discussing the evolution of man from the egg and whether conception could take place outside the womb, concerning "the chemical homunculus of Paracelsus."³⁰

Finally, even fictional literature adopted the trend. The greatest celebration of the theme occurs in a book that we are all supposed to know—Laurence Sterne's *The Life and Opinions of Tristram Shandy, Gentleman.*

First published in nine volumes between 1759 and 1768, the epistolary novel alerts us, right at the beginning of Book 1, that something is not totally normal about Tristram Shandy. Chapter 6 starts with the subtly threatening warning that "in the beginning of the last chapter I informed you exactly when I was born; but I did not inform you how." From here, although the secret is withheld in the hallucinogenic convolutions of the narrative, we can see it coming—especially since we have already been informed that his father was somewhat estranged from his mother around the time of his presumed conception. And also because chapter 2 suddenly interrupts the flow of the story to deliver a staunch defense of the homunculus's human rights.

> The Homunculus, Sir, in however low and ludicrous a light he may appear, in this age of levity, to the eye of folly or prejudice;—to the eye of reason in scientific research, he stands confessed—a Being guarded and circumscribed with rights. The minutest philosophers . . . shew us incontestably, that the Homunculus is created in the same hand, engendered in the same course of nature, endowed with the same locomotive powers and faculties with us:—That he consists, as we do, of skin, hair, fat, flesh, veins, arteries, ligaments, nerves, cartilages, bones, marrow, brains, glands, genitals, humors, and articulations;—is a Being of as much activity, and, in all senses of the word, as much and as truly our fellow creature as my Lord Chancellor of England. He may be benefited, he may be injured, he may obtain redress, in a word, he has all the claims and rights of humanity.

Yes, Tristram Shandy may well have been a homunculus. But it is rather apparent from his letters that he did not live a happy life.

THE BACKLASH

Like Tristram Shandy's, the life of the homuncular concept was not exactly peaceful and gay. Although Paracelsus's creature had some occasional admirers, and even gained the right to become a leading character in works of fiction, the academic reception of the recipe and its epistemological implications was not exactly heart-warming. Werner Rolfink, in his 1661 *Chimia in Artis Formam Redacta Sex Libris Comprehensa,* strongly refuted the homunculus of Paracelsus. The author said in the preface that his work "comes to life after having feared the light for the space of some years."[31] Some contemporary writers, including Thorndike, perceive in Rolfink's contribution the ultimate goal of initiating a movement away from alchemy toward chemistry, away from magic toward science.[32] The sixth and last volume is dedicated to the refutation of nonexistent chemical effects, including the homunculus—as ridiculous, in Rolfink's opinion, as the claim of gold being generated in the human body, as in the episode reported in 1593 of a Silesian boy who grew a gold tooth.

The *Destillatoria Curiosa,* attributed to different authors but probably written by George Kirsten and published in 1674, opens with a list of nine false entities, one of them being Paracelsus's homunculus.[33] Antoine le Grand's *Curious Scrutinizer of Hidden Things of Nature,* probably from 1676, considers that "the statement of Paracelsus that a homunculus was generated in a glass phial" is as absurd as the belief that the beaver castrates itself when hunted.[34] Julius Caesar Baricellus's *Hortulus Genialus,* published in 1620, dismisses the homunculus as both ridiculous and an abomination.[35]

To make matters worse, while some scholars thought of the homunculus as merely silly, others perceived it as a definite heresy. In 1664, the influential Jesuit Athanasius Kircher (see chapter 4), in *Mundus Subterraneus,* referred to this "creation" as an impious act.[36] In 1612 Joanes Bickerus published *Hermes Redivivus,* in which he disapproved of making a homunculus to attract and avert all sorcery by magnetic force, advising the reader to turn instead to prayers to God if he wanted to seek aid against incantations.[37] At this point, the little man is apparently in the process of gaining the imaginary powers of a voodoo doll. Taking a closer look, we find that such powers are not even new. A similar approach is described in Raphael Patai's 1994 *The Jewish Alchemists: A History and Source Book,* in which the author reports that the Medieval "pseudo-maimonides" produced recipes for "divination with the help of a homunculus created by melting certain metals at a certain astrologically determined hour."

The association of any theory with the term "homunculus" now looks potentially more and more damaging, especially since these aforementioned accusations are only further examples of the reactions against man-made creations of life that have occurred through history. The Golems seem so embarrassing to some that a distinguished Jewish scholar once said that he hoped the original manuscripts would never emerge "from their well deserved oblivion."[38] Pico della Mirandola, writing against astrology at the end of the fifteenth century, bemoaned the fact that the *Liber vaccae* had been ascribed to such a great man as Plato, for it was "stuffed with execrable dreams and figments." And William of Auvergne said of the *Liber vaccae* that "they call it the laws of Plato because it is against the laws of Nature." [39]

By this analysis we see that the homunculus has many faces, but none seems reassuring. So, by the time the spermists first started drawing their tentative little man inside the sperm head, it should have been more than obvious that branding their theory with such a word would not make for the best of public relations.

FURY AND SHAME

Among the spermists' contemporaries who expressed disgust or distrust toward the entire homuncular concept, Henry More stands out for his vibrant warning. More might have been sympathetic. As one of the most mystical Cambridge Platonists, he believed in ghosts, witches, and pacts with the Devil. But he still considered reason to be "a participation of the Divine Reason in God," and certainly tried to distance himself as much as possible from those dissenting sects that would abandon reason for the easy certitude of "inner light." This preoccupation resulted in the writing of *Enthusiasmus Triumphatus,* arguably the best seventeenth-century attack on Enthusiasm and clearly the work of a man with a mission. Like many liberal Anglicans of his day, More sought to see all Christians reunited once again, and viewed Enthusiasm (which many blamed for the English civil war) as an emphasis on the private and divisive, which, carried to its logical extreme, might end up in a cry for "one man, one religion." More claimed to have no desire "to incense the Minds of any against Enthusiasts as to persecute them," perceiving them as victims of a mental disorder, madmen suffering from a "misconceit of being inspired," and wishing only to dissuade others from following their teachings.[40] One of his main targets, was, inevitably, Paracelsus.

Arguing that "Paracelsus has given occasion to the wildest Philosophick Enthusiasms that ever were yet on foot," and that "Paracelsus and his Philosophy, though he himself intended it or not, is one of the latest sanctuaries for the Atheist and the very prop of ancient Paganism," he attacked Paracelsus's beliefs in "Gnomi, Nymphs, Lemures and Penates, Spirits endued with Understanding as much as Men, and yet wholly mortal, not having so much as an immortal Soul in them"; "that Giants, Nymphs, Gnomi and Pygmics were the conceptions and births of the Imagination power of the influence of the Stars upon Matter prepared by them, and that they have no souls; as it is most likely the Inhabitants of the more remote parts of the world have not, as not being the offspring of Adam", and most importantly, that there is an artificial way of making an Homunculus, and that the Fairies of the woods, Nymphs and Giants themselves had some such origin, and that Homunculi thus made will know all manner of secrets and mysteries of art, themselves receiving their lives, bodies, flesh, bone and blood from an artificial principle.

When writing his *A demonstration of the Existence and Providence of God,* first printed in 1696, the Calvinist Bachelor of Divinity John Edwards, fellow of Saint John's College in Cambridge, took the attack to an even broader, if necessarily less balanced, level. In the preface to his two-volume effort, he stated that his goal was to illustrate the errors of the atheists. Although he was not really expecting to convince them, since this was "perhaps next to blanching an Ethiopian" in its ultimate impossibility, he still hoped that "I shall do something towards preventing the spreading of this infectious disease they are the authors of."

In order to do so, Edwards proceeded to argue for the unmistakable imprimatur of God in all of the known natural facts of his time, which led him to dedicate Book 2 entirely to the "coming forth of the Foetus" and its "owing to marvelous Care of the Almighty, to the particular Midwifery of Heaven." From this midwifery is born "the exact symmetry of all parts when taken together," so perfect that, as other "Writers of the Church" had previously noted, "there were the same Proportions in the Fabrik of the Ark that there are in the Body of a Man . . . that is, his Longitude was sixfold to its Latitude, and tenfold to its Profundity" so that "there is such a Harmony and Symmetry of the Members, that they all have an exact Reference to each other." He cited this symmetry as proof that "there is something divine in the Disposition of the Parts of Man's Body." From this the author unleashed his most flamboyant bit of rhetoric, in a direct derivation from the Vespusian figures, made popular in our times mainly through the drawings of Leonardo:

> The Height of a Man is the same with his Bredth, i.e. the Space between Head and Feet, and between the Hand stretched out is alike . . . So Man is a quadrate Figure; and yet, if you place him thus with his Arms and Hands stretched out, you'll find that the figure of the Body makes a perfect Circle, the center whereof is his Navel. Here, we may say, we have found the Quadrature of a Circle. This is no Workmanship of Humane Skill, here is no Automaton made by Art, no Daedalus' walking Venus, no Archytas' Dove, no Regiomontamus' Eagle and Fly. Here is no Albertus Magnus or Frier Bacon's Speaking Head, or Paracelsus' Artificial Homuncle.

If the fury of the religious men were not enough to make the spermists proceed with caution when choosing a name for their encapsulated creature,

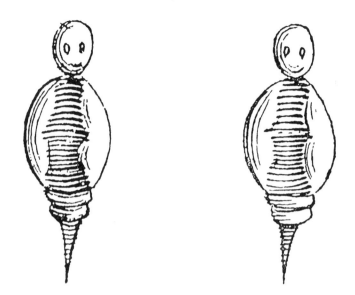

Fig. 44. Redi's drawings of parasites from the gut of the octopus. F. J. Cole's caption to this figure reads "Cestode parasites exhibiting the homuncular form, after Redi." (From Cole, *Early Theories of Sexual Generation,* 1930).

the irritation of their own peers would have been the icing on the cake. For the same attack on the homunculus was repeated by Francesco Redi himself, in the very same *Experiments on the Generation of Insects* that, in 1688, emerged as a milestone in the dismissal of spontaneous generation (see chapter 3). In Redi's words, Paracelsus was nothing but "a charlatan" who "impiously would have us believe that there is a way to create manikins in the retorts of alchemists." Redi then went on to add that

> I am still more scandalized at the assertion of others, who make these lies a foundation for conjecture concerning the greatest mystery of the Christian faith, namely, the resurrection of the body at the end of the world.

Never mind that, ironically, Redi's earlier drawings of two parasites from the gut of a female octopus were reproduced in the twentieth century with the caption "parasites exhibiting homuncular form" and pointed out as a case of "the tendency to detect the human form in similar animals—a tendency which was responsible for the seminal homunculi of later authors."[41] With the aforementioned passage from *The Generation of Insects,* the damage was definitely done: although Redi aimed most of his anger against those "others," the link was now tightly established between challenging "the greatest mystery of the Christian faith" and the homunculus.

The echoes of all these voices combined help us to perceive the reason for all the fear and suspicion surrounding the word "homunculus" in the seventeenth century. Interestingly, browsing through modern literature, we sense that its negative connotations were by no means erased with the passing of the centuries. The homunculus, it appears, is still considered a somewhat shameful lapse of reason. This subliminal shame is well illustrated by the numerous works of writers from our own decades who have attempted to rehabilitate Paracelsus as a great doctor and philosopher. We can read book after book on his life and doings without ever encountering a single mention of the infamous recipe. And, if such a reference appears, it will most likely be in an apologetic light, as we can first glimpse in this passage from "Paracelse et sa Posterite" by Georges Cattaui:

> Perceiving in human words an echo of the Divine Verb and in man, the temporal microcosm, a correspondence with the eternal macrocosm, many Renaissance humanists, inspired by the ideas of Empedocles and Pythagoras, considered the supreme cosmogonic principle to be a demiurge, whom Paracelsus named Hylaster. This material principle was conjugated with an "anima mundi" that involved all things. Thus Paracelsus's "nature" is full of witches, demons, gnomes, mermaids, sirens, and elves.[42]

The need for a redeeming explanation is even more transparent in Serge Hutin's "Les doctrines secretes":

> In "Prognostique," Paracelsus gives the recipe for preparing the homunculus. Is it only the Promethean dream of the alchemist becoming able to create life himself? This recipe has to have a symbolic meaning: the making of the homunculus is the analogy for the rebirth, the regeneration sought by the alchemic initiation, able . . . to transport the initiated from the corruption of the old man to the new, regenerated man.[43]

WHODUNIT?

So, to come back to our original question: why on earth would the pioneer spermists choose the term "homunculus" to refer to the little man inside the sperm head? Bartholomaei Castelli's *Lexicom Medicum Graeco-Latinum* of 1713—the major medical dictionary circulating half a century after the dates of Hartsoeker's and Leeuwenhoek's first drawings—has only a very short entry for "homunculus," equating it to alchemical experiments and diabolical doings. And in a 1720 treatise dedicated to spermatology, Martino Schurigio's *Spermatologia historico-medica,* the only association between sperm and homunculi again refers solely to the dangerous undertakings of Paracelsus's kitchen.

Given such a background, the microscopists who first aimed their lenses at the human sperm, and admitted the possibility of having seen a little man nestled inside the head of each one of these newly found "animals," should have anticipated that certain choices of terms would amount to shooting themselves in the foot. If they really wanted to sell their idea successfully to the learned world, they should have known better than to christen their little man with a name so loaded with negative and derogatory connotations.

And, apparently, they did. Nicolas Hartsoeker, the man who drew the well-known illustration first published in 1694, wrote his *Essai de dioptrique* in French and called his little person either "le petit animal" or "l'enfant." Leeuwenhoek stuck to the term "animalcules" for the spermatozoa and to "small man" for the presumed person-to-be encased inside. True, it can be claimed that "homunculus" is nothing but the Latin term for "small man." And Leeuwenhoek published the official versions of all of his letters in Latin. But you can read through all of them, in their original text, as carefully as you please. Just like Hartsoeker, he never once makes use of the H word.[44] Apparently, Hartsoeker was not fluent in French, and Leeuwenhoek did not have a good mastery of Latin, which made them both resort to the help of friends for the completion of their published works.[45] This could have added an extra factor of confusion to the translations, resulting in unfortunate phrases. However, to the contrary, everybody's choice of terms seems to have been rather careful.

THE FAIRY TALE

It is uncertain when Leeuwenhoek first referred to the famous drawings of little people with tails. We know that these drawings were not his own doing. History has it that they were presented to him by Dalenpatius, a pseudonym of the French aristocrat François de Plantade, whose name survives mainly in association with the satirical and iconoclastic novel *Le conte des fées du Mont des Pucelles.* But Plantade, educated by the Jesuits in Montpelier, was not your average man of letters. He also entertained a solid scientific curiosity, which made him travel to England and Holland with the goal of visiting famous laboratories and learning new techniques. Upon returning to his native city, he eventually became renowned as a meteorologist and astronomer. He died in 1741 while climbing the Pic du Midi during a geographic survey.[46] His interest in lenses could have led him to look at human sperm, and to "discover" minute human beings swimming in each sample. The drawings he supposedly gave to Leeuwenhoek are the closest we know to a perfect adult, shaped exactly like any one of us, hat and all—except for that little tail that makes all the difference.

"I detected certain animalcules, of almost the same shape as the young of frogs," reads the letter from the Frenchman to the Dutchman.

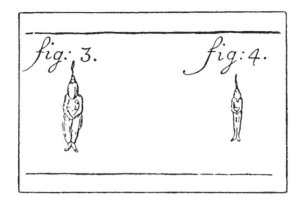

Fig. 45. The drawings sent to Leeuwenhoek by François de Plantade ("Dalenpatius"), depicting little men who became visible under the microscope when the "animalcules" in samples of human semen shed their skin.

> Their bodies scarcely exceed in size a grain of corn . . . whilst their tails are four or five times the length of their bodies. They move with extraordinary agility, and by the lashings of their tails they produce and agitate the wavelets in which they swim. Who would have believed that in them was a human body? But I have seen this thing with my own eyes. For while I was examining them . . . one appeared . . . and sloughed off the skin in which he had been enclosed and clearly revealed . . . both its shins, its legs, its breast, and two arms, whilst the cast skin, when pulled further up, enveloped the head after the manner of a cowl. It was impossible to distinguish sexual characters on account of its small size, and it died in the act of uncovering itself.[47]

That Monsieur de Plantade liked a good joke is obvious enough from the contents of *Le conte des fées*. But, considering his seriousness when undertaking scientific missions, and bearing in mind the uncensored pleasure of repeating Nature's jokes through jokes of knowledge discussed in chapter 4, to claim that this illustrated report was nothing but "an hoax" or "a deliberate fraud," as some modern writers have stated,[48] seems to be rather far-fetched.

The appearance of these figures in the literature has been debated in excruciating detail, from the real date of their first publication (1678 or 1699) to the possibility that Plantade's plate was bound by mistake with a wrong letter by Leeuwenhoek;[49] but the question of whether anybody, at that time, talked of "homunculi" has never even been raised. It seems interesting that apparently Leeuwenhoek himself was, at first, rather suspicious of these miniatures with hats and tails; and even that, supposedly misled by the translation of the letter provided by a Dutch medical friend, he accused Dalenpatius of ascribing a blood system and circulation to the spermatozoa, immediately proceeding to write about the errors of observation that can occur in microscopic work.[50] But, whatever twists of plot really occurred in the seventeenth century, it is unquestionable that Leeuwenhoek, in the version of his letters that has reached us, only goes as far as saying that the figures represent "generis animalcula, at mortua."[51] Yet the legend to the same figures in today's writings frequently reads something like "Leeuwenhoek's drawings of the imaginary homunculus."

In all of his other letters, the Dutch microscopist consistently used expressions such as "animalia," "figura animalculorum," "animalcula seminalia," "corpore animalculi," or "interiores corporis partes." In the most respected English translation of his writings,[52] we find him claiming that "in a certain book it is laid to my charge that I proclaimed that a human being will originate from an animalcule in the sperm, although I have on the contrary never expressed an opinion on this subject." In other passages the translation speaks of "the figure of a human being," "living creatures," or "Man . . . already furnished with all of his members" contained in the "little animals" or "animalcules." Yet the word "homunculus" insists on cropping up throughout the numerous modern accounts of his work.

THE MAN IN THE EGG

It is also intriguing that, according to the reports presented to us by modern writers, nobody in the egg camp produced similar terminology. On this side of the barricades, the closest we come to any homuncular suggestion are the bizarre drawings of Theodore Kerckring, dated from 1670 and intended to support Harvey's postulate, "all that is alive comes from the egg."[53] Now, if flights of imagination went forth during this period, this example should definitely be the one to remember. Here we can see "eggs of different bigness, as Dr. Kerkringius affirms to have found in the testicles of a woman"; immediately followed by "a bigger Egg," drawn as a perfect avian-like oval in absolute dissimilarity to any mammalian gamete; and then by "smaller eggs from the testicles of a cow"—all of these presumably Graafian follicles, and highly unlikely to have actually been observed by the author in the way he describes them.

The following figure represents

an Egg which Dr. Kerkringius affirms to have opened 3 or 4 days after it was fallen into the Matrix of a woman, and in which he saw the little embryon . . . whereof he found that the head began to be distinguished from the body, yet without a distinct perception of the organs.

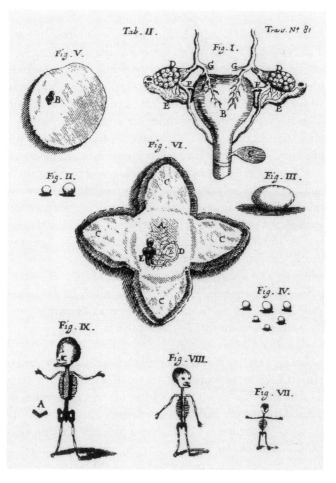

Fig. 46. Theodore Kerckring's drawing of human eggs and embryos, first published in 1670 in defense of Harvey's thesis. Figs. II and III represent "eggs of different bigness" from the "testicles" of a woman; whereas fig. IV represents "smaller eggs from the testicles of a cow." Figs. V and VI represent the "human embryon" inside the egg, at 4 and 14 days of gestation, respectively. Figs. VII, VIII and IX represent "the sceletons of Infants 3 weeks, 4 weeks and 6 weeks after conception." (As reprinted by Needham in *A History of Embryology,* 1934.)

The "little embryon," sitting in the upper left side of a large circle, in the place where the site of sperm entry would appear in the egg of the chicken (then much better known), has a head, a body, and a small tail. Anyone could have called it a "homunculus," but no one ever did. The resemblance between this image and the early stages of development in birds is even more striking in the following figure, "a bigger egg, opened a fortnight after conception," in which limb buds seem to be present, the tail is extended to look like an umbilical cord, and a network of blood vessels surrounds the whole structure, now positioned at the center of the egg. Nothing could be so far removed from true appearances during the early days of human development. And certainly nothing that the spermists ever drew comes closer to wishful thinking than the three following figures, representing "the sceletons of Infants 3 weeks, 4 weeks and 6 weeks after conception." These "sceletons," as reduced copies of the skeleton of a newborn, could easily have fallen prey to modern perception as caricaturing some sort of eerie homuncular osteology; yet no historian has treated them as such.

Another curious case of sin by omission has to do with a man we now tend to consider one of the founding fathers of ovism, Marcello Malpighi himself (see chapter 1). In his excellent scientific biography of the Italian microscopist, Howard Adelmann quotes from Malpighi's notes on a public lecture on anatomy given by Manzi and Sbaraglia at Bologna in 1676. If we are to believe Malpighi, Sbaraglia contended that the egg descends to the uterus through what he called a *vas deferens* (in reality, the ovarian ligaments), and that this egg already contains a "little man"—*"esse parvum homunculum,"* in the original notes.

Since Sbaraglia was a ruthless enemy of Malpighi (see chapter 1), Malpighi may have used the term "homunculus" just to deride him. Or he could just be using the Latin term for "small man." In any event, these notes were taken even before the sperm cells had been discovered. This seems to suggest that, even if the term "homunculus" really had a place in preformationist terminology, it was introduced by the ovists, not by the spermists. Yet, in modern accounts, the spermists are the ones who bear all the blame.

THE ROTUNDO-CONCAVO-CONVEX MACHINE

So it is not possible to blame the spermists for letting their imagination run wilder than that of the ovists. We have verified that the founding fathers of spermism consistently shied away from any association with the homunculus. It is impossible to evaluate exactly how conscious they might have been of the potential danger hidden in the word itself. But we can definitely establish that they never used that word at all.

Well then, who did it? Some smart ovist trying to discredit the enemy? Or perhaps an epigeneticist with the same goal in mind? Or some careless writer of the following century, when the debate over preformation had clearly shifted toward the egg and invoked much more sophisticated explanatory models? None of these hypotheses can be backed by a review of the original sources. A minor problem could arise with a comment made by Martin Lister in 1709. Pursuing his repeated attacks on Leeuwenhoek's beliefs, Lister conceded that the male semen was crowded with "vermiculi," but maintained that they functioned only to incite the male to perform the sexual act. He dismissed as absurd the view that such creatures could develop into men with the words *"Homunculi isti quanti sint, cum cogito, haec res agetur, allis, mihi certe fabula."*[54] Here he seems to be using the word merely as the Latin for "small men," and apparently nobody took much notice of its tentative emergence.

What's more, even those writers who overtly sought to satirize preformation did not employ the perilous term when they had the chance to do so. Take Sir John Hill's *Lucina sine concubitu,* in which he mockingly addresses the Royal Society by pretending to have invented a machine for trapping the seminal animalcules borne on the west wind. The text is obviously a joke on the idea, advanced by some early spermists, that human seed floated everywhere in the air.

> Accordingly after much Exercise of my Invention, I contrived a wonderful cylindrical, catoptrical, rotundo-concavo-convex Machine . . . which, being hermetically sealed at one End, and electrified according to the nicest Laws of Electricity, I erected in a convenient Attitude to the West, as a kind of Trap to intercept the floating Animalculae in that prolific quarter of the Heavens . . . When I had caught a sufficient number of these small original unexpanded Minims of Existence, I spread them out carefully . . . and then applying my best Microscope plainly discerned them to be little Men and Women, exact in all their Lineaments and Limbs, and ready to offer themselves little Candidates for Life, whenever they should happen to be imbibed with Air or Nutriment, and conveyed down into the Vessels of Generation.[55]

So, here is an eighteenth-century detractor of preformation castigating it with maximal irony. Still, when talking about the encapsulated generations, he offers a broad variety of terms—"Animalculae," "Minims of Existence," "little Men and Women," "little Candidates for Life," indeed, a colorful repertoire. But the H word never appears.

As for other contemporary antagonists of preformation, Needham attacked "the numerous absurdities which exist in the opinion of pre-existing germs."[56] Maupertuis reasoned that "both the system of the egg and that of spermatic animalcules are incompatible with the way in which Harvey actually saw the embryo to be formed."[57] The less prominent James Handley added that "we dissent in some things both from Leeuwenhoek and Harvey . . . Both semen and ova . . . we believe to be the causa sine qua non in every generation."[58] "Pre-existing germs," "the system of the egg," "spermatic animal-cules," and "semen and ova" never spelled "homunculus."

THE FATEFUL 1930S

We seem to be left with but one conclusion. The perpetrator of this mis-understanding must have done it very recently—and he had to be either strongly respected by the scientific community, or part of an assertive revision-istic episode. Or both. Since a large corpus of works on embryology and on the history of embryology, written mainly by embryologists, appeared around the 1930s, the origin of the trouble could be located here—for, thereafter, the term "homunculus" enters general use whenever spermism is discussed. Con-sciously or not, the scientists involved in these publications could have falsified their own history.

William Locy's *The Story of Biology,* published in 1925, does not use the word. Neither does Erik Nordenskiold's *The History of Biology: A Survey* in 1928. The same goes for Aute Richards's *Outline of Comparative Embryology,* from 1931. The term is still absent from Wells's *Patterns of Life* and from Joseph Needham's *A History of Embryology,* both from 1934; and it remains absent as late as 1949, in the first edition of Lester Barth's *Embryology,* and even in C. H. Waddington's *Principles of Embryology,* published in 1956.

But something happened in the meantime.

When writing his *Early Theories of Sexual Generation,* published in 1930, F. J. Cole had a precise goal in mind. "I have attempted to put the complete story of the Preformation Doctrine before the reader," he claims in the preface, "and to avoid the common mistake of ignoring all but the more salient fea-tures." He certainly pursued this goal to its most painstaking limit; but, while doing so, consciously or not, he also paved the way for the ensuing confusion. On the very first page, he mentions Paracelsus. He gives us once again an abridged version of the sixteenth-century recipe. He traces the origin of the word back to Cicero, and mentions its use in *Tristram Shandy.* After that, the term "homunculus," or one of its derivatives, appears at least sixteen more times—almost always in the author's own words, only once in a quote from an original source. It is through Cole that Redi's parasites with human faces be-come homuncular visions. And don't we have a feeling that somewhere—where was it?—we have seen a hilarious drawing of a homunculus with a moustache? Our instinctive association is Cole's doing again. In the middle of his detailed account of attacks on and defenses of spermism at the onset of the eighteenth century, all of a sudden he refers to a drawing of an unidentified organism that Joblot produced in 1718. It is a mean-looking mask, with a moustache, six "legs," and a "tail." In Joblot's time, secondary sources did

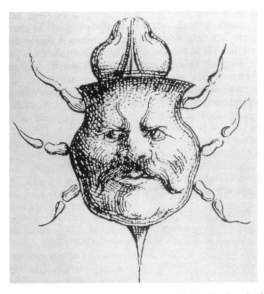

Fig. 47. Joblot's microscopic observation of "an animal that had in the back the face of a satyr," seen in an infusion of lemon. F. J. Cole's caption to this illustration reads "Homunculus in an aquatic animal, after Joblot." (From Cole, *Early Theories of Sexual Generation,* 1930).

refer to this drawing with words such as Abbé Regley's, in the introduction to Spallanzani's *Nouvelles recherches sur les découvertes microscopiques*: "[Joblot] observed under his microscope infusions of roses, anemones, jasmines, basils, teas, and mushrooms. The infusion of lemon gave him an animal that had in the back the face of a satyr." But, in Cole's book, the legend reads, "Homunculus in an aquatic animal, after Joblot."

After his opening paragraphs on Paracelsus and *Tristram Shandy,* does it make sense to claim that Cole was just thinking of "small men" whenever he wrote "homunculi?" The analogy may have been an unconscious one, but it surely was effective—even more so since the use of the term appears curiously biased against spermism, consistently sparing the fabrications of the ovists. In 1671, William Croone produced a manuscript paper on the development of the chick, containing an illustration claiming to represent the preformed embryo, although it seems clear that he drew only a fragment of vitelline membrane accidentally resembling the features of a bird. When mentioning this, Cole says that Croone's paper is important "not on account of its merits, which are negligible, but because it is the first reasoned attempt, based on observation and illustration, to establish the existence of a preformed foetus in the unincubated egg." Even when dealing with such a gross misinterpretation, the dangerous homuncular label is avoided, and the much smoother expression "preformed foetus" is used instead.

1930 was an ill-fated year for the reputation of spermism. The prolific Lamarckian E. S. Russell is now best remembered for *Form and Function,* one of the best and most comprehensive accounts of the history of comparative animal morphology. But, in 1930, he also published *The Interpretation of Development and Heredity.* The book opens with a revealing quote from Jonathan Swift:

> He said that new systems of nature were but new fashions, which would vary in every age; and even those who pretend to demonstrate them from mathematical principles would flourish but a short period of time, and be out of vogue when that was determined.

Russell announces, in his introductory chapter, that

> it will be necessary to treat of the matter in some extent historically, in order to be able to envisage modern theories in their proper perspective, to understand their mode of origin, and generally to follow the filiation of ideas. We shall find that, in spite of the vast accumulation of detailed knowledge . . . there is much less difference than one would expect between the fundamental hypothesis or modes of explanation adopted, say, by the Greeks, and those in vogue at the present day. This is because there are—apparently—only one or two possible ways of interpreting development open to the human intelligence, and these few alternative methods tend to recur again and again throughout the whole history of biological science.

Russell is absolutely right—and he himself falls prey to this binary condition of the human mind a couple of chapters later, when stating that epigenesis and preformation are the perfect expression of such recurrent dichotomy.

> The epigenetic view is dynamic, vitalistic, physiological; the preformationist is static, deterministic, morphological. The one stresses time or process, the other space and momentary state—the one emphasizes function, the other concentrates on form.

Guess which one is doomed to interpretation as one of those "systems of nature" that in the end "were but new fashions," as in his warning quote from Swift? "The preformationists contributed nothing of value to the understanding of our problems."

Maybe it is just a coincidence, but three paragraphs later, in a section dedicated to rescuing Bonnet from the rest of the preformationist bunch, the H word appears:

Some earlier preformationists thought of the germ as an adult in miniature, and imaginative souls saw in the spermatozoon a tiny homunculus with head and arms and feet.

From this point on, there is no going back. Although, as we have seen, the homunculus still took a couple of decades more to invade the embryological texts, the seeds planted in 1930 eventually blossomed all over the field. *The Elements of Experimental Embryology,* first published in 1934 by Julian Huxley and G. R. de Beer, opens with a chapter dedicated to a "Historical Introduction to the Problem of Differentiation." And here we are told once more about "the crude idea" that

> the preformation in the egg was spatially identical with the arrangement of parts in the adult and fully developed animal, or that the "homunculus" in the sperm, with the head, trunk, arms and legs which it was supposed to have (and which certain over-enthusiastic observers claimed to have seen through their microscopes . . .) only required to increase in size, as if inflated by a pump, in order to produce development.

Here, as in other modern works discussed earlier in this section, the term "homunculus" appears between quotation marks. These, like the use of italics adopted by some authors, are perhaps meant to suggest that this is not the exact word, only an approximate abbreviation of the concept. But do we ever stop to ponder such subtleties?

The final proof of our modern unrestrained correspondence between the homunculus and the little man in the sperm head lies where one would least expect it. Or maybe not. It appears inside a small rectangular box, containing eighty cards. Here is the note printed on the outside:

> Aleister Crowley was born on October 12, 1875, and he joined the Hermetic Order of the Golden Dawn in 1898, rising rapidly through the grades of the order. Crowley poured the entire contents of his magical mind into his tarot deck. Neither Crowley nor the artist of the Thoth Tarot deck, Lady Frieda Harris, managed to publish the tarot deck during their lifetimes. Crowley died in 1947 and Lady Harris died in 1962. The Thoth Tarot deck remained unpublished until 1969.

We do not need to look further than card IX, The Hermit. It stands under Virgo and Yod. And here is the description of the drawing:

> Here we have, in the hand of the Hermit, the Lamp of Sacred Wisdom. It contains the Sun, which is hidden beneath the surrounding darkness to fructify the earth. The Hermit is looking at the Egg (universe) which is surrounded by

Fig. 48. Card IX, "The Hermit," from Aleister Crowley and Frieda Harris's *Thoth Tarot Deck* © 1944, 1971 Ordo Templi Orientis. Used with permission. The "homunculus" is represented by Hartsoeker's drawing of the spermatozoon.

a snake, a symbol of life. The hounds of hell endeavor to snatch the sacred light and the little Homunculus. The wheat is in the masonic tradition.

It is perturbing enough to encounter the Egg and the Homunculus reunited in the same figure of a tarot card. But more amazing yet is the form assumed by the Homunculus. Escaping from the three-headed black dog that represents the "hounds of hell" is none other than . . . an exact reproduction of Hartsoeker's drawing of the spermatozoon! The circle has been closed.

Were Cole and Russell doing this on purpose? But why would they? Neither of them thought very highly of the seventeenth-century preformationists, but then again, such ideas had fallen from grace long before these men wrote their books. In 1930, there was no need to beat a dead horse. Cole and Russell were not even allies in their respective scientific concerns: Cole was one of those zoologists who agreed with the genetic theory of development. If, as discussed in the beginning of this chapter, we can establish some sort of equivalence between homunculi and genomes as different materializations of the soul, his feelings about both the word and the concept should have been nothing short of sympathetic (as we shall see in the Epilogue, the early geneticists were often called "preformationists" by their skeptical colleagues). Russell, on the other hand, adhered to the epigenetic embryological approach, and was thus one of those classic embryologists who sincerely disliked genetics. If anything, these two authors were rivals. Yet they both unwittingly initiated the same historiographic confusion—and exactly in the same year!

It was the strange fate of *emboitement* to rise from Swammerdam's misunderstanding of the meaning behind his observations of the imaginal disks of insects (see chapter 1), only to expire in our global misunderstanding of the meaning behind the real name of the little man inside the sperm. Cole and Russell, most likely, did what any one of us would have done. By equating preformation with nonsense, and spermism with an amusing detour along the path of knowledge, we may just as well, without even noticing, have come to equate the man inside the sperm with all those other tales of occult beliefs and bizarre experiments to be regarded as nothing but the follies of a benighted past. The H word may have entered the secondary literature without being part of anyone's explicit agenda. We just came to regard preformationism as so silly that we no longer bothered to verify the sources. And, in so doing, we stamped the verdict "to be dismissed" all over spermism, with all the rest of preformation in tow.

The Music of the Spheres

From the wells of disappointment where the women kneel to pray
LEONARD COHEN

HIS CHAPTER ANALYZES a strong philosophical advantage held by the ovists—and a strong philosophical disadvantage that cursed them.

On the bright side, the ovists based their system on the egg. Eggs are spherical. As we shall see, the sphere represents the perfect shape. This was certainly very exalting for those who embraced the egg credo. All the details matched. According to the theory of preformation, God had created all humans at once. He had created man to mirror His own image. As Malebranche would say, He had created the universe to enjoy Himself in the contemplation of its splendor. Therefore, what could be more logical than to assume that He had encased His creatures within corpuscles bearing the quintessential shape of perfection? To think like this was not only logical. It was also very rewarding.

But, on the dark side, eggs come from females. And females were considered imperfect creatures. This was certainly a cruel embarrassment to those who had to accept that their idea of generation made all lineages of animals unfold from the very entrails of inferiority. If ovism was the true system of reproduction, God was sending a mixed message. He had locked us inside perfection. And then He had locked perfection inside imperfection. So much for the reward.

Let us analyze the bright side first.

According to M. Vandermonde in his 1751 *Essai sur la manière de perfectionner l'espèce humaine,*

> the beauty of men is but a reflection of the beauty that the Creator has dispersed all through the Universe. The order, the arrangement, the proportions, the symmetry, are in all of His works . . . Proportion and symmetry are the first requisites of beauty , , , It was following this rule of beautiful nature that the greatest architects built the most sublime buildings.

This is already a strong statement in defense of the egg. Eggs, basically, are spheres. Even if the well-studied eggs of birds did not follow the rule, one could always argue that bird embryos develop in the yolk, as had been described from the ancients to Malpighi—and the yolk *is* spherical. If, as Vandermonde believed, symmetry and proportion are the main prerequisites of beauty, nothing is more symmetrical and perfectly proportioned than the sphere. This old belief entered the Scientific Revolution as powerful as ever. Galileo (1564–1642) may have shaken the basic premises of the church by postulating that the earth had movement. But, concerning the intrinsic mechanisms of this movement, he was still as Aristotelian as his clerical judges. He assumed that inertia led to a circular movement, making any body abandoned to its own devices move in a uniform fashion around a circumference.[1]

The immense symbolic power of the sphere is well expressed in the vast number of cosmogonies that represent both the universe and the world with this shape, the only means through which all the points at the surface can be equidistant from the point at the center—an achievement of such astonishing transcendence that, according to Chevalier and Gheerbrant's *Dictionnaire des symboles,* "if a being is conceived as perfect, it has to be symbolically imagined in the form of a sphere."

THE UNIVERSAL SPHERE

When the great geometer Pythagoras (ca. 560–ca. 480 B.C.) set out to explain the universe, he held that everything must be spherical and circular, because those are the shapes of perfection, and the universe was meant to be perfect. It had been conceived and executed by a geometrizing deity, and it was infused, invigorated, and protected by the benevolent influence of this deity's power. In the beginning of the seventeenth century, this concept remained one of Johannes Kepler's (1571–1630) main obsessions. Just as the German astronomer was about to change the universe forever, he still held dearly to the old Pythagorean theme, a theme that Descartes and Malebranche reemphasized time and again right after him: the notion of God as a geometer, creating the world according to a mathematical plan.

As S. K. Heninger Jr. wrote in "Pythagorean Cosmology and the Triumph of Heliocentrism,"

> The Pythagoreans pointed out that the sphere has a perfect shape: all points on its surface are equidistant from the center; it therefore has no beginning or end; consequently, it possesses a distinct, integral, self-sufficient identity. It is the appropriate form for a cosmos set apart from infinity . . . Similar arguments proved that the circle is the perfect plane figure: all points in the circumference are equi-

distant from its center; it likewise has no beginning or end—indeed, it is a common symbol of eternity.[2]

During the fourth to third centuries B.C., this cosmology was taken over by Plato (427–347 B.C.), whose writings on the shape of the universe carry such a strong Pythagorean undertone that a rumor was born, reaching as far as the Renaissance, that Plato's *Timaeus* was nothing but a plagiarism from a lost Pythagorean fragment. Assuming that the Creator had estimated that "there is a thousand times more beauty in the similar than in the dissimilar," Plato repeated in *Timaeus* that the universe had the form of a sphere.[3] This cosmology was passed from Plato to his disciple Aristotle, and from Aristotle to Ptolemy (ca. 100–170), in Alexandria and Canopus.

After this long string of revisions, the resulting theory of the universe depicted a succession of ten concentric spheres. The whole physical world comprised a solid sphere of vast, but not infinite, radius, of which Earth itself, spherical and motionless, occupied the center. Then came seven concentric spheres, each with its astral bodies in their assumed order: Moon, Mercury, Venus, Sun, Mars, Jupiter, and Saturn. The sphere next to Earth carried the Moon, and enclosed within itself the central portion of the universe, filled with the four corruptible elements: *earth* (forming dry land), *water* (forming the oceans), the atmospheric *air,* and an upper layer of *fire*.[4] These four elements moved naturally in straight lines, earth and water down toward the center of the universe and air and fire up toward the Moon sphere, each stopping its movement when it reached its natural place. From the Moon to the other planets the space was filled with *ether,* the natural motion of which was revolution in a circle around the center of the universe. After the spheres of the planets came the *starry firmament,* which contained the fixed stars and completed one revolution every thousand years. Then came the *crystalline orb,* containing the signs of the Zodiac and completing one revolution every 49,000 years. The whole was enclosed within the *primum mobile,* which revolved every 24 hours, but in the opposite direction.[5]

Ptolemy, who, unlike Aristotle, was a real and serious astronomer and an influential geographer, was the first to address the obvious problem that the times of revolution of the planets were not as regular and predictable as such a system would entail. To solve this discrepancy, he developed an extremely complicated set of geometric propositions, described in detail in his famous *Almagest*.[6] The book included all of Ptolemy's relentless observations and plottings of planetary motions, recorded from A.D. 127 to 151. Here, he held that each planet described a small circle (an *epicycle*) about its mean position, while the mean position revolved in a larger circle (a *deferent*) about Earth. This accounted for the displacement of the planet's center of revolution vis à vis Earth, forming an *eccentric circle*.[7] All these combined motions matched the observed rotations of the planets to a satisfactory degree of accuracy. More importantly, all the shapes involved were still perfectly round.

Toward the end of the twelfth century, the scientific writings of Aristotle began to be introduced into Western Christendom, and were combined with Ptolemy's data. In the thirteenth century, the Dominican Saint Thomas Aquinas (ca. 1225–1274) achieved a partial synthesis of the Aristotelian doctrines with Christian theology, creating the system of thought known as Scholasticism. As the blanks of the ancient impersonal cosmos were filled in with Christian imagery, the outermost sphere of heaven became the *Empyreal* abode of God and the saints, while the Devil and his dark angels dwelt in the heart of the earth. The Scholastics referred to several passages from Scripture to reinforce their view of a geocentric universe, and claimed that "for the safety of faith, the opinion that the earth does not rest at the center of the universe cannot be tolerated."[8]

This system was challenged in 1543, when Nicolaus Copernicus (1473–1543), shortly before his death,[9] published his treatise on the revolution of the celestial orbs,[10] in which he postulated that the Sun, not Earth, was positioned at the center of the universe. But, in truth, Copernicus was reprising an old Pythagorean idea, early disputed and almost forgotten, according to which there was a sphere of fire in the center of the created world around which Earth revolved. Thus, in proposing his new system, Copernicus was going back to the oldest systems created by Western civilization. Later, when Johannes Kepler started his own astronomical pursuits, the Pythagorean ideas were still very much in his mind. "What if the Pythagoreans taught the same thing I do," Kepler wrote in *Harmonia Mundi,* "covering the meaning in an envelope of words?" One more circle was closing itself.

The perfection of the sphere still held such persuasive power over sixteenth-century astronomical thought that Copernicus retained Ptolemy's concepts of eccentrics and epicycles, although he changed their combinations to accommodate the new place held by the Sun. Both Ptolemy and Copernicus were extremely skilled and dedicated astronomers. They both studied the motions of the planets for decades and compiled amazing amounts of data. Yet they both preferred to construct their theories based on layer upon layer of circles within circles rather than consider another shape for the celestial orbs. Over the course of twenty-two centuries, from Pythagoras to Copernicus, to deny the sphere remained all but unthinkable.

However, the displacement of Earth from the center of the system, the point of supreme perfection, was still uncomfortable for most of Copernicus's contemporaries—so much so that it allowed for the appearance of yet another alternative, proposed in 1588 by the Danish astronomer Tycho Brahe (1546–1601). This alternative was mathematically equivalent to the Copernican scheme, but here, the Sun revolved around an Earth again made central and stationary, while the five planets and the Moon revolved about the Sun. The whole revolved daily about Earth, and so did the stars. Brahe worked out this scheme in the course of twenty years of systematic and refined observations of the planets, using instruments of a previously unapproached excellence. Fearing persecution at home, Brahe eventually moved to Prague, where he was later joined by his young German colleague Kepler, who obtained all the Dane's observations after his premature death, reportedly from having eaten too much at dinner.[11]

Although he was strongly influenced by the Pythagorean postulates and by the appeal of the ever-geometrizing God of Plato, Kepler was a pivotal figure in combining purely *geometric* astronomy with inquiry into the *physical* cause of planetary motions. In the face of all the evidence he himself had computed, and measured against Tycho Brahe's enormous database, Kepler still tried to work with circles, eccentrics, and epicycles. But, finally, he was overcome by his own observations. As a result, he had to postulate (in the *Astronomia nova* of 1609) his famous so-called first law: the planetary orbits are *ellipses,* not circles, with the Sun at one focus. As we shall see below, he himself was dismayed with this conclusion, very much imposed on him by long years of study. The ellipse was the denial of the perfection held by the sphere. Now, that perfection was gone forever.

THE ROOF OF VOYAGING

Needless to say, the sphere did not cast its seductive spell only in the West. In the Islamic tradition, we find a tale concerning the original creation of water as a white pearl with the combined dimensions of earth and sky. In A.D. 950, the philosopher al-Farabi described Creation as a successive emanation of spheres from a lineage of ten intellects.[12] The Islamic philosopher and physician Avicenna (980–1037) reprised the essentials of this theory, which considered the universe to be composed of concentric spheres, from the peripheral sphere to the sphere hidden at the center of the earth.[13]

Things were not much different at the antipodes: in Polynesian astronomy, the earth had over it a number of arched heavens, consisting of concentric hemispheres of solid matter. Each island considered itself to be the *pito,* or navel, of the universe, and therefore assumed that its location was the exact center of this system of superimposed spheres, separated from the other islands by the celestial cupolas that rested on the earth—which meant that a journey to another island, situated within another zone of the earth's surface, was the equivalent of a journey to another heaven.[14] This universe was certainly magnificent, but it could also be cozy. For the inhabitants of the Gilbert Islands, the night sky was the vast roof of a house running north and south, the observer being the central pillar. The celestial sphere was called, literally, "the roof of voyaging."[15]

China offers another interesting case of reverence for the sphere. As Kiyosi Yabuti points out in his 1973 "Chinese Astronomy: Development and Limiting Factors," the Chinese were apt astronomers and experts in algebra from the onset of their culture, and kept perfecting their systems so that "before the fifteenth century there was not much disparity in the level of scientific achievements between China and the West. Chinese inventions contributed to the awakening of European civilization in the Renaissance."

Seeking a way to spread the Christian faith through this immense country, the Jesuits quickly understood that one of the best ways to win Chinese admiration was through the display of their own astronomical and mathematical knowledge.[16] In 1610, less than two years before his death, the celebrated Italian Jesuit Matteo Ricci (see chapter 8) wrote that "especially due to the books and maps which we have printed, to our Mathematics and to all the many novelties on this matter . . . they . . . treat us with extraordinary respect. What impresses them the most, because it is something never before heard or registered by memory among them, is that to China came foreigners that could be their superiors in such sciences."[17]

Ricci kept asking the Company of Jesus to send him "learned men," and most of all "a good astronomer," to see him through his mission of converting the Chinese to Catholicism. One of the many aides sent from Europe for this purpose was the Portuguese Manuel Dias, who arrived in China in 1610 and died there in 1659. Searching for an astronomical gate to the Chinese heart and soul, he tried to perceive what would most efficiently trigger their admiration—and promptly went to work to translate into Chinese the enormously popular John of Sacrobosco's thirteenth-century treatise *The Sphere,* which thus became *Tien wen lio.*[18]

RINGS OF STONES AND WHEELS OF LIFE

Our ancestors certainly sensed something immensely powerful in the magic shape of the circle, since they left behind all those circles of stones that puzzle archaeologists, titillate "gods came from outer space" aficionados, and generally impress all of us commoners.[19] We may never know exactly what the driving inspiration behind the construction of those circles was, but we know for certain that circles, afterlives, and the heavenly spheres above have been connected in man's spiritual configurations from very early on—and were so deeply rooted that they still emerged in unchanged forms after Christianity had taken over Europe. In any event, Christendom changed only the size and the color, not the shape: Jesus passed around pieces of bread during the Last Supper, but when this moment is ritually reenacted, at the climactic point in the Catholic Mass, the host, penetrated by the spirit of God, is no longer a random-shaped piece of bread: it is now a perfect white circle.

In this symbolic reshaping, Christendom was only—consciously or not—reprising an old theme. The early Greek sign for sphere, or globe, was a circle with a cross inside. This representation appeared as a symbol at the onset of the Bronze Age, becoming abundant in rock carvings and in ancient Egypt, in China, in pre-Columbian America, and in the Near East. This symbol was included in the early ideographic writing systems used by the Egyptians, Hittites, Cretans, Greeks, Etruscans, and Romans. It was consistently associated with spiritual powers, and so the Christian church incorporated it among its symbols as the *consecration cross* or the *inaugural cross,* used by the bishop, after immersion in blessed water or oil, to touch twelve different places on the church's walls. It also became the halo of the saints, representing the spiritual powers or energy that holy people emanate. According to Ad de Vries's 1974 *Dictionary of Symbols and Imagery,* the sphere represents *perfection* itself, evoking the form of the deities, the creative motion, and the wheel of life.

The filled circle, another representation of the globe, appears not only in old rock carvings but also in Egyptian hieroglyphs, in Japanese Buddhist symbolism, and in India (where, placed over the brows of women, it indicates their married state). The empty circle is also one of the oldest ideograms, spread over walls of prehistoric caves and rock faces. Among most primitive peoples, it was a representation of both the Sun and the Moon. In astrology, it represents the *eternal,* the *endless, without beginning or end, life itself;* and, in modern Western ideography, it also stands for *all possibilities* within a given system. According to the *Dictionary of Symbols and Imagery,* evoking eternity and Heaven, the infinity of the universe and the cycle of existence, the phenomenal world, and all that is precise and regular, the all-encompassing properties of the circle represent *perfection,* just like the sphere.

Interestingly, the ellipse is the symbolic opposite of the circle, meaning *nothing, zero,* and *absence.* Does it seem strange, then, that Kepler's ellipses first appeared so unpalatable as replacements for the beautiful circles of the planets' orbits, so eternal, so endless, so deeply associated with the perfection of life? Kepler himself did not like the idea, although he had clear evidence in its support. Elliptical orbits were profoundly contrary to his preoccupation with circularity and uniform motion. He gave in only when his much-desired uniformity was restored by his "second law": although the orbit of a planet is elliptical, the imaginary line connecting each planet to the Sun will sweep over equal areas in equal times.[20]

Celestial orbs were another symbol of perfection, and thus they had to be circles—a feeling enhanced by the perception that all that stands outside the circle represents *chaos*.[21] The Italian Dominican philosopher Giordano Bruno (1548–1600) tried to teach that stars were suns with galaxies of their own, thus standing outside our own system of concentric circles. The Catholics did not like him. The Lutherans did not like him. The Calvinists did not like him. In 1600, after having suffered excommunication and seven years in prison, he was burned at the stake. Before Kepler broke the dogma in spite of his own aesthetic preferences, perfect spheres followed perfect circles. Giving proof of their utmost perfection, they made celestial music.

THE CURVED LYRE

The music of the spheres is the most eloquent testimony to the perfection embodied by all things shaped like eggs. Revolving in concentric circles, the planets produced a hum in the air, the Celestial Music that only the gods could hear. Other traditions claimed that not only gods, but also "specially gifted persons," could "at times" hear that music, as the idea appears outside the Western realm with slightly modified tones. Job speaks of the morning stars singing in harmony. The Psalmist refers to it poetically two times (xix. 2–4 and xlii. 9). The Talmud speaks of the noise made by the Sun in its rotation through space, cutting its way through like a saw cutting through timber and producing sawdust, which is seen as a strong beam.[22]

Every planet had a different pitch in the curved strings of the circular lyre. In the West, this vision was enhanced by the Pythagorean school. Pythagoras assumed that all things have innate numerical relationships, like the notes of the musical scale, because the universe had been harmoniously arranged by the Creator. The music of the spheres symbolically expressed this unifying system of created beings. Earth to Moon: one tone. Moon to Mercury: semitone. Mercury to Venus: semitone. Sun to Mars: one tone. Mars to Jupiter: semitone. Jupiter to Saturn: semitone. Saturn to the fixed stars: minor third. In Pythagoras's eight-stringed lyre, Earth produced the lowest note, and Saturn the highest one. This idea became commonplace enough to be included by Pliny in his *Natural History* (II, xx), but the Pythagorean metaphysical implications gradually faded in the process of popularization. In its initial formulation, the music of the spheres formed an all-inclusive scale, a unity composed of eight parts, a cosmos delimited by its internal relationships through the unity of sound.[23]

In the seventh chapter of book 5 of *Harmonia mundi,* Kepler stated very clearly what he meant by "music":

> The heavenly motions are nothing but a continuous song for several voices, to be perceived by the intellect, not by the ear; a music which, through discordant ten-

sions, through syncopations and cadenzas as it were, progresses toward certain pre-designed six-voiced cadences, and thereby sets landmarks in the immeasurable flow of time.

However, the idea of perfection represented by celestial music played by spheres was so old and so seductive in its stunning beauty that, although several modern scholars insist with J. L. E. Dreyer that "the harmony is to Kepler only a mathematical conception; he does not imagine that there is any music of the spheres,"[24] he is still often described in popular literature as having really imagined it so. From the evidence discussed above, it could equally be argued that Pythagoras was speaking of music in a strictly metaphoric sense. But we still prefer to think about real music coming from real spheres, to the point of producing numerous poetic offerings on the subject and even attempting to reproduce the polyphony of the planets in a modern sound recording.[25] Such is the power of form. One century after Kepler had replaced spheres with ellipses in the design of planetary orbits, Newton's declaration that Earth was not a perfect sphere, but rather one flattened at the poles, still raised rage and mockery in many learned circles.[26]

THE CURSE OF THE LEFT TESTICLE

Alas, the power of form is not the only representative of the powers that be. The power of place is equally strong. And, in this case, the latter could just as well have neutralized the former. In their symmetrical opposition, both were symmetrically charged with symbolism. Thus far, we have looked at the *positive* charge. For the remainder of this chapter, we shall deal with the *negative* charge. Being an ovist implied the belief that all mankind was encased inside the ovaries of the first woman. But, as several different signs indicate, even as ovism was taking shape, women were still perceived as imperfect men.

One of those signs is the curse of the left testicle.

According to Will Durant in *Our Oriental Heritage,* it was common belief among the descendants of Moses that boys were produced by the right testicle and girls by the left, since the latter was smaller and weaker. In his 1994 *Obstetrics and Gynecology,* Harold Speert states that Hippocrates and Galen also believed that "male infants were a product of the right ovary; females, of the left." Aristotle was more conservative, remarking that animals with "no testes," such as "the footless animals . . . the classes of fishes and serpents," were also able to produce offspring of both sexes. Therefore, he warned us that "to put it in this way [that the left generates females and the right generates males] is to seek for the cause from too remote a starting-point."[27] However, he is frequently quoted by modern writers as having embraced the theory of the sex-determining difference between right and left testicles as promptly as his predecessors. Which goes to show that this belief, by all accounts an enduring and pervasive one, was not left behind as time marched on. Popular and academic books circulating during the time of the Scientific Revolution were still carrying this tale, as can be illustrated by the following examples:

From the pseudo-Scotean *The secrets of Nature Revealed,* for sale in 1754: the male child is conceived on the right side of the womb, and the female on the left.

I knew a soldier who having lost his left testicle in an hospital in Antwerp, had afterwards sixteen children, all boys. And, being willing to make an experiment of the contrary kind, I cut the right testicle of a dog, and making him after line bitches, I saw that the puppies were afterwards all of the feminine gender.

In Nicolas Andry's *Orthopaedia* (see chapters 2 and 3), translated into English in 1743: the author quotes Abbé Quillet's *Callipaedia* regarding precautions to be taken by spouses at the time of conception: "the Fifth rule is directed to the wife, whom he advises, if she would have a Son, to lie upon her right Side during the time of Conception. The Sixth Rule is directed to the Husband; and he tells him, that if he wants a Son, he must take care that only the right Testicle perform its office, and for this reason, he must tie the left one pretty tight with a String."

In *The secrets of Nature Revealed:* "the woman breeds a boy easier and with less pain than girls, and carries her burden not so heavily." Assuming, as was then still frequent, that "the seed only comes from one vessel at each copulation" (for "if it came from the two it would spread through the womb and there may be at once several children within the mother"), the author is certain that "the pleasure is a great deal more poignant when the seed flows from the right vessel of generation than when it issues from the left." As a method for testing the gender of the progeny to come, he then suggests the following: "let [the pregnant woman] milk a drop of her milk in a basin of fair water, and if it sinks to the bottom . . . it is a girl . . . but if it be a boy, it will spread and swim at the top. This I have often tried, and never failed."

In the pseudo-Aristotelian *The Complete Master-Piece,* for sale in 1741, the anonymous author claims that, if the woman is pregnant with a boy, "she feels it first on the right side." Again, according to this source, "[a woman] breeds a Boy easier, and with less Pain than a Girl and carries her burden not so heavily." Also, if the marks under the eyes, "which are of a wan blue Color," are more apparent under the right eye, "she is with Child with a Boy; if the Mark be more apparent in her Left Eye, she is with Child of a Girl." Then comes another variant of the test for the gender of the fetus: "let her drop a Drop of her milk in a Basin of fair water, if it sinks to the Bottom as it drops in, round in a Drop, it is a Girl; if it be a Boy, it will spread and swim to the Top."

This distinction between right and left is scarcely innocuous in the image it casts of the female gender. It springs from the same source as the Aristotelian idea that the sex of the progeny is determined by the heat of the male partner during copulation: the more heated the passion, the greater the chances of obtaining male offspring. Thus old men, Aristotle noted, should abstain from sexual contacts during the cold months: their bodies being colder, the chances of producing females substantially increase under such conditions. We can see a pattern starting to take shape. Right is good. Warm is good. Male is good. Left is bad. Cold is bad. Female is bad. The "sinister side" has not been named by chance. *Sinister* means "left" in Latin.

The dichotomy between right and left as a reflection of the dichotomy between the positive and negative sides of the universe is a concept as old as Western culture. After Judgment Day, according to the Christian tradition, good people will sit at the right of God, and bad people will have to sit at the left: right is the direction of Paradise, and left the direction of Hell. In *Dictionnaire des symboles,* Chevalier and Gheerbrant point out that, in some rabbinical commentaries, Adam was initially created not just as an androgynous figure (see below), but one in which the right side was male and the left side was female.

According to the same source, the Greeks had already defined the right side as the side of the arm that holds the spear, and saw all their favorable presages appearing at the right: the right symbolized force, precision, success. Incorporating a line of thought derived straight from the Aristotelian positive-negative axis, the philosophers of the Christian Middle Ages asserted that the left, together with being feminine, was also satanical and nocturnal, as opposed to the divine and diurnal qualities of the masculine right. Conveniently ignoring in these matters that the left is the side of the heart, the Western Christian tradition turned the right into the active side and the left into the passive, meaning that the right symbolizes the future and the left the past, where man can no longer interfere. Thus the right acquires all the beneficial powers, with the destructive powers assigned to the left.

The rituals of black masses include the sign of the cross made with the left hand, and all the elements of black magic favor the left as their trademark, requiring an entrance with the left foot first or the turning of the left side to the fire. Similarly, the Devil marks the children consecrated to him at the left eye, with the tip of one of his horns.

In the 1995 edition of his *La magie des nombres,* Jacques Languirand argues that it is unlikely that Westerners read from the left to the right by chance: all things, in our culture, are supposed to proceed in this direction, and thus the right becomes further reaffirmed as the direction of the future.

> At the left is the woman: her roots reach back to the past, to the origins. [The left] comprises the unconscious, the mother, the passive principle. At the right is the man . . . [the right] comprises the conscious, the father, the active principle.

And, for Carl Jung, the psychological equivalents are clear:

> Left is the past, the sinister, the repressed, the involution, the abnormal, the illegitimate; in short, the affective life. Right is the future, the happiness, the openness, the evolution, the normal, the legitimate; in short, the social life.[28]

Languirand adds that, according to the Hindu doctrine,

> the left hand corresponds to the moon and the left eye looks toward the past, symbolized by the moon. The right hand corresponds to the sun, and the right eye looks toward the future, symbolized by the sun.

This last quote reveals that the West is by no means the only "dextrocractic" culture. Maybe because the vast majority of humankind is right-handed (being left-handed was still considered a serious deviation when I attended elementary school—I, for, one, was forced to change my natural deviance), the same association appears in worldwide mythologies. For the African Bambaras, the number four, the feminine number, is synonymous with left, whereas three, the masculine number, is synonymous with right. The right hand symbolizes order, expertise, work, and fidelity. The left hand symbolizes disorder, uncertainty, all the variations of human conscience. The funerary customs of the Dogon prescribed that the dead should be buried turned to the right if they were men and to the left if they were women. In the rituals of India, to turn from left to right was favorable, but to turn from right to left was morbid: this last turn was practiced only in funerary ceremonies.

Even when this dichotomy is reversed in favor of the left side, as happens in the Far East, women are still at a loss. For the Chinese, the left is the honorable side: it represents the sky, hence the *Yang*. The right represents the earth, hence the *Yin*—and, within this context, it is the right that becomes the feminine side; a side so bad that houses should never be expanded in its direction. The left is now male, the symbol of strength, the side where the general stands during battle. The left hand is the hand that gives, the right hand the hand that receives. In the Japanese tradition, left is the side of wisdom, of faith, and of instinct; it is the side of the Sun, the male element. The left is always above the right, which symbolizes the Moon, the water, the female element.

Menstrual Evil

If authors repeated and paraphrased themselves freely and abundantly on the subject of the left testicle, they were even more emphatic in demonizing menstrual blood. They certainly had a long mythological heritage to back these beliefs, since the strictest of primitive taboos was laid upon menstruating women: any man or thing that touched them at such times lost virtue or usefulness.[29] The Macusi of British Guiana forbade women to bathe during their periods lest they poison the waters; and they forbade them to go into the forests so that they would not be bitten by "enamored snakes."[30] Tales of lizards and snakes seeking menstruating women at four-way crossroads, or climbing into their beds at night, are extremely frequent in Portuguese traditional culture. According to a quote in Arthur William Meyer's *The Rise of Embryology,*

> the Siamese, who imagine that evil spirits swarm in the air, believe that it is these who enjoy the first fruits of their girls and who cause the "wound" which renews itself every month, a "wound" of which the menstrual blood is the result and proof. It is contact with this blood of which the Maori male is so afraid . . . the Maoris . . . identify menstrual blood with an evil spirit, Kahukahu.

Another type of malice ascribed to menstrual blood appears in the legends of the Brazilian Indians, as recounted by Freitas Mourão in his 1984 *Astronomia do Macunaíma*:

> the sun, Vei, and the moon, Capei, used to be good friends and were always together. Capei was a strong young Indian, with a very beautiful and clean face. One day he fell in love with one of the daughters of Vei and played with her during the night. Vei did not want them to marry, and thus he asked his daughter to wet Capei's face with her monthly blood. From them on, Vei and Capei hated each other, and that is why they always appear in the two extremes of the sky, as far apart from each other as they can. This is also the reason why the moon, Capei, has dark spots on his once so handsome face.

All over the world, sexual relations have been traditionally forbidden during the woman's menstrual periods (see chapter 4), and the spell of impurity reaches out to childbirth, which is frequently considered unclean, thus subjecting women to a whole series of rites of subsequent purification.

The author of *The secrets of Nature Revealed,* in which the "monthly terms" are more often called "flowers," explained that the menses were regulated by the cycle of the Moon and its influence over the sea—not so innocent a statement, since it implicitly puts women directly under the control of the hidden powers of nature, and those powers always carry with them a fair amount of danger. Such danger is clearly illustrated in one of the author's warnings: if the womb has too many "flowers," the child conceived in it will be leprous. Furthermore, he warns the reader,

> the discharge of the monthly terms is of such pernicious quality that if a dog should chance to lick them he would immediately become delirious and mangy; and if any green plant is sprinkled with them, it will immediately become withered and die.

The author's explanation for all these nasty effects is that "these monthly terms are superfluities of their substance, which becomes a kind of imperfect sperm or seed."

Likewise, the much more academic Andry, in the same book that gave birth to modern orthopedic science, does not hesitate in repeating Quillet's warning to husbands:

> In his third Rule he recommends it to Husbands not to touch their Wives, during the time of their Menses; for if they do he assures them, and not without Reason, that their Children will be deformed.

He also repeats Quillet's descriptions of how menstrual blood drives dogs mad and kills green plants, in terms identical to those used in *The secrets of Nature Revealed,* although he remarks in the end that this is "by no means true"— false, yes, but just too good to pass up.

Another case illustrates these authors' delight in repeating folk ideas even if they were too learned and cultivated to believe in them themselves. The most vivid description of menstrual evils is definitely the one contained in Jean Palfyn's *Description Anatomique des Parties de la Femme qui servent à la Génération,* also from the eighteenth century. Although the author states in the end that he does not agree with these popular beliefs, he indulges in a complete description of them:

> Several authors say, with Pliny, that there is nothing more monstrous than this blood. Its vapor, or its touch, are enough to sour the new wine; to make the seeds sterile; to kill the blossoms in the trees and make the dried fruits fall. They say that the glass in the mirrors becomes tarnished just by its presence, the sharpness of iron becomes feeble, the beauty of ivory disappears, the bees die, the copper and the iron rust, even the air becomes infected. Others say that a woman put some of her menstrual blood inside a cake that she gave to a man, hoping that it would work as a love potion; but the man died of the poison. Some peasants believe that menstruating women can make young animals die just by looking at them, and that by these means women can kill even the basilisk.

Ironically enough, this same demonic menstrual blood was considered by numerous classical philosophers, including Aristotle, as a serious candidate for the female contribution to the formation of the fetus: the active nature of the male semen would wake up this passive mass and shape it into its specific animal form. In his *Natural History* (VII, 15), Pliny explained the generalized consensus with an interesting example of noncausal correlation:

> This discharge, which is productive of such great and singular effects, occurs in women every thirty days, and in a greater degree every three months. In some individuals it occurs oftener than once a month, and in others, again, it never takes place. Women of this nature, however, are not capable of having children, because it is of this substance that the infant is formed. The seed of the male, acting as a sort of leaven, causes it to unite and assume a form, and in due time it acquires life, and assumes a bodily shape.

This concept survived in numerous writings until the end of the Renaissance. That nobody stopped to consider the apparent contradiction of all living animals being generated from a material with such devastating powers demonstrates even more clearly how the evil nature of menstruation had become an undeniable truth throughout endless centuries.

THE CANONICAL HERMAPHRODITE

The attitudes discussed above spring naturally from one simple, long-held belief: women are not truly a separate gender, but rather a lesser, inverted, more imperfect form of masculinity. According to Aristotle, women were men whose development was arrested too early: "mutilated males," unable to reach full bloom because the coldness of the mother's womb was stronger than the heat of the father's semen. In this widely accepted picture, women were naturally colder and more passive than men, and their sexual organs had not matured to the point of being able to produce active seeds. The Catholic Church seems to have been pleased to adopt this description, and Galen, who launched the anatomical ideas that were to prevail for more than a thousand years in the West, gave the concept an even more solid framework by writing, around 200 A.D.:

> Just as mankind is the most perfect of animals, so within mankind, the man is more perfect than the woman, and the reason for this perfection is the excess heat, for heat is Nature's primary instrument . . . the woman is less perfect than the man in respect to the generative parts. For the parts were formed within her when she was still a fetus, but could not because of the defect in heat emerge and project on the outside.[31]

And, as with all things inverted and imperfect (see the discussion on monsters in chapter 4), dark powers inhabit their bodies, so that their mere presence is a threat.

The modern reader can sense a sort of morbid pleasure in texts from the period of the Scientific Revolution that explore these mysteries. Very often the authors offer heartfelt apologies for touching on such delicate issues, explaining that their work is necessary to shed light over obscurity—and then proceed to write hundreds of pages on the subject they obviously cherish so much. A good example is the initial apology presented by the author of *The Complete Master-Piece:*

> Though the Instruments or Parts of Generation in all creatures, with respect to their outward form, are not perhaps the more comely, yet, in compensation of that, Nature has put upon them a more abundant and far greater Honor than in other parts, in that it has ordained them to be the means by which every species of being is continued from one generation to another . . . And this methinks should be sufficient to show the great Honor Nature has put upon them. And therefore, since it is our duty to be acquainted with ourselves, I need not make any Apology for anatomizing the secret Parts of Generation.

Furthermore, the author adds, it was very important for women to be informed on the innermost secrets of their bodies: they would then be less likely to die when giving birth.

Having said this, the author proceeds to inform the reader of several different forms of detecting and confirming "true virginity" in women. Having dispensed this practical advice, he finally gets down to the details of female genital anatomy. The subtext on women's inferiority starts to emerge as soon as he explains that the "external part" is generally called "Pudenda," from "the Shamefacedness that is in Women to have them seen." He next describes the clitoris in these terms:

> it is like the Yard in Situation, Substance, Composition and Erection, growing sometimes out of the body two Inches, but that happens not but upon some extraordinary accident . . . The Head of it being covered with a tender Skin, having a Hole like the Yard of a Man, but not through; in which, and in the Bigness of it, it only differs from it.

Further supporting the parallel between masculine and feminine genitalia, he remarks that women also have "spermatic vessels," although they are shorter than those of men because "the Stones of a Woman lye within the Belly, but those of Man outside."

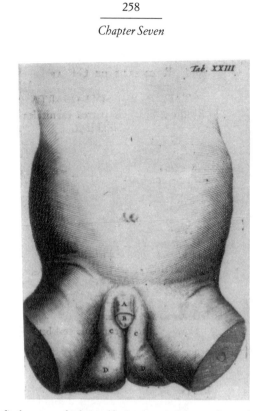

Fig. 49. de Graaf's drawing of "the malformed genital parts of an infant," with "the clitoris looking like a male penis." Illustrations of this kind could have fueled the enduring popular belief in correspondence between the clitoris and the penis, and inspired later tales of the clitoris often protruding two inches outside the vagina. (From *De mulierum organis generationi inservientibus,* 1672.)

In *The secrets of Nature Revealed,* we find the same ideas, often put almost in the same words:

> The Clitoris is a sinew and hard body, full of spongy and black matter within; and in form it represents the Yard of a man, and is subject to erection and falling as that does . . . The Clitoris sometimes grows out of the body by 2 inches.

Also,

> testicles in women are within, they are smaller then those of men, they are rough and uneven, whereas in men they are smooth.

Here, the author explores the inferiority of the female organs through yet another path. He is obviously in agreement with the old idea of the double semen, produced by both men and women during copulation and producing an embryo as the result of the mingling of the two seeds. As he explains it, "from the vessels wherein the seed of the woman is contained her seed is ejected, and by the exuberance of it she is moved to copulation. This seed is called by the Greeks Sperma, and receives its original form from the radical moisture of the whole body . . . The moisture from whence the seed is derived into every nerve, vein and artery, is the most refined and noblest part of the human frame; containing in itself the whole nature and complexion of every part of the Body, or, in other words, being the very essence of man." But let us not get carried away. Although man and woman both contribute this seed, the author alerts us, "you must take notice that man is the agent and woman the patient." Besides, "let me observe that the seed of a man is more strong and hot than the seed of a woman."

Jean Palfyn follows closely in these footsteps, explaining that the external parts of the woman can also be called "Shameful," since "they make women experience *pudeur* and an honest fear whenever they are unwillingly exposed"; and, to strengthen his claims, he points out that "the ancients already considered them as such, and that is why God told Moses that women should never climb to the altar but by degrees, so that no one could see what honesty tells us to keep hidden." As for the clitoris, the author describes it as the place where "God has hidden the source of all feminine pleasure"—but not, mind you, because God wanted women to have pleasure. According to Palfyn, this pleasure is necessary to persuade women to "expose themselves to bear for nine months such a heavy load, suffer the pains of delivery, and endure all the care they must take in bringing up their children."

Like the other writers of the same period mentioned above, Palfyn dedicates eight pages to the discussion of methods suitable to establish whether a woman is a virgin, which he considers necessary "for the reputation of those who judge the honor of girls, for the satisfaction of the husband and for the rest of human society." Among those methods considered "uncertain" are such prescriptions as: "if you throw seeds of Pourpier over burning charcoal and a girl who has been deflowered breathes the fumes, she sees marvelous things: if the girl is chaste she will not see anything extraordinary"; or "as long as Virginity has not been challenged, a Virgin will be immune to the stings of the most irritated bees." The more "certain" proofs include the presence of the hymen and the pain experienced by the woman on her wedding night.

Even when the tone becomes more sympathetic toward women, the idea of their condition as a lesser mirror image of the perfection of men still prevails. This is well illustrated by the final conclusions on the matter in *The Complete Master-Piece*:

> Woman, next to Man, the noblest piece of the Creation, is Bone of his Bone, and Flesh of his Flesh, a sort of second self. And in a married state are accounted but one; for, as the Poet says,

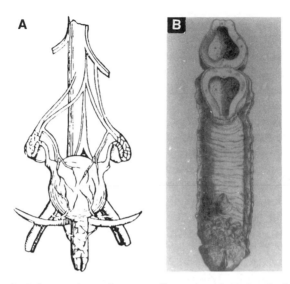

Fig. 50. Vesalius's famous sixteenth-century illustrations depicting the female genital anatomy as an internal development of the male's external genitalia and their immediate repercussions: *(A)* vagina as penis from Vesalius's *De Humani Corporis Fabrica* and *(B)* vagina and uterus as depicted by Vidus Vidius in *De anatome corporis humani,* 1611. (As reprinted by Gilbert in *Developmental Biology,* 1994.)

> *Man and his Wife are but one right*
> *Canonical Hermaphrodite.*

These ideas had been around for a very long time. Echoing Hippocrates, Galen had also talked of a female semen, originating in vessels surrounding the female testes, becoming separated in those testes, and then passed down through the same tubes into the uterus, where it produced a frothy coagulum when coming into contact with the male semen, thus producing the embryo. During the Middle Ages and the Renaissance, several illustrations of the genital anatomy of women appeared in which the ovaries and the womb were depicted as an inverted, inward imitation of the penis and testis. Interestingly, one of the most famous and frequently reproduced of these illustrations was produced in 1543 by Andreas Vesalius (1514–1564), the famous anatomist from Padua who overturned many of Galen's anatomical teachings, and—at least according to the legend—in the process risked the retaliation of the church for arguing that man and woman have the same number of ribs. In Vesalius's two major works, *De Humani Corporis Fabrica* and *Tabulae Sex,* the portraits of female genital anatomy depict an internal development of the male's external genitalia.

The scholarly community was by no means immune to the appeal of such configurations. As soon as Trembley discovered regeneration in the polyp (see chapter 4), John Turberville Needham (see chapters 4 and 5) immediately proceeded to proclaim that Eve had been born from the rib of Adam in exactly the same manner:

> The body of the first woman was not formed from earth, like the body of her husband, but she was rather generated from him through an accelerated vegetative process, nourishing on his substance during his sleep, until she separated in a state of perfection, like what is observed among the young polyps and the organized bodies of the same kind.[32]

Summing it all up, the author of *The Complete Master-Piece* decided to make it rhyme:

> Thus the Women's secrets I have surveyed
> And let them see how curiously they're made:
> And that, tho' they of different sexes be,
> Yet in the Whole they are the same as we:
> For those that have the strictest Searchers been,
> Find Women are but Men turned Out side in:
> And Men, if they but cast their Eyes about,
> May find they're Women, with their Inside out.

THE FEMALE INSIDE

Amazingly, this idea of the two sexes united in one is by no means ridiculous in the light of modern developmental biology. Sex determination in animals is still a gigantic unsolved puzzle. Natural hermaphroditism as a way of life and reproduction is common in organisms ranging from worms to snails. The capacity for sex change in numerous species is widely documented and almost dazzling. Severing one simple gland in male toads makes them become perfect females, complete with ovaries and wombs and as fertile as any other. In fishes that swim in groups led by a male, the death of that male can prompt one of the females to acquire all the masculine traits and take up the lead. Some fishes are so invariant in this practice that cytogeneticists gave up trying to classify them according to the usual terms and had to create the new term "unisexual." These cases are far from being all uncovered, and continue to provide scientists with unexpected surprises. But, in the case of mammals, it now appears beyond question that all embryos start their lives following the developmental program of only one of the sexes. As compared with what our predecessors believed, the tables were turned in just one little detail: in the beginning, all mammals are females.

In the early steps of our development, we acquire the rudiments of sexual organs (the *gonads*) in an undifferentiated form, able to become either ovaries or testes.[33] In the absence of a Y chromosome (typically the male-determining chromosome in mammals), these gonads become primitive ovaries, and the hormones released by these ovaries trigger the development of an earlier structure known as the *Müllerian ducts* into the oviducts, the uterus, the cervix, and the vagina. It is only after the Müllerian ducts have formed that, if a Y chromosome is present in the cells, factors released into the blood trigger the primordial gonads to become testes. The testes then segregate a male-specific hormone, the *anti-Müllerian duct hormone,* which destroys the Müllerian ducts and substitutes another structure, the *Wolffian ducts.* At this point the testes produce a second hormone, the famous *testosterone,* which stimulates the development of the Wolffian ducts to form the *vas deferens,* the tube through which the semen passes into the urethra and out of the body. Testosterone further masculinizes the embryo by triggering the formation of the penis, the scrotum, and other components of the male genitals, all while repressing the development of the mammary glands. Thus, the mammalian body is initially programmed to express a female phenotype, unless the Y chromosome changes the rules after the beginning of the game.[34]

Confirming these patterns, recent studies in rabbits show that if the undifferentiated gonads are removed from embryos, these embryos will develop with a female phenotype. The same happens in humans born without functional gonads. Also, if a chromosomal accident occurs such that the affected individual has only one X chromosome, but no Y chromosome (a condition known as *XO*), ovaries start to develop but then atrophy before birth, and all the germ cells die before puberty. However, under the influence of ovarian hormones known as *estrogens* derived from the mother and from the placenta, these infants are born with a female genital tract.[35] If the male-determining gene in the Y chromosome of mice is absent or fails to be expressed, or is expressed too late or in too small a number, some gonadal cells differentiate as primordia of ovarian follicles and development proceeds according to the female pathway—regardless of whether the embryo's genotype is XX (female) or XY (male).[36]

More interesting yet, the *primordial germ cells,* destined to migrate from different embryonic regions to the ovaries or the testes, where they become eggs or sperm depending on the sex of the embryo, also seem to be initially programmed according to female instructions. Most of these studies have been carried out in a species of prosimians from Madagascar, *Galago crassicaudatus crassicaudatus,* in which significant numbers of primordial germ cells frequently follow a wrong path and wind up in a wrong, *ectopic* location,[37] in numbers often greater than those that actually reach the developing gonads. The most common host for these lost travelers is the adrenal gland, located above the gonads, but they can also be found in the walls of the aorta or in the mesonephric tissue, destined to give rise to the kidneys.[38] The same phenomenon was later detected in certain strains of laboratory mice. In either species, whenever this ectopic location of primordial germ cells occurs, those cells that should have become spermatozoa (because they are contained within male bodies) start to develop as eggs. They do so in perfect synchrony with eggs developing in ovaries, while their companions that managed to find their way to the testes develop normally as spermatozoa. This phenomenon stops only when the support of ovarian follicular cells becomes necessary for the continuing maturation of the egg, and then the eggs that should have been sperm finally die.[39] Were it not for the secondary intervention of the Y chromosome, the mammalian world would contain nothing but females.[40]

WHEN THE TWO WERE ONE

Obviously, when the myths of Creation started to take shape, none of this was known. But some sense of a common origin must have bathed our primeval self-perception, since, in numerous mythologies, the two sexes were initially created as one. Sexual separation generally appears as an afterthought of the Creator—and, bowing once more to the old concept of feminine inferiority, it is the female that constitutes the afterthought.

In Roman mythology, Athena was born from the head of Zeus. We also find this pattern in the Bible. In the best-known passage, God creates Adam and takes him to the Garden of Eden. It is only at this point that God thinks, "it is not good for the man to be alone. I will provide a partner for him"; and, while Adam is sleeping in a trance, fashions Eve out of the man's rib. Next (Gen. 2:23),

> He brought her to the man, and the man said: *Now this at last—bone from my bones, flesh from my flesh—this shall be called woman, for from man it was taken.*

But in an earlier, more obscure passage (Gen. 1:27–28), Adam and Eve seem to be created at once as a single body:

> So God created man in his own image; in the image of God He created *him* and blessed him and called him Adam; male and female He created *them.*

Christ Himself, when He came to earth as God incarnated, has been explained by several theologians as being simultaneously masculine and feminine.[41]

The possibilities underscored by this unisexual beginning were particularly enticing for successive generations of philosophers. The putative original plasticity allowed by this context still echoed in Nicolas Andry's *Orthopaedia,* in which he notes that

> in former Days, the Navel connected the two Sexes, so that they made but one body, though they were really two; but this String coming at last to break, each body had its liberty.

This passage could have been directly derived from the Persian or Talmudic myths of Creation, in which the first human being is a person with two bodies and two sexes, male and female, united at the back like Siamese twins; the separation of the two bodies occurs only as an afterthought of the Creator.

The same motif appears in one of the versions of Creation contained in the Indian Upanishads:

> The procreator had no delight. One alone has no delight. He desired a second. He was as large as a woman and a man embraced. He caused that self to fall into two pieces. Therefore arose a husband and a wife. Therefore one's self is like a half fragment. Therefore this space is filled by a wife. He copulated with her. Therefore human beings were produced. But she asked herself—how now does he copulate with me once he has produced me from himself? Come, let me hide myself.

So she becomes all kinds of female animals, and he becomes the male animal of that species, and at every turn they copulate, so that all kinds of animals are produced, even down to the ants.[42]

If, in this version, the position of women already appears as secondary, things become even bleaker in an old Hindu legend, in which the female is nothing but a composite of the Creator's leftovers: when Twashtri, the Divine Artificer, came to the creation of woman, he found that he had exhausted his materials in the making of man, and had no solid element left. To overcome this problem, he simply put her together out of the odds and ends of Creation:

> He took the rotundity of the moon, and the curves of the creepers, and the cling-ing of tendrils, and the trembling of glass, and the slenderness of the reed, and the bloom of flowers, and the lightness of leaves, and the tapering of the elephant's trunk, and the glances of deer, and the clustering of rows of bees, and the joyous gaiety of sunbeams, and the weeping of clouds, and the fickleness of the winds, and the timidity of the hare, and the vanity of the peacock, and the softness of the parrot's bosom, and the hardness of adamant, and the sweetness of honey, and the cruelty of the tiger, and the warm glow of fire, and the coldness of the snow, and the chattering of jays, and the cooing of the *kokila,* and the hypocrisy of the crane, and the fidelity of the *chakravaka;* and compounding all these together he made woman, and gave her to man.[43]

THE SOURCE OF DISHONOR

These primeval beliefs had implications reaching well beyond making women fare poorly in comparison to men. As we have discussed above, being inferior can be perceived as being in touch with dark, subterranean forces that the inferior person's mind is not even able to tame. Thus inferiority can be easily equated with danger. In its early stages, Christian theology openly equated women with sin, since it was through the first woman that original sin had impinged upon mankind. In India, where original sin was not part of the universal creed, the Code of Manu described the female condition in strikingly similar terms: "the source of dishonor is woman; the source of strife is woman; the source of earthly existence is woman; therefore avoid woman"—and do not forget that "a female is able to draw from the right path in this life not a fool only but even a sage, and can lead him in subjection to desire or to wrath." [44] Following parallel lines of thought, most great cultures barred women from learning to read or being instructed in any kind of philosophical methodology (or even, in the case of the Chinese, from playing the flute), since acquiring such tools would make their potential danger all the more evident— not to mention that a woman equipped with knowledge would become depraved, or at least acquire a perilous resistance to subjection.

One of the legends of the Brazilian Indians tells the story of the young virgin Ceiuci, who bore a son, Juripari, from the juice of a local fruit that the girl had carelessly spilled over her belly. This juice was animated by the Sun, Coarici, who saw in Ceiuci the perfect maiden to produce a hero charged with changing and correcting all the evils that were damaging the earth, particularly the dominance of women over men. After eliminating the power of the female sex, Juripari established a number of ritual sacred celebrations that women were forbidden to attend. Albeit aware of that danger, Ceiuci disobeyed her son and tried to attend one of those festivities; and from that heresy she died. Juripari could not bring her back to life, but was allowed to carry her into the heavens, where she became the Pleiades. The Indians refer to that cluster of stars as Ceiuci. [45]

Never to be left unattended, women all over the globe were ascribed the destiny of being the property first of their fathers, then of their husbands, and eventually of their sons. Paracelsus made this claim via a truly bizarre path in his tract "De homunculi." [46] According to the author, any experience of lust was enough to trigger the production of seed, and if not expelled from the body, this seed would become "polluted sperm," the substance from which soulless homunculi and monsters could arise. Women, unlike men, always retained this lust-generated seed inside the body, where it putrefied and caused diseases such as monstrous growths that mocked pregnancy. As we have seen in chapter 2, Paracelsus suggested that men should solve this problem either by marrying or by castrating themselves. Since women could not be castrated, Paracelsus's argument followed, their parents should see to it that they got married as early as possible, placing them under the rulership of men and thus ensuring that all of their seed was properly used for the production of ensouled offspring.

The custom of keeping married women in seclusion appeared in Islamic as well as in Hindu traditions—and went as far as stipulating that a doctor could address a married woman only through a thick curtain. Although we tend to associate the burning of wives on their husband's funeral pyres with old tales from India, according to Herodotus this practice was already common among the ancient Scythians and Thracians. African tribes observed similar rites, and, to this day, burying alive the wife and daughters in the grave of the dead husband is still occasionally practiced in remote regions of the Andes.

The Westerners who explored the sophisticated Japanese society of the early decades of our own century were delighted by their discovery of the code of the "Three Obediences." A woman was to obey, in sequence, her father, husband, and son; and, except for etiquette, no other education was required for her. As in many other places, if the husband caught the wife in adultery, he was allowed to kill her and her lover at once. If he spared the lover, several Japanese philosophers agreed, the husband himself should be put to death. The philosopher Ekken advised men to divorce their wives if they talked too loudly or too long. But, if a husband happened to be dissolute and brutal, his wife should treat him with double kindness and gentleness. Were the Westerners shocked by such backwardness? Of course not. Much to the contrary, they marveled at a system that had produced such wonderful wives, and considered the possibility of introducing similar rules back home.[47]

We all know of the recent problem created in India by the use of sonograms to determine the sex of a fetus early enough for an abortion to be carried out if it should be a girl. We have all heard stories of the Chinese throwing their newborn daughters into the river. Once again, these Eastern practices also existed in the West. In the early days of Rome (508–202 B.C.), if a child was born deformed or of the female sex, the father was permitted to expose it. The same husband could act as a direct judge of his wife, condemning her to death with no further ado, for infidelity or for stealing the keys to his wine.[48] And, needless to say, the Greeks developed the paradigm of all democracies— but the thought of women having anything to do with government seems to have not even occurred to them.

True, this is all old news. But, interestingly enough, the venerable age of these beliefs never really entails their demise. Right up to our own days, as discoveries have accumulated and knowledge has progressed, every newly uncovered aspect of biology has been promptly converted into yet another proof of women's inferior status. As John Farley argues in his 1982 *Gametes and Spores,*

> As a result of the close association made in the nineteenth century between ova and asexual cells, and the realization that asexual reproduction was a widespread phenomenon, the question arose, what were the reasons for sexual reproduction in the first place? The answer . . . reflects Victorian social attitudes toward sex. Sex was simply the means of procreation that had, by the division of labor, placed the task of procreation into the hands (or, more appositely, the ovaries) of a special individual, the female. Socially and biologically, the female existed solely to bear and raise offspring. The nineteenth-century biological attitude toward sex mirrored that century's social attitude toward women. The status of women was therefore a law of nature; to argue otherwise was to threaten the social and biological fabric of the race.

We may have spread several layers of modern polish over the surface. Underneath, however, the universal perception of female inferiority and subjection to the male continues to lurk.[49] Louis Aragon may have written that "woman is the future of man," and he certainly meant well. But this overquoted sentence still vaguely suggests that both sexes cannot possibly inhabit the same realm.

Venus and Mars

A deeper understanding of the connotations instinctively associated with the female and male genders can be derived from the symbols we now currently use for each sex. The symbols ♀ and ♂ may have become casual household items, but configurations thousands of years old hide deep inside these drawings. We may know that ♀ also stands for Venus, both the goddess and the planet; and that ♂ is the symbol of the god and the planet Mars. We may also know that the alchemists used ♀ for copper and ♂ for iron. Yet we need to delve deeper, for symbols are never innocent; and certainly not when so many regulatory powers over life on earth have been ascribed to the planets, and so many active powers have been associated with the metals. Let us bear in mind that one of the greatest players at the onset of the Scientific Revolution, the physician Hermann Boerhaave (see chapter 2), wrote in his *Essay on the virtue and efficient cause of magnetical cures* that

> the Sympathy and antipathy, so notable in Animals, Minerals and Vegetables, is . . . to nothing else to be ascribed, than the Influence of the Celestial Bodies;

and their Impression according to the Constitution of each Species and individual Being, either in general or sometimes according to the singular Fabric of one Species in particular.

According to Carl G. Liungman in his 1991 *Dictionary of Symbols,*

the planet ♀ was worshiped by several peoples and cultures of antiquity as the *divinity of fertility,* the *goddess of war, beauty, and love.* In its role as the goddess of war and fertility it is associated with the Morning star. In its role as the divinity of sexual love it is associated with the Evening star. In the cradle of our culture, ancient Greece, it was not realized until around 500 B.C. that these two stars were one and the same.

The astrologers consider ♀ to be a benefactor, and, as a horoscope symbol, Venus represents the drive for togetherness, the aesthetic drive, and sensuality, together with the instinct to seek the company of other humans. The day of Venus is Friday, the day of Freja, the fertility goddess of the Nordic countries. According to Boerhaave,

the image of Venus is manifold, but the following is more frequently engraved in stones: a Woman with a long robe, and in her Hand a Bay Leaf. The Virtue thereof is to give an Agility in performing business, to bring everything to a good Success, to free from the fear of being drowned, and to procure Honors from Women.

It seems only logical, then, that the planet Venus has long been linked in the West with women, and with the representation of the female in biology, botany, medicine, and other natural sciences. This association became so enduring that, in the middle decades of the twentieth century, ♀ was also appropriated by the women's liberation movement. But what about the association with copper, an association that goes back to antiquity and is so prevalent as to include the use of ♀ in cartography to indicate a copper mine? According to Liungman,

the background for this is that the island of Cyprus in antiquity was a great exporter of copper. It also happened to be one place where the goddess Astarte or Ischtar [godly equivalents of Venus] was especially worshipped. Consequently, when the sign for the planet associated with this goddess of fertility was established as ♀ it also became associated to this *island of copper* (*cuprum* in Latin, from which the island's name, Cyprus, is derived).

In our times, the mining company Stora, which during the first centuries of its existence dealt only with the extraction of copper, adopted ♀ as its logo. To close the circle, the astrological age of Taurus occurred between 4500 and

2350 B.C. and was believed to have been influenced by Venus, the ruling planet of Taurus. Through copper, which the same symbol represents, the Bronze Age was brought into existence during this period. As populations increased, some cultures ceased focusing mainly on hunting and herding and turned their attention to agriculture—the feminine realm *par excellence* during primitive times.

The use of ♂ for Mars is contemporary with the use of ♀ for Venus; both symbols appear to have emerged about 2000 B.C. The symbol ♂ originally referred only to the god of war and aggression (one of the Roman Empire's most important divinities, right after Jupiter), and was later applied to the red planet. Since Mars was the warrior god, the connection of this symbol with iron is easy to understand: iron was the metal used for weapons of war.[50] Since Mars takes a little less than two years to circle the Sun, ♂ was also one of the signs adopted in botany to represent plants with a two-year growing cycle. But let us get to the heart of the matter.

In psychological astrology, according to Liungman,

> Mars represents such characters as *self-assertion, aggressivity,* and *the ego,* i.e., the urge to distinguish oneself as a unique individual . . . The planet is associated with the astrological element of fire . . . In its aspect of fire, it symbolizes the *zest for life.* A human being needs an adequate amount of it to have *courage, physical energy,* and *independence of mind and actions* . . . When . . . its influence is too strong . . . the worst characteristics of the Martian principle are embodied: *cruelty, selfishness, lust for war,* and a *desire for total dominance.* This planet, however, is not only a symbol for the *warrior,* the *murderer,* and the *slave owner,* but also for the *surgeon* who heals and the *hunter* who brings home the food. On the physiological level Mars represents the *blood,* the *outer sex organs,* the *adrenal glands,* the *muscle tissue* and the *nose.* The professions ruled by this planet are the military ones and those which have to do with cutting and piercing, such as *butcher, surgeon* and *carpenter* . . . With the circle filled, [the symbol] is used in military contexts to denote a *grenade thrower* or *mortar.* Among the weekdays, Mars represents Tuesday, the day of Mars in Latin countries.

According to Boerhaave,

> the image of Mars is diversely figured, sometimes with a Banner or Flag in his Hand, sometimes with a lance or any other warlike instrument, but always armed, and sometimes on Horseback. The Virtue of a Stone wherein such a Figure is graved, is to make him who wears it Victorious, bold and valiant.

Of the twelve houses circling the globe in conjunction with the signs of the Zodiac, the first represents the self and those interests that are self-centered, and it is linked to Aries and ♂. In old astrological traditions, Mars was also the ruler of the Zodiac sign Scorpio, with the pointed arrow signifying the scorpion's tail. In modern Western astrology, Scorpio was transferred to the influence of Pluto, the last planet to be discovered in the solar system, in 1930—and the symbol for this planet means *death and rebirth, total transformation, atomic energy, ruthlessness, dictatorships, large organizations,* and *competition.* The interaction between Mars and Pluto has yet other symbolic ties: the American astrologer Alan Oken suggests that Pluto is merely a higher octave of Mars. Mars breaks down the form, but it is Pluto (from which the name of the now so dreadful *plutonium* is derived) that transforms its atomic energy. The planet Mars represents the aggressive and emotional energy between enemies in wars, whereas the planet Pluto represents the ultimate destruction, the *atom bomb.* Mars signifies passion in sexuality; Pluto represents the orgasm. Mars is the anger that makes the soldier pull the trigger. Pluto is the force that separates the soul from the body. Nice company.

In light of this picture, it seems only logical that ♂ ended up representing the male gender. The pointed arrow had always had phallic symbolism and, in computer language, it means *exponentialization* (multiplying the sum by itself), whereas in physics it sometimes denotes *centers of gravitation.* Masculine enough, and even more so if we add one more mundane detail to the whole: in its early days, the Swedish carmaker Volvo used ♂, the symbol of Mars, iron, and self-assertion, for its logo.

The Mark of Dracula

None of this is meaningless in terms of the symbolic roles ascribed to males and females—because, as a matter of fact, none of it is merely symbolic. As we have seen, ♀ supposedly symbolizes sexuality, seduction, pleasure in being surrounded by other people, and fertility. But, in any account of the meaning of symbols, references to fertility largely override all the others. Languirand notes that the cross at the bottom of ♀ has nothing to do with the symbols of Christendom—the sign precedes Christianity by at least two millennia—but rather with the sign +, for sum, implicitly meaning coupling and fertility. It is good for a soil to be fertile, and it is certainly good for a woman to be fertile. But this apparent goodness is not without its dangers.

In the first place, this association places all the responsibility for fertility upon women; and this is where the blessing becomes a burden. At the most prosaic and immediate level, fertility entails child-rearing. For all of our modern well-intentioned efforts to counter a centuries-old trend, this beautiful act still has the flip side of restricting a woman's freedom and barring her from entering the workplace on a basis of true equality with her male counterparts. More seriously yet, if George Washington had no children, why does everyone blame it on Martha? Although gynecologists agree that male sterility is very common and is actually dangerously on the rise in our days, when a couple is sterile, women tend to take all the blame. And, as any practitioner in a sterility clinic will tell you, women perceive themselves as guilty when they are the ones affected, but think of their mates as innocent victims, deserving to be showered with all kinds of consolations, if the problem lies with them.

Also, it comes as no surprise that the male planet Mars should act upon the bile, as Boerhaave and his contemporaries would have had it. This bears proof of how accustomed we have become in our civilization to understand, and forgive, any man's short temper—while women themselves tend to be uncomfortable with anger in other women. Anger comes naturally with the masculine territory, but when women express it, the "phenomenon" is unconsciously perceived as an undue intrusion. On the other hand, Venus acts upon the kidneys. Again, this is not innocent, and loops back into the blind knot of fertility: the kidneys are the most sensitive definers of the lumbar region, marking the area where pregnancy takes place. It is interesting to note that, in his *Essay on the virtue and efficient cause of magnetical cures,* Hermann Boerhaave considered Mars a strong agent against "Timidity" and "Pusillanimity," but considered Venus, "holding in her hand Flowers and Apples," as having the sole powers "for procuring Mirth, Beauty, and Strength to the motherly Body."

There is also another implication, often claimed and just as often contested on the grounds of insufficient documentation, but nonetheless recurrent in popular perception. According to Liungman,

> the planet Venus, ♀, was in earlier times associated with a common goddess of fertility, war, sex and peace. This was true of nearly all known ancient cultures around the Mediterranean. But from about 2400 until 250 B.C. this Venus goddess was gradually replaced by masculine divinities.

The fateful 2400 B.C. marks the probable period during which such wildly grandiose and self-serving monuments as the pyramids of Gizeh were built. It is certainly not a coincidence that the time between 2350 and 200 B.C. was the Age of Aries, ruled by Mars—a time, notes Liungman, "marked by endless wars, a time when new trade routes were discovered, and long voyages into uncharted regions of the world were embarked upon; a time of struggle and daring escapades; an age dominated by man, not woman." The goddess ♀, who had ruled over the Age of Taurus, was now superseded by the male ♂, as iron wills took over from agricultural skills.

Moreover, the vertical line pointing upward (as in ♂) is *active,* whereas the horizontal line crossing a vertical line pointing downward (as in ♀) is *passive.* Languirand notes that, in astrology, any arrow turned upward means impulsivity, and when tilted to the right (as in ♂), it becomes *impulsivity geared toward action.* Moreover, the arrow also means *active eroticism,* since the triangle at the top is the symbol of *fire.* Thus, when ♂ overcomes ♀, the ensuing change in people's lives reaches up and down to all levels, and spells out a very clear message. Power and victory displace sensuality and seduction. Yet the fading of the ♀ light does not stop here.

The ♀ Venus associated with the symbol for Nirvana signifies artistic creativity. This sounds good, but only until we learn that the symbol for Nirvana is also the symbol for *self-denial,* for *defeated ego,* and for *deceitful and secretive activities,* such as *poisoning.* In this light, ♀ appears to have a strong affinity to any contemporary country music production for female vocalists, or to all the likes of *Arsenic and Old Lace.* It is fascinating to see all the clichés fall into place.

And the clichés are not all necessarily pathetic in their expression of what a woman truly represents. Venus was often depicted as ♀ in Western symbologies, but both African and Middle Eastern populations also frequently represented the planet (and the associated goddess) as a five- or eight-pointed star. This represents the Morning or Evening star, certainly, but not just that. The five-pointed star appears in the headgear of the Apache chief Geronimo, who lived during the nineteenth century. And Liungman promptly adds: "compare with the eight-pointed star on the headgear of the cruel Vlad, Count Dracula, on a fifteenth-century painting. In both cases the star associates to ♀ and its role as a war divinity." We all know that the ♂ Mars was the quintessential war divinity. Yet Dracula did not use it. Let us not soil masculinity with morbid connotations, please. Evil has to be ♀, as we have verified over and over.

If both genders had a warrior component, they seemed to be well aware of it in the mythological realm, and to profoundly despise each other's territories. The two Zodiac signs ruled by Venus lie directly opposite those ruled by Mars. For the astrologists, the ♀ Venus does not function smoothly under Aries, which is ruled by Mars—and the ♂ Mars fares poorly when operating under Taurus, which is ruled by Venus. The ♀ Venus also rules Libra—and, as it happens, neither the Sun nor the ♂ Mars has a good relationship with Libra. The ♀ Venus, symbol of the female principle, may be the Moon's daughter, and this sounds romantic enough. But the ♂ Mars, symbol of the male principle, is the Sun's son—and the Sun is always a notch above the Moon in many civilizations' values, except for the scary realms of the Occult. After all, given the choice, would you rather be driving a ♂ Volvo or working as a miner in a ♀ copper mine? In the end, even the mining company Stora, represented by the logo ♀, eventually abandoned the extraction of copper and turned its attention solely to iron—the very ♂.

So the ovists could derive all the mythological support they needed from the perfection of the sphere, the quintessential shape of the egg. But, as perfect as eggs might be in their form, the theory that the ovists had to defend could not avoid the disturbing fact that God had chosen to put His most perfect creation inside the body of a lesser being. The notion of women as imperfect men was certainly fading away among the scholars of the Scientific Revolution, but, as is clear from all the popular texts quoted in this chapter, the concept was still tenaciously alive in the collective perception of their times. The spermists may have proposed the unthinkable by holding that mankind was carried inside worms; but, judging from all the material we have just analyzed, it is hard to tell which would be more unpleasant to the general public during the days of preformation: to be hatched from a worm that at least came from a body made in God's own image, or to accept that God had played a nasty trick on mankind by submitting us to the humiliation of being encased within a perfect form—but one that was hidden inside women. Measuring the power of myth against the power of man's pride, this certainly was a dramatically even match, tied to the last punch.

Magical Numbers

Like the other, this game is infinite

JORGE LUIS BORGES

HE BOOK OF NATURE is written in mathematical characters and can only be understood by those who understand mathematics." Galileo's famous statement can now be seen as the battle cry for mathematical hegemony in the Scientific Revolution, an idea that evolved from a provocative challenge to a quasi-religious credo, eventually gaining the wide acceptance that marked the days of Isaac Newton. In the frenzy of quantification that followed,[1] the respectable configurations of mathematics acted as a legitimizing incentive for generation to evolve by numbers. These numbers certainly worked like a charm for the triumph of preformation. But, as quantitative studies became a crucial part of the equation, ovism certainly gained an edge over spermism, as we shall demonstrate in the first part of this chapter.

TWO HUNDRED THOUSAND MILLION

It seems undeniable, when analyzing the writings on reproduction from the eighteenth century, that quantification had become the fashionable thing to do. In *A History of Embryology,* Needham quotes extensively from Albrecht von Haller, in what quickly becomes an orgy of numbers, proportions, measurements, and fractions:

> We cannot examine in this case [the growth of the chicken egg] the size of
> the egg at the moment when it is put to incubate, but it cannot be more than
> $4/100$ inches long, for if it were, it would be visible, and yet 25 days later it is
> 4 inches long. Its relation is therefore 64 to 64 millions or 1 to 1 million. This
> growth takes place in a singular manner, it is very rapid in the beginning and con-
> tinually diminishes in speed. The growth on the first day is from 1 to $91\frac{1}{8}$ and
> what Swammerdam calls a worm grows in one day from one twentieth to one
> thirtieth of a grain to seven grains, i.e. it increases its weight by 140 to 240 times.
> On the second day the growth of the chick is from 1 to 5, on the third day, 1 to

not quite 4, on the fifth day from 1 to something less than 3. Then from the sixth to the twelfth day, the growth each day is hardly from 4 to 5, and on the twenty-first day it is about from 5 to 6. After the chick has hatched, it grows each day for the first 40 days at an approximately constant rate, from 20 to 21 on each day. The increase of the first twentyfour hours is therefore in relation to that of the last 24 hrs as 546¾ to 5 or 145 to 1. Now as the total increase in weight in the egg is to that of the whole growth period (up to the adult) as 2 to 24 ozs. all the post-embryonic growth is as 1 to 12, i.e. it is to the growth of one day alone early in incubation as 1 to 7½ . . . The growth of man, like that of the chick, decreases in rapidity as it advances. Let us suppose that a man, at the instant of conception, weighs a hundred-thousandth of a grain and that a one-month old embryo weighs 30 grains; then the man will have acquired in that time more than 300,000 times the weight that he had to begin with. But if a foetus of the second month weighs 3 ozs. as it approximately does, he will only now have acquired 48 times the weight he had at the beginning of the period. This is a prodigious decrease in speed, and at the end of the ninth month he will not weigh more than about 105 ozs. which is not more than an average increase of 15 per month. A child three years old is about half the size of an adult. If then the adult weighs 2250 ozs. the three-year old child only weighs 281 ozs. which is an eighth of the adult weight. Now from birth to 3 years he will grow from 105 to 281 or as 5 to 14, but on the following 22 years he will only accumulate 2250 ozs. or eight times what he had at three years. The growth of a man will therefore be in the first month of intra-uterine life as to 1 to 300,000, in the second 1 to 48, in each of the others as 1 to 15. In the first 3 years of extra-uterine life his growth will be from 164 to 281 and in the succeeding 22 years from 281 to 384, and the growth of the first month to the last will be as 300,000 to ²⁸⁄₄₅₆ or 136,800,000 to 28, or 4,885,717 to 1. The whole growth of man will consequently be as 108,000,000 to 1.

Haller is enjoying himself so much with all these complicated numbers that we almost miss the pioneering point he makes in this passage: that the rates of embryonic growth are much greater than those of life after birth, with both tending to decline from a remarkable initial speed to slower and slower rates as development progresses—and that this phenomenon occurs both in the embryo and in the infant. Haller's excitement with microscopic numbers was most likely fueled by the way microscopy served his fervent religiosity (see chapter 1), apparently confirming that whatever may be hidden to the human eye is yet disclosed to the divine vision. Also, this type of approach was most certainly related to his admiration for mathematics, as Hochdoerfer nicely summarizes in her essay *The Conflict Between the Religious and the Scientific Views of Albrecht von Haller*:

The mathematical method of investigation appeals to Haller as sound, and satisfies his sense of accuracy. He believes that this method was taught to men to approach the truth more slowly, yet with more certainty, and that man has been taught by mathematics the difficult lesson that he can not believe anything which cannot be proved . . . In view of the fact that exact mathematical tests may not be applied to many problems in natural science, Haller defends the reasonable use of hypotheses as a means of approaching the truth. He emphasizes the fact, however . . . that hypotheses should never be confused with actual truths. They may merely be regarded as a temporary contrivance, a sort of scaffolding to aid the student of nature in his ascent to a position from which he can view truth more clearly. By resorting to the use of hypotheses Haller feels that he is not giving offense to the spirit of higher mathematics. He calls attention to the experience of Newton, who, though he was bent upon destroying all arbitrary opinions and stated that he hoped to banish the use of all hypotheses from scientific investigation, yet was not able to get on without using an hypothesis himself.

So Haller leans against this hypothetical scaffold and uses it to reach deep into the magic of numbers, digging out the key to the greatest mystery of them all: exactly how many people God intended to bring to life. Buried in the eighth volume of Haller's monumental *Elements de physiologie,* the chapter dedicated to reproduction includes a brief passage in which the numbers shine through as a definitive advantage to the ovists.

In what concerns preformation, it is simple to avoid the objection that, both in the ovary of Eve or in the testis of Adam, there was no size and space enough for all the seeds of all men, those who existed before and those who shall exist during the duration of the world, to be contained in such minute particles. I can deduce that, if now live on earth around a thousand million men, and if each generation represents thirty years, being that the age of the earth is six thousand years, we would have a total of two hundred generations, and the total number of humans should be therefore around two hundred thousand million. This number is by no means surprising . . . We just have to consider that there was only one egg contained in the ovary of the first daughter of Eve, instead of two, leading to all the eggs liberated in the subsequent generations . . . Each mother is the envelope of a foetus, and from millions of such envelopes even more millions result.

TEN THOUSAND ZEROS

Here, the statistics are clear: God placed two hundred thousand million people inside the ovary of Eve, and these are the humans destined to live from Creation to Judgment Day. If Haller felt compelled to imagine Eve and her first daughter both having only one ovary, this was most likely due to the fact that preformation, as a doctrine, demanded that all life, for each species, start from only *one* organism: somehow, postulating a single ovary seems to have been easier for the Swiss physiologist than conceiving of the primordial woman with two ovaries, one of them eggless. Haller may also have reasoned that Eve was created, rather than generated: this would explain why her daughter, the first in the human lineage to result from reproduction, also had a single ovary, with yet a single egg. From there on, ovaries could start coming in pairs, and the generational unfolding of lives could proceed according to the doctrine. Ovism had to face this primordial dilemma, but once it was resolved, quantification did not pose any extravagant problems.

Within the scope of preformation, ovists were the only ones that could reason along these lines. Spermists, on the other hand, were at a total loss on the quantitative front: sperm cells were just too numerous to count, and their colossal numbers were, from the beginning, the main factor in spermism's doom (see chapter 2). Hartsoeker started his career as a leading spermist, and the discomfort caused by excessive numbers was one of his first signs of distress. "If all animals of the same species had been enclosed inside the first female, as is held by Swammerdam, Father Malebranche, and several others," he wrote in his *Cours de Physique,* right before disavowing preformation altogether,

> or in the first male of the same species, according to several other Philosophers, those who now inhabit the earth would have been, at the time of Creation, of a smallness truly infinite and incomprehensible. If, for instance, all the rabbits that have been produced since the time of Creation had been enclosed inside the first male rabbit, any of those rabbits that live now on earth would have been, compared with the first one, not what unity is to unity followed by sixty zeros, which is more or less what a single grain of sand is to all the visible world; but what unity is to unity followed by more than a hundred thousand zeros. In order to reach this number, I suppose that the small animal from the seed of the male rabbit is but ten thousand million times smaller than the living rabbit, since God created the first male and female rabbit six thousand years ago and these animals can reproduce after six months." [2]

Hartsoeker's stand on the issue of these numbers seems rather tentative, making it unclear whether he is actually endorsing this view or discoursing on the model to expose its ultimate impossibility. It is reasonable to argue that in this passage Hartsoeker was just engaging in a demonstration, by reductio ad absurdum, of the shortcomings of the preformationist doctrine. This possibility is even more likely when we bear in mind that Hartsoeker was the only distinguished preformationist to actually make public that he was abandoning the doctrine that he had first endorsed so strongly, although, as we have seen in chapter 4, it was the phenomenon of regeneration in the crayfish that apparently prompted him to do so. But a single apparent disparity between one's beliefs and nature's resources may not be enough to discourage a hard-headed man so radically. Maybe Hartsoeker was already feeling uncomfortable with his own theory before the crayfish emerged as the last straw. Maybe the extravagance of preformation's numbers had something to do with the growth of his unrest. The crayfish could have been a red herring. Numbers could have been the real problem, and in that case Hartsoeker's considerations of rabbit generation would be a poignant testimony to the spermists' tribulations on the numerical front.

However, the secondary sources have been prompt to pronounce the aforementioned quote a true spermist statement, thus once again (as in the case of Hartsoeker's infamous "homunculus," analyzed in chapter 6) bringing disgrace upon preformation. In *Early Theories of Sexual Generation,* Cole sets the tone by writing that

> [According to Hartsoeker] if, for example, all the rabbits which have been produced since the beginning had been enclosed in the first male rabbit, a rabbit existing to-day would have been at the time of the Creation, not as unity is to unity followed by 60 zeros, which is about what a grain of sand is to the whole earth, but as unity is to unity followed by more than 100,000 zeros. To reach this truly appalling result [Hartsoeker] assumes that a seminal animalcule of the male rabbit is at least ten thousand million times smaller than the rabbit itself, that the creation of the first rabbit took place about six thousand years ago, and that rabbits begin to reproduce at the age of six months.

The author uses this example to illustrate the basic flaws of preformation, further strengthening his own viewpoint by immediately adding, without even the pause of a simple paragraph, the fallacious argument (see chapter 4) that "preformation does not explain the existence of monstrosities . . . nor can preformation explain the regeneration of lost parts."

Later in the book, Cole cites L. Bourget's 1729 "Lettres philosophiques sur la génération & le mechanisme organique des plantes et des animaux" to point out that this ovist disputed Hartsoeker's calculations, "the object of which was to reduce the theory of emboitement to absurdity." Bourget's own numbers were nonetheless equally breathtaking: according to his computations, the egg that had been created in the first year of the world would be to the last and smallest to appear at the time of his writing as 630,720,000 is to 1. However, Cole notes, claiming to repeat the words of "the eager and flamboyant Bonnet," that "this was very small compared with the terrifying mathematics of Hartsoeker."

Similarly, according to Meyer, in *The Rise of Embryology,*

Hartsoeker . . . estimated that the size relationship of the first grain of wheat to such a grain after six thousand years would be inversely as unity is to unity followed by thirty thousand zeros, and thus brought ridicule upon the idea. He similarly estimated that if all rabbits born since the beginning of time had been enclosed in a rabbit living in his day, the size relationship of the former to the latter would be as unity is to unity followed by more than one thousand zeros! Such calculations helped make *emboîtement* ridiculous.

It is true that the numbers entailed by spermism were prone to criticism even by the spermists' contemporaries (see chapter 2 for Lyonet's comments on how such multitudes of encased lives would cause any male to burst). Like a considerable number of his contemporaries, Maupertuis (see chapter 4) addressed the problem of all the lives that would be wasted per ejaculation if spermism were the true model of reproduction:

> This little worm, swimming in the seminal fluid, contains an infinity of generations, from father to father. And each [preformed creature inside the sperm] has his seminal fluid, full of swimming animals so much smaller than himself . . . And what prodigiousness when we consider the number and tiny size of these animals. One man calculated that a single pike fish, in one generation, could produce more pikes than there are men on earth, even assuming that all the earth is as densely populated as Holland.[3]

Even the hundreds of eggs laid by frogs, or the thousands of eggs shed by sea urchins, were easier to quantify than the millions of spermatozoa present in any ejaculate of any species. Besides, eggs do not move. But sperm cells swim, and often swim fast, in exasperatingly tight herds.[4] Modern biologists have trouble counting them now, even after the cells have been paralyzed through the use of fixatives, while the observers have access to excellent microscopic tools, and hold in their hands a hemocytometer, a handy all-purpose device used mostly to count red blood cells. Now picture yourself in the seventeenth century, with rudimentary laboratory equipment and only your brain available to count these creatures that will not stop moving. What do you do? How tenacious do you have to be in order not to give up? Still, in spite of these frustrating limitations, and in spite of the ovists' unfair advantage concerning initial numbers, most spermists never gave up. If Hartsoeker did, maybe we should consider his choice an excellent example of one man's intelligence probing the fashionable trends of his day.

674 BONES

In truth, regardless of one's particular stand on generation, from the mid-seventeenth century onward, the fascination with numbers weighed heavily on naturalists' approaches to their subject matter. Some examples include Leeu-wenhoek's effort to calculate the probable size of blood vessels and muscles in the "smallest animals" (probably bacteria) that he had observed in an infusion of ginger; and Spallanzani's carefully noting, in the midst of his experiments on animal regeneration (see chapters 1 and 4), that if all *four* legs and the tail of a salamander were cut off *six* times during the *three* summer months, a *single* animal would generate 674 new bones in the new parts.[5]

Life and nature were in the process of becoming mathematized and geometrized. A quantitative fever was sweeping the learned world, like a tidal wave that would stop at nothing. Once it got started, it stretched in no time from mathematics to philosophy, engulfing all of the life sciences in the process. And even after this long run, the wave was still so powerful that it spread its numeric waters over sociology and politics, eventually reaching as far as the shores of faith and religion.

"Almost from its inception in the mid-seventeenth century . . . probability theory seemed to promise a calculus of moral sciences," notes Lorraine Daston.[6] If we bear in mind all the endless musings and ruthless polemics, unveiled throughout this book, about Nature's morals and Nature's goals, then it is easy to imagine the life sciences smoothly merging with the vaster realm of "moral sciences."

As Louis I. Bredvold points out,[7]

> Already in the seventeenth century . . . the imagination of mankind was pro-foundly impressed by two aspects of mathematics: first, its infallibility as a form of reasoning, in striking contrast to the inconclusivity, not to say futility, of most other forms of philosophizing, in which the continuous argument seems only so much evidence of the insolubility of the problem; second, the successful applica-tion of mathematics to the phenomena of motion . . . what could be more natural than to expand the application of this admirable instrument of the mind to the study of all the realms of human experience?

Thus, according to legend, on the night of 10 November 1619, René Descartes had a dream. He woke up with a self-assigned mission, that of applying the infallible method of mathematics to all the phenomena of the universe and to the hiddenmost corners of human thought, in a complete and homogeneous metaphysical system that would include not only physics proper, but also medicine and ethics. Through Descartes's initial impulse, although he died leaving behind only dispersed fragments of a grandiose plan, even human behavior entered the realm of those deductive and infallible sciences that were to shape the face of a new era.

Independently of Descartes, his English contemporary Thomas Hobbes also held that "geometry is the only science that it hath pleased God hitherto to bestow on mankind,"[8] and developed an entire ethical and political philosophy under the guiding light of mathematics. The Dutch philosopher and maker of lenses Spinoza, who died in 1677, left in manuscript another great ethical treatise based on geometry, "Ethica More Geometrico Demonstrata." Even Leibniz, who was prompt to denounce the inadequacies of Descartes and felt deeply unsatisfied by the "Euclidean ethics" developed by his professor Erhard Weigel,[9] was haunted by a dream of a General or Universal Science, one that would make all aspects of life quantifiable under the global umbrella of calculus—including, as we have seen in chapter 2, the ever-elusive concept of the soul.[10]

As mathematics expanded to religious and moral matters in its aim to explain the entire functioning of society, even the degrees of spreading of original sin were quantified in the process. In 1699, the Scottish philosopher John Craig produced a calculus of moral evidence and its decay in time, entitled *Theologiae Christianae Principia Mathematica.* His figures showed that, at the date of publication of the book, the evidence in favor of the truth of the Gospel was equivalent to the testimony of twenty-eight contemporary disciples, but that it would diminish to zero in the year 3144.[11] As this *furor mathematicus* entered the eighteenth century, the works of the English poet Alexander Pope emerged as the most memorable landmark in the organization of human behavior within a realm permeated both by poetry and by the new mathematized logic of the West. Even as other voices were raised to protest that "Strict logicians are licensed visionaries,"[12] many Utopians dreamed of becoming the Newton of Ethics or the Newton of Politics.

In a context in which even the mind and the soul were expected to become explainable through equations and extrapolations, it was quite natural for the life sciences to have embarked on this kind of quest. Even in the ever-unpredictable domains of physiological processes, numbers started to rule over the chaotic nature of life.

Reproduction, with its enduring high status within the realm of the life sciences as "the mystery of mysteries," was not to be left out of this numerical wave swept ashore by the hand of God. According to Daston,

> 18th-century collectors of statistical regularities opposed them to the workings of probability as a signal evidence of the hand of divine providence in human affairs. A typical argument of this sort . . . concerned the ratio of male-to-female births. John Arbuthnot [1710] . . . argued that if this regularity [the probability of either male or female equaling ½] were due to what he called "mere chance" . . . the probability of the observed ratio over a long period was astronomically small. Ar-

buthnot concluded that this was palpable evidence of design; namely, the divine provision for an equal number of men and women of marriageable age to insure propagation of the race *via* monogamy. Due to the greater "wastage" of young men due to their more hazardous lives, it was prudent to begin with a small surplus. No less a mathematician than De Moivre praised this statistical version of the argument from design revealing the "wise, useful, and beneficent purposes," inherent in "the steadfast Order of the Universe" . . . statistical regularities remained a favorite weapon of 18th-century theologians intent upon finding the signs of an "all-wise, all-powerful and good" agent in demographic data.

As a reproductive model, preformation was ideal for the part. Unlike epigenesis, it definitely resorted to the "all-wise, all-powerful and good" work of God in the organization of generation. And then again, also unlike epigenesis, it allowed for actual calculations of numbers of generations, and numbers of years, and proportions relative to sizes. The ball was in the preformationists' court. They seized it and played on.

25,088 Eyes

Statistics and infinitesimal calculus both emerged in the age of preformation, and the coincidence was so fortunate that most likely it was not a coincidence at all. If you are going to postulate encasement of generations, the more scientific foundations for the concept of infinite smallness, the better. And do you need a little help from technology? Here. As it happens, the microscope has just been invented.

Ah, the universe suddenly opened up by the microscope. A universe so much bigger than what the scholars of previous generations had estimated that it dazzled the imagination with the concept of minimal infinities. Infinities that, needless to say, matched the idea of preformation as a perfectly fitted glove. What a blessed combination. "We can even conceive that with stronger microscopes," wrote Jean Sénébier in 1777, in his preface to one of Spallanzani's books,[13] "we could see the Present pregnant of all the Future to come, in its infinite succession of Organized Beings."

Early comments on the microscope repeat again and again the same excitement over this smallness, permeated by incredible, brand new numerical wonders. The thread of the young spider is four hundred times smaller than the thread produced by an adult of the same species. The sting of the wasp has eight spikes. The venom of the viper contains crystals of an unbelievable smallness. The eyes of the insects are breathtaking: 6,236 for the silkworm, 1,400 for the louse, 25,088 for the dragonfly. The powder on the wings of the butterflies is in reality an array of small feathers regularly disposed and of an infinite variety. The skin of humans is covered with millions of minute scales, and we can count our own pores: 144,000,000 in a square foot, 2,016,000,000 over the entire body surface. The acorn encloses the oak and the entire forest. The snowflakes are composed of small stars, small flowers, and small hexagons. Infinity revealed. At last.

In a certain sense, Western knowledge had spent thousands of years awaiting this moment. Lynceus, the demigod companion of the Argonauts, from whom the lynx got its name due to its exceptionally sharp vision, appears in classical mythology as the first suggestion in Western culture that smallness invisible to the common eye could actually be seen, given the necessary tools. Quite appropriately, the first academic society focused on the microscope was the Academy of the Lynx, founded in Rome in 1601 by the duke Frederico Cesi. The aim of the academy's members was deliberately to achieve the keenness of vision characteristic of the animal they had chosen as a symbol. Galileo joined the ranks in 1609, and by 1610 already had his own microscope. This was certainly a bizarre piece by modern standards, since it had a concave and a convex lens measuring several "brasses," and one brasse is the equivalent of five feet. We can only wonder how anybody managed to use such an object, but apparently Galileo achieved some moderate success. In his 1967 biography of Swammerdam, A. Schrierbeek quotes an anonymous Englishman who traveled to Italy and reported the following: "I heard Galileo himself say that with his optical instruments he could distinguish the organs of sense and movement in small animals. In particular, he had found that in a certain insect each eye is covered by a thick membrane pierced with holes, like the iron visor of a warrior's helmet, by means of which he has access to the images of things visible."

It is not meaningless that Galileo himself took time off from the faces of Venus and the moons of Jupiter to play with this new "optical instrument." In spite of all the initial frustrations,[14] the "instrument" was becoming more and more fashionable every year. Not only did it reveal new worlds, but the vagueness of what those worlds meant made it flexible enough to suit any agenda: it could be seen as a lesson on the endless repetition of forms just as well as an opposite lesson on the reduction of the initial form to a more elementary one. Both the proponents of pan-animism and the atomistic materialists made use of the "instrument," and both parties claimed that the "instrument" supported their views.

Finally, with its growing importance within the debates of natural philosophy, the "instrument" needed a name. Not just any name: a respectable terminology became necessary. In 1625, another member of the Academy of the Lynx, Johan Faber, proposed the name "microscope" for the visionary new object, which up to that point had gone by designations such as "conspicilium muscarum" or "conspicilium pulicarum." These terms mean, respectively, "fly-glass" and "flea-glass"; their choice reveals how tight the connection was between the early use of the microscope and the study of insects. Now, dignified by a more encompassing designation, the glass was destined to expand from insects to all of the hidden manifestations of life.

Lynceus had worked his marvels by means of tools that were superhuman. Now the tools existed at a human level, and at first they seemed to promise an endless feast of possibilities.

"As Small as She Pleases"

It is easy now to deride "mathematical" calculations such as Hartsoeker's, or to smile at Haller's account of the two hundred thousand million humans hidden inside the ovary of Eve. But our present reasoning is framed by self-evident boundaries that were nonexistent during the Scientific Revolution. Preformation took shape two centuries before the introduction of achromatic lenses, which would make microscopic images much clearer,[15] hence much less open to flights of imagination. Moreover—and this is certainly the key point—preformation emerged two centuries before the postulation of the cell theory, which would impose a lower limit on size for the forms of organized life.[16] The preformationist body of work, with all the numbers that sprung from it, developed during a favorable interval when a new door had just been opened and there were no walls enclosing the new room thus revealed.

"Nature can work on sizes as small as she pleases," wrote Bonnet in 1764, insisting that we should not restrict the infinite possibilities of nature with the strict limits of our senses and our instruments. "We cannot establish the moment in which creatures begin to exist as the moment in which they become visible."[17] On the same note, Jean Sénébier wrote in 1777 that "the impossibility of perceiving Animals smaller than those revealed by the Microscope does not make their existence unlikely . . . it is easy to imagine Animalcules smaller and smaller, or at least there is nothing wrong with this supposition . . . there could even be some Liquors which are inhabited by nothing but these rigorously invisible Beings." As for the ones that the microscope had already made visible, Sénébier stressed the amazing fact that "the faculty that the germs of Animalcules have of resisting considerable fire and cold allows them to live and to subsist in all the parts of the earth," and added that "it is not hard to imagine Animalcules living in the Flame, in the Burning Bodies, perhaps even within Light itself."

Moreover, at the birth of preformation there was no clear concept of gametes, or sexual cells. Eggs were known to hatch, but even mammalian eggs remained a mystery. There was no confirmed assessment of the function of the sperm, no descriptions of a real fertilization taking place. More precise microscopes and the original cellular concepts appeared only in the 1830s. It was not until 1824 that Prévost and Dumas noted the universal existence of sperm in sexually mature males and their absence in immature and aged individuals, together with the absence of sperm in the infertile mule, establishing for good—but not without much initial resistance—that "there exists an intimate relation between the presence of sperm in the organs and the fecundating capacity of the animal." Carl Ernst von Baer first described the mammalian ovum in 1826, and immediately after him scientists like Coste (1833) and Bernhardt (1834) described the germinal vesicles, the huge nuclei of immature eggs, in a number of mammals. However, even after all this had been observed, the idea of the egg as a cell and the germinal vesicle as a nucleus still took several decades to develop. Finally, Oscar Hertwig observed fertilization in the sea urchin in 1875, establishing for good the participation of both gametes in the process, and the way in which their interaction functions.

But the seventeenth century was free of all these constraints. And, in the time that followed, the microscope gained more and more adepts, to the point of becoming a popular tool of entertainment. When we find Nicolas Andry arguing, in his 1743 *Orthopaedia,* that "children amuse themselves frequently by looking at Gnats, and other small insects, through Microscopes; this obliges them to shut one Eye, which, by being frequently repeated, may give them this Deformity we call the Monopia, or one Eye less than the other," we get a clear idea of how widespread this type of microscopy had become.

And children were by no means the only humans peeking through magnifying glasses and letting their imagination follow its course over the elusive forms thus discerned. As Jean Sénébier wisely pointed out,

> this knowledge that we watch grow and perfect itself every day should make us aware of the imperfection of our ideas about the universe as a whole and of the danger of trying to establish too broad laws for it. [When using the microscope] we are like a Man who has never seen a watch and to whom are offered separately all the pieces that make up the watch.[18]

Naturalists had microscopes, and they were then allowed to have dreams.

138240000000000 ANIMALCULES

Early microscopes, with their rudimentary imaging and their technical limitations, worked as liberating devices, which at first seemed to allow infinite reduction in size. All that their users knew for sure was that things existed that the eye could not see, and that these things could be incredibly small. Listen to Leeuwenhoek:

I usually judge that three or four hundred of the smallest animalcules, laid out one against another, would reach to the length of an axis of a common grain of sand; and taking only the least number, to wit 300, then

300 × 300 = 90000

90000 × 300 = 27000000 animalcules together are as big as a sand-grain.

Let us assume that such a sand-grain is so big, that 80 of them, lying against the other, would make up the length of one inch.

80 sand-grains in the length of one inch × 80 = 6400 sand-grains in a square inch.

6400 × 80 = 512000 sand-grains in a cubic inch.

512000 × 27000000 animalcules which make up the bigness of a sand-grain = 138240000000000 animalcules in a cubic inch.

This number of animalcules is so great, that if one had as many sand-grains, of the bigness aforesaid, then one could lade them more than 108 of our ordinary sand-lighters, that is, reckoning one *schagt* of sand (which is 144 cubic feet) to every lighter.

I have let my thought run likewise on the very little vessels that are in our bodies, and have judged that they are above a thousand times thinner than a hair of one's head; and I have therefore put the proportion of the very little vessels thus in relation to the body, in order to arrive afterwards at the proportion of the vessels in the little animals.

First of all, I sought to know how many hairbreadths are equal to the length of one inch; and having by me a copper rule, whereon the inches are divided into 3 parts, and each of these again into 10 parts (thus altogether, an inch divided into 30 parts), I laid hair from my periwig upon these divisions; and observing them thus through a microscope, I judged that 20 hairbreadths are equal to $\frac{1}{30}$ of an inch. Consequently, there are 600 hairbreadths in the length of one inch.

Further, I measured, roughly, the thickness of my body above the hips, and judged (taking one thing with another) that the diameter of my body was 8 inches.

And finally, lots of multiplications and divisions later . . .

To sum up, then:

As a sixth of a hairbreadth is:

To a length of 5400 miles [the circumference of the earth, against which Leeuwenhoek is now measuring the diameter of the hair, since he found that through other methods the number of vessels in the little animals would be "exceeding great"]

So is one of the smallest vessels in the smallest animalcules:

To the thickness of a sand-grain (of such size that 80 thereof, lying one against another, equal a length of one inch).

> Sir, you have here the wonderful proportions that I conceive to exist in the secret parts of Nature: and from this appeared also, that we all have yet discovered but a trifle, in comparison of what still lies hid in the great treasury of Nature; and how small must be those particles of water which, to all appearance, pass many a time through such tiny vessels.[19]

This excitement with numbers was more than feverish; in its restlessness, it was almost religious. On his deathbed, Swammerdam still dreamed of his unfinished work on bees, "since the wisdom of God is so mathematically proven therein." And an amateur entomologist, the Dutch painter Johannes Goedaert, published a book of assorted microscopic observations in 1660 whose frontispiece claimed "The Almighty God is at His greatest in the smallest things." The microscope could be more than a tool of knowledge: very much in tune with some basic principles of Christendom, it could become a tool of modesty, showing us how perfect are the works of nature compared with the works of man when both are put side by side under the lenses—how perfect is the leg of the bee, and how shabby is the edge of the razor! As Hermann Boerhaave would say in his *Orations,*

> what particle of the human body is believed to be more simple than a hair? But this same hair, if it is diligently studied, displays the beauty of a most elaborate structure; so that the whole of human lifetime and all possible zeal may not suffice to produce in the mind a complete understanding of it.

Moreover, microscopy could also become a tool of redemption, as in the case of the louse, "which horrifies our eyes but becomes much embellished under the microscope"; or, even more so, in the case of the spider, "an animal that we consider very vile," or in the case of the "insignificant fly":

> [The spider] was better treated than us by Nature in what concerns a very delicate organ. Some spiders have six eyes, others four, others yet up to ten . . . by these means, the spider can see simultaneously what is below and what is around [spiders had long been assumed not to have any eyes at all] . . . The fly presents under the microscope riches that amaze, luxuries that dazzle. Her head is covered by diamonds; her body is all covered by brilliant blades; she displays long silk gowns and ravishing feathers. A circle of silver surrounds her eyes.[20]

By the mid-eighteenth century it had become customary for churchmen to write introductions for treatises in the natural sciences, and these essays often gladly engaged in exploring the transcendental side of the microscope. As Jean Sénébier wrote:

> Hypotheses are the resort of the Physician who cannot be instructed by observation alone: Nature almost forces him, with her obscurity, to imagine what she in-

sists on hiding. Of all the challenges that the Philosopher can face, none is harder than to determine the relationships between the microscopical Beings and the Universe . . . The study of Nature is difficult to the infinite; she is covered by very thick clouds; she is surrounded by steep abysses; the road that leads to truth is unique; this road is closed by a multitude of natural obstacles; and we augment the darkness that hides this road with our hasty conclusions. We can only wish, then, that those BONNET, those SPALLANZANI, those TREMBLEY, those HALLER, those DE SAUSSURE, and all those other Men whose genius and talent has rendered them worthy of being Nature's confidants, dedicate themselves to study her, become her voices and her painters, and secure themselves at the same time the precious title of Benefactors of the Human Species, increasing the happiness and the wisdom of Man and rectifying Man's judgment.

And then, bringing the grandiose exhortation back to the righteous grounds of preformation:

Having established the pre-existence of germs, we can easily see the regularity within each species derived from it, with the same germs always providing the same foundations for the Organized Beings. The Microscope has strongly fortified this deduction, showing us within the same species of Plants the same organs of generation, little particles with the same configuration and Bodies to be developed that all resemble each other. The Eggs enclosing the Tadpoles before fecundation are similar to the same Eggs after Fecundation. Even Crystals of the same Salts always resemble each other.

STEP BY STEP

Any simple comparative chronology will make clear how the inventive microscopy of the seventeenth century, and all the numbers that sprung from it, builds up to the heyday of preformation.

The first publication in this new field was by Francesco Stelluti, consisting of a single plate representing microscopic features of the bee[21] at a magnification of five diameters, issued in 1625 under the title "Melissographia," or "Description of Bees." Using more refined instruments, Swammerdam described red blood cells, the "blood corpuscles," in 1650. The big boom started in 1665 with the publication of *Micrographia: Some physiological descriptions of minute bodies by magnifying glasses,* by the first curator of the English Royal Society, Robert Hooke. *Micrographia,* with its excellent and abundant illustrations, clearly demonstrated how much the microscope could do for the biological sciences. In propelling this instrument to the role of star tool in biological observations, Hooke found an excellent partner in another curator of the Society, with whom he took turns sharing the only microscope owned by the institution. This man was Nehemiah Grew, and his whole work was dedicated to plant structure. Grew's publications on this subject, dating from 1672, 1673, and 1682, contain in all over a hundred engravings from microscopic drawings—and many of them are so precise and accurate that they still could easily figure in any modern botanical textbook.

But the curators of the English Royal Society, for all their importance in popularizing the use of the microscope, were obviously not the only men using the new machine to take a fresh peek into the mysteries of life. By 1661, Malpighi had already published the first work in microscopic anatomy. At the same time, Swammerdam was using the microscope, in conjunction with dissections of animals, to refute the concepts that insects had no internal structure and originated by spontaneous generation. So the microscope was already well on the way to becoming a powerful tool in explorations of reproduction when Leeuwenhoek started publishing in 1673.

Leeuwenhoek used microscopes allowing for magnifications of several hundred diameters, although to this day nobody knows for sure how he managed to come up with such clear and precise pictures. Some authors, like Hughes in *A History of Cytology,* contended that Leeuwenhoek could just have been "a person of abnormal visual acuity." Maybe we will never know, for in any event Leeuwenhoek kept the details of his methods jealously to himself—and the resulting sense of a vision enshrined in mystery must have provoked the curiosity of his contemporaries even further. His exciting data and regular communications, which he kept producing almost until the time of his death, popularized even more the use of such devices.[22] It was by using them that he established that microorganisms were in fact "small animals," and finally proceeded to describe the "animals" swimming in the semen. As we have seen, Leeuwenhoek's work is not only full of small creatures: it is also full of detailed calculations of the sizes of those creatures and estimates of their numbers per drop of liquid, expressed as fractions, subdecimal values, and equations. So the notion of smallness, its importance in life, and its role in fertilization mechanisms were all established in close parallel with the use of microscopic techniques.

Considering the dates of advances in microscopy and the lives of those involved in the preformationist debate, it is hard to imagine that one would not have influenced the other. Let us not forget that Harvey's *Exercitationes de generatione animalium,* with the famous statement that "all that is alive comes from the egg," was published in 1651. In the same year, the English anatomist Nathaniel Highmore published *The History of Generation,* containing the earliest microscopic images ever printed in England. Harvey and Highmore met after King Charles I lost the battle of Edgehill and fled to Oxford in 1642, bringing with him a number of royal physicians, including his first physician, William Harvey. Highmore's interests seem to have drawn him to the famous anatomist, and in the years that followed (from 1642 until Oxford fell to Parliament in June 1646), the two men developed a friendship that led Highmore to become a staunch proponent of Harvey's model of the circulation of the blood. During the same period, the two naturalists worked jointly and separately on the problem of generation, and it is often claimed that Harvey based many of his epigenetic ideas on Highmore's microscopic drawings—but, as we have seen in chapter 1, Highmore himself did not escape the attraction of the preformationist model, and hinted at it often enough in his book. Finally, thanks to the microscope, in 1673 Malpighi published the first detailed study on the fine anatomy of chick development, *De formatione pulli in ovo.* At this point, the combination of microscopy and reproductive theories had already produced a considerable body of work. Then Malpighi's data was appropriated by the ovists (see chapter 1), and the link with preformation tightened even further.

As we have seen in chapters 1 and 3, Charles Bonnet used the microscope avidly in his youth, and scrutinized with great zeal the activities of minuscule creatures for endless days and nights. After all this initial effort, he was practically blind by the time he was twenty. However, his mind continued to wonder at the world of smallness so recently discovered, and his long life was marked by the constant production of an ever-growing paean to the excitement of calculating sizes and applying numbers to the vision of life made possible by the new machines and the resulting new theories.

"The Sun is a million times bigger than the earth," he wrote in 1762, in the opening pages of his *Considérations sur les corps organisés,*

> and has at its edge a globule of light, of which several thousands enter the eye of the Animal twenty-seven millions of times smaller than a mite. But Reason sees even further. From this Globule of Light she sees the coming of another Universe, that has his Sun, his Planets, his Plants, his Animals, and, among those, an animalcule, which is to this world that I have just mentioned what we are to the world where we live.

Bonnet's final conclusions on organized bodies are also the final synthesis of the revelations of the microscope, the mandatory resort to geometric and numerical approaches, and their combination at the core of preformation:

> we should not suppose an infinite *Emboîtement,* which would be absurd. The divisibility of Matter to the infinite, through which we could support this type of *Emboîtement,* is a geometric truth and a physical error. Any Body is necessarily

finite; all of its Parts are necessarily determined: but this determination is unknown to us. We absolutely do not know what are the last terms of Matter's division; and it is this very ignorance that must prevent us from regarding the encasement of the Germs inside each other as impossible. We have only to open our eyes, and to walk around, to see that Matter is prodigiously divided . . . We have been shown that one ounce of gold can be quite sub-divided through human art to produce a string of 80 to 100 leagues of length: we have been shown in the Microscope Animals of which several thousands do not make up together the size of a small grain of sand . . . We discover in a bulb of Hyacinth his lineage up to the fourth Generation . . . Let us not judge Matter solely through the more or less close relationships that She has with our Body.

FIAT LUX

When microscopes revealed an invisible world, and prompted their users to undertake all kinds of Byzantine calculations to estimate the size or number of the creatures that inhabited it, they also provided a direct link with one of the most persistent questions in human thought, one that occupied much of Aristotle's energies and was endlessly reprised by his medieval Scholastic heirs: how small could matter be?

This question, obviously, entailed all kinds of related problems that established the content of numerous treatises. Do minimum parts have sizes? Are they the same as points? Does divisibility have a limit? Do all things have a maximum and a minimum? Whenever a mixture occurs, are the minimal parts generated all at once, or one after another? Summarizing it all in the second part of the fourteenth century, the Scholastic John of Jandun held that

> of any natural thing there is some quantity so small that its form cannot be preserved in a lesser quantity . . . The matter of water can be brought to such a quantity that the form of water cannot exist in it; similarly air by condensation can be so diminished that it is corrupted . . . When minimum parts are divided, they take the form of what surrounds them: if they are in the air, they take the form of air, and so on.[23]

Reaching forward to the future, this question was bound to be at the core of the preoccupations expressed by the earliest commentators on microscopy. As Sénébier wrote,

> It is probable that the Animalcules are the combinators of the Elements. Maybe even smaller Animalcules recover all the dispersed Elements resulting from putrefaction . . . I suppose there are as many microscopic Species as there are combinations to be performed: the first combinations would be performed by

those beings who combine solely two Elements, and in different proportions; other [Animalcules] would combine three Elements in a similar fashion, etc. They perform these combinations like the Plants and Animals that surround us, and they prepare these combinations for the Beings of the Visible World.[24]

Jandun's proposition was also clearly continuous with the Aristotelian posture. Drawing his conclusions mainly from biological observations, Aristotle had postulated that the smallest particles of matter should be "the minimum quantity of flesh [from which] no body can be separated out; for the flesh left would be less than the minimum of flesh." [25] And Aristotle himself was already writing mainly as a commentator on the ideas proposed by some of his most illustrious Greek forerunners; namely, those who, like Anaxagoras of Clazomenae, postulated that there was an infinity of substances within the primordial mixture, and that nothing here changed its qualities—everything is in everything, a theme that was later to be reprised with the aphorism attributed to Paracelsus, "the whole is hidden in the whole."

In setting the stage for the birth of preformation, this concept becomes even more relevant when we note that Anaxagoras held that there are infinite numbers of things, and that none of them is the smallest thing: each part of each thing still encompasses a portion of everything. As explained in a fragment from Simplicius:

> since the portions of the great and of the small are equal in number, so too all things would be in everything. Nor is it possible to say that they should exist apart, but all things have a portion of everything. Since it is not possible that there should be a smallest part . . . as things were originally so they must be now too, all together. In all things there are many ingredients, equal in number in the greater and in the smaller of things.[26]

This noble ancestry contributed more to the credibility of preformation than just the concept that the minutest thing can contain everything: it actually linked the idea directly to the problem of seeds and their offspring, thus introducing reproduction into the picture with the archetypical preformationist leitmotif: everything is in the seed as it shall be in the offspring. Still, according to Anaxagoras, at the microscopic level, the original mixture was not uniform. Matter, infinitely divisible though it is, was from the first coagulated into particles or "seeds," providing a natural unit from which cosmogony could begin. As this philosopher would ask, "how could hair come from what is not hair and flesh from what is not flesh?" [27] And his answer to this question is again totally akin to the future concept of preformation: flesh contains a portion of everything, including primordial particles of hair—it expresses itself as flesh only because it contains more particles of flesh than of all the other things hidden in the mixture.

Ever since Aristotle, this quest for the *minima naturalia* had appeared particularly important in cases in which new substances appeared and old ones disappeared. These problems applied to events such as corruption or chemical combination. And, obviously—once again to preformation's benefit—they also applied to generation. The Stoics in the century after Aristotle compared generation with yet another category of these phenomena: total fusion from which the constituents could not be regained, as the compounding of drugs. When the Scholastics appropriated these questions as one their most widely discussed philosophical subjects, the central dilemma of generation still occupied a fulcral point: do the forms of miscible elements remain in the mix, and if they do, do they undergo strengthening or weakening?

In the thirteenth century, the controversial English Franciscan and natural philosopher Roger Bacon tried to perfect this theory by positing degrees between potentiality and actuality through which forms pass on their way from the elementary state to their union in the form of the mix. In Bacon's opinion, matter could be divided endlessly, but from a certain lower limit on, it would lose its fundamental properties, since the form of the basic element could no longer inhere in the matter: fire, for instance, would always be fire—but at a certain degree of smallness it would abruptly lose its warmth.[28]

In sketching the background of preformation, we note that generational undertones infiltrate Bacon's postulates, holding firm behind his main conclusion, as presented by Norma Emerton: "since God had created matter and form together, each species must have its own specific matter, as well as its own specific form"; thus—and here the soundtrack of *emboitement* emerges subtly in the background—"there should be a hierarchy of matters and of forms, the higher absorbing or perhaps building on the lower." Within this scenario, the ultimate point sounds rather familiar, and it would certainly not displease Bonnet: "the final form itself, as well as the lower forms composing it, was in some sense present from the beginning, since it was the perfection toward which the lower forms were tending."

To make the parallel with later controversies on reproduction even more striking, some of the Scholastics ended up debating whether a final form could be achieved without a constructing, regulating, figure-giving element, and, if such an element existed, how it acted upon the whole mix—a problem that suggests all the troubles with "organizing principles" and "vital forces" that later constituted one of the main struggles of epigenesis (see chapter 4).

Further paving the way to the reasoning of the Scientific Revolution, geometric approaches soon joined the arithmetic concerns, filling the gap left by Aristotle with ideas derived from Plato. And it is highly arguable that, of all the disciplines of life, morphogenesis remains, to this day, in name and method, the most indebted to Platonic idealism—and, through it, to the merits and demerits of preformation. Its mathematical bent finds its most eloquent contemporary synthesis in the singular vision of D'Arcy Wentworth Thompson, who offers in his momentous *On Growth and Form* the following well-tempered appraisal:

> the celebrated doctrine of "preformation" implied on the one hand a clear recognition of what growth can do throughout the several stages of development, by hastening the increase in size of one part, hindering that of the other, changing their relative magnitude and positions, and so altering their forms; while on the other hand it betrayed a failure (inevitable in those days) to recognize the essential difference between these movements of masses and the molecular processes which precede and accompany them, and which are characteristic of another order of magnitude.

Another interesting reappraisal of ancient and medieval theories on smallness and form, contemporary with the triumph of preformation, appears in the writings of the Belgian "philosopher by fire," or chemical philosopher, Jan Baptista van Helmont (1579–1644), the man to whom we owe the coining of the word "gas."[29] Helmont's approach to the smallness of matter is truly fascinating as a contemporary idea of *emboitement* developed in the realm of chemistry—an analogy made even more impressive by the author's choice of words. Just as Haller calculated how many eggs were encased inside eggs in the ovary of Eve, so Helmont calculated how much matter was contained inside matter for any given corpuscle. And, just like Haller, he actually produced a final number.

According to the Helmontian concept, "particles" were the smallest constituent of matter, but within each particle was another smaller particle, over another yet smaller particle, and so on until one reached the center. This precise point would represent $\frac{1}{8200}$ of the size of the particle, and it was the location of the *"sperma,"* or "spark of light"—the ferment that Helmont presumed to be at the core of all corpuscles of matter, transmuting elemental water into the myriad substances that we encounter in the phenomenal world.[30]

A Hundred Thousand Angels

It becomes clear, then, that quantification did not emerge by spontaneous generation at the onset of the Enlightenment. Western culture had long displayed an insatiable appetite for it. The Pythagoreans are generally cited as the main agents in introducing the idea that numerical correlations are manifest everywhere in Nature,[31] embedding in the West the idea that all things are measurable through numbers. As Aristotle would later summarize it, not without hinting at a considerable amount of skepticism,

> the Pythagoreans . . . took up mathematics; they were the first to advance this study, and . . . they thought its principles were the principles of all things. Since of these principles numbers are by nature the first, and in numbers they thought they saw everything that exists and comes into being . . . ; since, again, they saw that the attributes and the ratios of the attunements were expressible in numbers; since, then, all other things in the whole of nature seemed to be modelled after numbers, and numbers seemed to be the first things in the whole of nature, they supposed the elements of numbers to be the elements of all things, and the whole heaven to be an attunement and a number.[32]

The continuity between the *furor mathematicus* of the Scientific Revolution and the transcendental powers attributed to numbers long before the seventeenth century certainly made the mathematical approach a comfortable path for the evolution of thinking. Magical numbers are manifold and spring from many different legacies. For the Western train of thought, one of those legacies was certainly the esoteric Hebrew literature produced from the first century A.D. onward, as Paul Johnson points out in *A History of the Jews:* since the Torah was holy, letters were holy—and so were numbers, since each letter had a numerical value. If one could find the key, secret knowledge could be obtained. Thus the fifth verse of Psalm 147, "Great is our Lord and of great power," was used as a lettered figure code for $236 \times 10{,}000{,}000$ celestial leagues to provide the basic measurements of the head and limbs of the Creator[33] and uncover some of His secret names. These names could function as passwords, allowing the celestial doorkeepers to admit the ascending soul into a fantastic series of eight palaces leading to Paradise. And 8 was used for a reason: it was a magical number taken from the Greek Gnostics. Another magical number was 22, the total number of letters in the Hebrew alphabet:[34] Creation itself had been enacted through combinations of Hebrew letters, and, when discovered, these codings revealed the secrets of the universe. The numerical interpretations of Hebrew numbers produced entities such as the Tetragrammaticon, $10^2 + 5^2 + 6^2 + 5^2 = 186$, one of the code names for God. These musings eventually became the core of the Jewish cabala, and, as we have seen in chapter 6, the driving forces behind efforts of creation ex nihilo, such as the Golems.

Keeping up with these numerological pursuits, the Scholastics of the Middle Ages trained their students in the application of numbers to the most esoteric problems. According to Jonathan Spence in *The Memory Palace of Matteo Ricci,*

> the central role of mathematics in the thinking of the Catholic church had been spelled out by Thomas Aquinas in the thirteenth century. He had seen it as an admirable early topic of study for the young because of its methodology of proceeding in a straightforward manner from a thing to its properties, so that it was the easiest and most certain of human sciences.

So, long before the Scientific Revolution made mathematics the ultimate state of the art, the church fathers had accumulated a long history of resorting to numbers to train the reasoning of their pupils.

One of the best-known examples of these *sophismata* is the popular question of how many angels could stand upon the head of a pin. This, argues Tobias Palmer in his 1995 *An Angel in My House,* is "one of the most popular questions asked about angels for centuries." In fact, angels (consistently drawn with wings and a halo from the fourth century onward) had represented a major focus of interest for medieval thinkers. In *A Book of Angels* (1994), Sophy Burnham lists some of the more pertinent questions raised on this subject: were the obscure "Sons of God" mentioned in Genesis (see chapter 3) simply angels? Did they, when mating with "the daughters of men," give rise to the race of the Nephilium? Are they born and do they die with the birth and death of stars?[35] Can two of them occupy the same space simultaneously? Does this mean that they are incorporeal? Can objects be incorporeal? When were they created? When the darkness was separated from light? When the oceans were separated from the earth? Thomas Aquinas, who in a single week held fifteen discussions about angels at the University of Paris, declared that they had been created when God created earth and sky.

Although Burnham argues that the famous issue of the head of the pin may very well be nothing but a myth, and modern scholars have been unable to find references to it in original medieval sources, it is often pointed out today as symbolizing the debate between a *finite* versus an *infinite* number of angels, which in turn symbolizes the intellectual ability, already present in the Middle Ages, to deal with the notion of infinite quantities, or at least with the assumption that those quantities could actually exist.

Palmer, for one, argues that the head of the pin has a real historical value. And, according to the same author, there is actually an answer for it—interestingly enough, an answer that carries subtle undertones of the visions of encasement brought forth by the doctrines of preformation. "One angel—one angel alone—can stand over the head of a pin," writes Palmer. "If an angel is already in this place and another angel comes, then the two angels will fuse to become one and the same. This is one of the favorite games of those beings, who can slide inside each other over the head of a pin in numbers that reach more than ten thousand. What is interesting is that, as the number of angels grows inside the shape of the single angel, the intensity of its beauty grows in the same proportion. One of the most beautiful visions that exists in the world is indeed the intensity of a hundred thousand angels united—over the head of a pin—in one single burst of light, amazing and overwhelming."

5,490 WOUNDS

The fascination with angels certainly did not fade under the inquiring light of the Enlightenment, since leading preformationists like Albrecht von Haller kept insisting that man's lack of appreciation for nature had prompted God to create beings that could have communion with Himself, beings to whom He could reveal all His glory—the angels, who were nearer to God in their innate impulses and thus essentially good.[36] And, long before the days of Haller, other sacred figures were made even more sacred by attaching precise numbers to the momentous events in their lives. Ludolfus of Saxonia, a monastic Carthusian and fourteenth-century devotional writer, who was to greatly influence the sixteenth-century founder of the Jesuits, Ignatius of Loyola, described the martyrdom of Christ in powerful terms meant to lead his readers to feel their Savior's pains as they read about them: "After all the nerves and veins had been strained, and the bones and joints dislocated by the violent extension, he was fastened to the cross. His hands and feet were rudely pierced and wounded by coarse, heavy nails that injured skin and flesh, nerves and veins, and also the ligaments of the bones."[37]

This detailed description of Christ's sufferings, however, could not be enough to inspire its readers. To obtain a maximal effect from his passage, Ludolfus felt compelled to tell us exactly how many wounds the Son of God received in his brutal fastening to the cross. There were, precisely, 5,490 wounds. Ludolfus even had a source to quote for the accuracy of this number: the devout Bridget of Sweden.

The magical spell of numbers is also obvious in several Renaissance medical treatises. The sixteenth-century Portuguese doctor Amato Lusitano explained that in cases of acute diseases, the worst episodes should occur on the seventh or fourteenth days, just as the best diagnoses could be obtained on the eighth and fifteenth days. In doing so, Amato drew upon a whole range of illustrious predecessors, from Galen to the fifteenth-century Italian philosopher Giovanni Pico della Mirandola (founder of the Christian cabala), to back this instruction:

The number 7 is dissonant with the number 1, just as the number 14, which is composed of 7 plus 7, so that, from these disagreements and dissonances, there grows a great struggle between disease and nature, and, as a consequence, there occur on these days extremely strong crises. And more. As in the musical instruments the eighth is harmonious and unisonic with its beginning (and so is the fifteenth) [38] as we can see from the first series, so also on the 8th and 15th days should judgments be made for perfect and complete health.

In fact, on the 7th and 14th days one sees a crisis of agitation and evacuation, but on the 8th or 15th day of the same crisis we see health emerging, triumphantly so according to the numbers . . . The number 7 is, therefore, the day of critical days. On this day we see all the convulsion promised at the onset of the disease . . . If on any other day a worsening of the disease occurs, it is because that day tastes to Nature as the 7th day, having some sort of link and connection with it . . . If the crisis occurs on day 3, it is because it tastes to Nature as half of 7. Now the days 14 and 20 should not be considered as even, because the 2nd week starts on the 8th day and these are 7 days plus seven days—14 and 20 are sequestered odd numbers. [39]

FIVE HUNDRED IDEOGRAMS — BACKWARDS!

In the Western heritage that brought about the Scientific Revolution, numbers were not only objects of philosophy, visions of Heaven, or keys to magical realms. They could also be useful tools for organizing one's brain or astonishing one's audience—or, preferably, combining both of these aspects in all their splendor. The Italian Jesuit Matteo Ricci took this intellectual baggage with him to China at the end of the sixteenth century, logically (in his mind) resorting to the teaching of mnemonic skills to train the memory as a means to impress the Chinese, so that they would let his congregation preach Christianity in this vast, religiously unconquered land.

As Spence notes, if Ricci expected to impress the Chinese through his mastery of memory skills, it was because he himself, like the other learned people of his time, must have been deeply impressed by these techniques. [40] Ricci must have made good use of the mnemonic devices implemented from the Middle Ages to the Renaissance, because he wrote casually in 1595 of running through a list of four to five hundred random Chinese ideograms and then repeating the list in reverse order—and these numbers certainly did impress the Chinese, who spread tales of Ricci's ability to recite volumes of the Chinese classics after scanning them only once. This was only natural to Ricci: back in Europe, his peers admired those men who had trained their minds so skillfully in the art of mnemonics that they could roam in seconds through hundreds of thousands of memory images, each one fixed in a particular place in their

minds. Mathematical knowledge fitted in well with the basic premises of mnemonic theory, since the mind found in the harmonious order of mathematics
something particularly easy to remember—as easy, for example, as the arrangement of Euclid's geometric propositions. And, in any event, God knows
what appealed the most to the Shakespearean audiences of commoners: all
those beautiful rhymes, all those twisted plots—or the magical ability of the
actors to memorize all those endless monologues?

To further stress the importance of numerical miracles achieved through
memory, it should be emphasized that other mnemonic systems, divergent
from the conventional rhetorical school, had appeared at different times in
Europe—and often based their function on geometric or magical aspects. The
memory system of the thirteenth-century Catalan philosopher Ramón Llull,
for instance, incorporated both algebra and geometry. As Frances Yates explains in her 1966 *The Art of Memory,*

> The figures of [Llull's] Art, on which its concepts are set out on the letter nota-
> tion, are not static but revolving. One of the figures consists of concentric circles,
> marked with the letter notations standing for the concepts, and when these
> wheels revolve, combinations of the concept are obtained. In another revolving
> figure, triangles within a circle pick up related concepts . . . Think of the great
> medieval encyclopedic schemes, with all knowledge arranged in static parts, made
> yet more static in the classical art by the memory buildings stocked with images.
> And then think of Llullism, with its algebraic notations . . . The first art is the
> more artistic, but the second is the more scientific.[41]

As for the magical element, the work of Giordano Bruno (see chapter 7)
is certainly the finest example from the sixteenth century, for, as Frances
Yates also explains in *The Art of Memory,* Bruno's art of mastering thousands
of memory megabytes was in fact permeated by magic.[42] As we now know,
magic did not save Bruno from the stake. But the magic of memory continued to dazzle Europe, and we did not put this spell to rest until we invented
computers.

The magical effect of astonishing numbers was also so strongly present in
Ricci's instincts that, even when he was functioning as a diplomat rather than
a churchman, his mind was still perfectly clear on the appeal of numerology.
In 1596 he wrote to the general of his order, Claudio Acquaviva, about the
crowds of Chinese literati who were now visiting his home, and listed the three
precise motives that brought them there: their eagerness to study the mnemonic system, their conviction that the Jesuits could turn mercury into pure
silver, and their desire to study Western mathematics. As Spence points out,

the list is completely believable in the context of the European intellectual and religious life of Ricci's time, when memory systems were combined with numerological skills and the arcane semiscientific world of alchemy to give the adept a power over his fate that mirrored the power of conventional religion.

As for the Chinese interest in discovering Western mathematics, Ricci tried to stimulate it even further by filling the picture with numbers and subdivisions. The great river of mathematics, as he explained to the Chinese, had four main branches: arithmetic, geometry, music, and astronomy-plus-chronology. "These four main branches," Ricci added, "subdivide into a hundred streams."

> One measures the magnitude of the universe . . . and also the heights of mountains and hills, lofty buildings of all kinds, the depth of pits and valleys, the mutual distance of two places . . . Another subdivision computes the sun's rays, so as to explain the sequence of the seasons, the varying length of day and night, and the hours of sunrise and sunset, and thus deduce latitude and longitude; the exact moment that years, months, and days commence; the equinoxes and solstices . . . Another subdivision constructs instruments such as spheres for observing the heavens and earth . . . for regulating the eight classes of musical instruments, and marking the passage of time . . . Again another subdivision regulates the arts which work in water, earth, wood and stone; builds cities, erects towers, terraces and palaces.

ALL YOU CAN EAT

All of the digressions into different realms of numerology described above shed light on one aspect of numbers directly connected to their long-presumed magical powers: during most of the making of our civilization, numbers were not yet circumscribed by the tight arithmetic limits in which we have enclosed them in our days. They were called everything from "poetic entities" to "the visible envelopes of beings," "the highest degrees of knowledge," or "the keys to all mysteries," and were the subject of such solemn pronouncements as "God is in all of us as the numbers are in the unity." The *Dictionnaire des symboles* states that "since, in the traditional mentality, there in no operation by chance, the number of things or facts acquires in itself an extreme importance and even allows us, in certain cases, to achieve a true understanding of things and events. Each number has its own personality, leading to the appearance of numerical interpretation as one of the most ancient symbological sciences."[43] In his 1982 "Science and Religion in the Thought of Nicolas Malebranche," Michael E. Hobart makes this point very efficiently in a brief and incisive passage:

The concept of number as we know it did not achieve explicit formulation until the late nineteenth and early twentieth centuries. In this period investigations resulting in the 'arithmetization of mathematics' demonstrated that, in order to understand mathematics in theory, one needed to grasp the principles of mathematical operations in their most elementary expression—arithmetic, the fundamentals of which presupposed and rested upon the idea of number.

So the preformationists were blessed once more. They could engage even more freely in dazzling calculations of generations and regressions of sizes—and therefore keep finely in tune with the quantifying spirit of their times—since they were dealing with numbers in a much more fluid, imprecise state than what current concepts allow. In the secondary revisions, none of this worked to their honor or benefit. But, at the beginning, all the right elements were in place to provide the preformationists with exciting materials to revel in. Reproduction. Microscopes. Mathematical infinities developing during a period when the time for life on earth was still supposed to be finite, according to Scripture. A brand new world. A feast of forms. A feast of numbers. A feast of possibilities. All in the context of the Scientific Revolution. You have to recognize an embarrassment of riches when you see one.

EPILOGUE

The Fat Lady Will Not Sing

We can only see what we have already seen

FERNANDO PESSOA

So, is it over? Have we heard the fat lady sing yet?

Predictably, we have not.

A century after the combined pursuits of Haller, Bonnet, and Spallanzani expired in a dead end, and after Caspar Wolff and the joint efforts of Blumenbach and Immanuel Kant failed to produce an epigenetic model free of invisible forces, we find the German biologist Oscar Hertwig (1849–1922), the man who first described fertilization and gamete nuclear fusion in the sea urchin, writing "What is development? Does it imply preformation or epigenesis? This perplexing question of biology has reappeared recently as a problem of the day." This passage dates back to 1894.

In the preface to the modern reprint of Hertwig's book (1977), Joseph A. Mazzeo states that this work is "a fine example of the last version of the battle between upholders of preformation . . . and of epigenesis." The book's title could not have been clearer as a statement of its central theme: *The Biological Problem of To-day: Preformation or Epigenesis? The Basis of a Theory of Organic Development.* But Mazzeo failed to anticipate that Hertwig's turn would by no means be "the last version" of the "battle." The battle is far from over. As with any other fundamental debate, the basic concepts involved never die. They just acquire modern incarnations and find their way back to the core of the crucial philosophical perplexities they represent. Nothing lost, nothing gained—only transformed. Two centuries after his beheading, Lavoisier would have been pleased.

It might be appropriate at this point to bring in some elements of modern pop culture to evaluate the present ramifications of the old controversy. Consider, if you will, the megahit movie *Jurassic Park.*[1] The central "scientific" elements of the plot, leading to the assumption that live dinosaurs could be cloned from fossilized dinosaur DNA, rest on two main misconceptions, at least as monumental as the millions of dollars amassed by the enterprise. The first misconception would have been the delight of any spermist. The makers of *Jurassic Park* seem to think that isolated DNA can generate a new organism, but they could not have been more wrong. It takes a true sperm cell to generate any animal.

Here is the problem: Isolated DNA is totally inactive and ineffective for giving rise to a new organism. In order to function properly as a nuclear component, DNA has to be organized into *chromatin:* it has to be coiled around basic proteins called *histones,* and this structure then has to be coiled again onto itself, until it reaches a structural level at which it is ready to receive and transmit instructions to the surrounding environment. This is true of any somatic cell. It is even more true of sperm cells, which become specialized during their maturation into swimming machines ready to deliver their genetic contents to the egg at fertilization. The DNA in the sperm head has only half the chromosome complement present in the somatic cells of the organism; the other half comes from the egg, restoring the original chromosome number. Moreover, sperm DNA is even more tightly coiled than other DNA, so much so that it dispenses with large proteins like histones and substitutes much smaller basic units, the *protamines.*[2] And DNA organization is not all that matters: unlike other cells, the sperm is practically devoid of cytoplasm. Also, when it enters the egg, it brings with it a fundamental piece of machinery, the *centrosome* (see chapter 5), which will organize the newly fertilized *zygote* to ensure correct cell division.[3] Scientists who try to clone vertebrates all over the world well know the aches and pains of trying to imitate a fertilization system without these contributions from the sperm. For once in his tormented life, Hartsoeker would have been pleased.

And Hartsoeker's opponents would have smiled, too. One of the most exasperating problems in modern cloning is how to artificially imitate the activation stimulus that the sperm gives to the egg when it binds to the egg's cell membrane at the beginning of fertilization. Modern laboratories have tried everything: heat, mechanical stimulation with needles, exposure to acid media, aging, electrical pulses that release repeated influxes of extracellular calcium to the egg's cytoplasm, which in turn releases more calcium from inner storage compartments. We know that sperm binding causes this effect, which awakens the egg from its previous dormant state and allows for all the events that follow. We try to imitate the sperm's action as best as we can, with more and more sophisticated instruments, amassing more and more basic knowledge. But the egg's sleeping beauty insists on being stubbornly exclusive in choosing the sperm as her prince: no matter what we try to do, the sperm does it better. After decades of research, our success rate in vertebrate cloning remains distressingly low. And we definitely do not know why. Something about the sperm-derived activation mechanism keeps escaping us. The means by which the sperm wakes up the egg are as elusive as the old *aura seminalis.* Swammerdam would have been pleased.

The second misconception in *Jurassic Park* is even more serious. Even if the park's scientists could have invented a way of transforming their isolated DNA into a functional sperm, this sperm still would have nowhere to go if it were not injected into a viable egg, with its chromosome component also reduced to half and arrested at the right stage of maturation to undergo successful fertilization. Yet *Jurassic Park* has plainly removed any eggs from the initial picture. Of course, even if the filmmakers were aware of this irrevocable condition for development, they would have been facing a tough call: how do you construct a believable scenario in which dinosaur eggs remain alive and viable over millions of years? To this day, with all the technology now available to us, no one has been able to successfully preserve living eggs.[4] So the whole rationale of the movie is made unsustainable by an old postulate: *Ex ovo omnia.* No way around it. Harvey would have been pleased.

But the problem goes far beyond pleasing a group of dead European white male naturalists with our modern shortcomings. Maybe the reader has the answer ready for the lack of dinosaur eggs: the creators of *Jurassic Park* could just have injected their newly created sperm into any other reptilian egg, could they not? And then—finally!—all those majestic brontosauruses, those nail-toting velociraptors, that ever so cruel tyrannosaur, could have hatched and roamed their island. Alas, this is not possible, either. Reproduction rests upon yet another irrevocable premise: to obtain a given species, the animal has to hatch from an egg of that same species, and that species alone. During the maturation period, eggs store in their cytoplasm a host of messages sent by the nucleus that are crucial in the organization of embryonic development. In most mammalian species, for instance, maternal (i.e., egg-derived) control over development ceases only after at least four cycles of cell division, when the embryo finally starts synthesizing its own RNA and becomes responsible for its own fate.[5] This pattern is even more impressive in numerous invertebrates, in which the different cytoplasmic components are specifically apportioned to the two cells resulting from the first cell division, thus determining the organogenic fate of all the cell lineages derived from them. This principle can be summarized in a very short sentence: the destiny of the embryo is already inscribed in the egg before fertilization.

Does this ring a bell? Certainly. Like any current debate related to reproduction and heredity, this brief analysis of the main flaws in the plot of a simple movie rings the old bell of preformation. Of course, preformation was bound by the limits of its own time. Of course, we all know now that DNA replicates,

The Fat Lady Will Not Sing

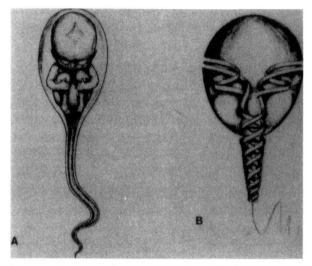

Fig. 51. The homunculus, then and now: Cartoon by embryologist Lauri Saxén, first published in 1973. (From Gilbert, "Commentary: Cytoplasmic Action in Development," 1991; reprinted from *Quarterly Review of Biology*.)

rather than unfolding from pre-existing DNA. But the final preformationist idea of the blueprint of the new organism contained inside the gamete was not all that far removed from reality, after all. Even modern cartoons have paid homage to this analogy (which is arguably merely coincidental, but fascinating nevertheless), depicting, as in the case of a drawing by embryologist Lauri Saxén, a double helix of DNA coiled inside the sperm head in the manner of Hartsoeker's famous drawing of the putative little man who should have been there. Besides, we have to admit that Kant's and Blumenbach's last conciliatory concept (see Prologue), in which epigenesis is directed by a set of preprogrammed instructions, is not, in its essence, all that far removed from our current views in developmental biology. And our views are still very much open to constant scrutiny and revision. That is why the fat lady cannot sing yet. Perhaps she never will. The plot is still unfolding. For all we know, it could unfold forever. As in the quote from Borges at the opening of chapter 8, this game is infinite. And even more so because Pessoa is devastatingly right about the limitations of our own minds: as much as we try to beat the odds as we strive toward knowledge, we can only see what we have already seen.

THE PREFORMED GERMPLASM

We should now rewind our story back to Oscar Hertwig's book, since it represents one of the first modern reprises of the preformation/epigenesis controversy. The roots of this work can be traced back to the period between 1875 and 1890, when biologists' conception of the cell was undergoing a radical transformation. Thanks to light microscopy and specific dyes, the nucleus had become a prominent cytological entity—and, inside the nucleus, the darkly staining chromosomes began to assume the chief role in providing a visible natural link between the cell and evolution. The physical continuity of the chromosomes throughout the cell cycle, the constancy of their numbers, the accuracy of their movements, and the longitudinal splitting of the chromatin threads, along with the fusion of male and female pronuclei in fertilization, all combined to give these structures an exceptional position in cytology.

Oscar Hertwig, together with researchers such as E. Strasburger and Hugo de Vries, was among the first scientists to sharply distinguish the nucleus from the cytoplasm of the cell. So was the man whom Hertwig would confront, his fellow German biologist August Weissmann (1834–1914). Hertwig considered himself an epigeneticist;[6] he wrote *The Biological Problem of To-day* mainly as a response to Weissmann's "germplasm" theory, which Hertwig perceived as a new version of preformation, since it stressed once again the notion of development proceeding by some sort of unfolding from a primordial complexity into a similarly complex end product.

Weissmann had made a pioneering point in drawing a radical distinction between germ cells and somatic cells, but stretched this distinction so far as to claim that germ cells were derived directly from germ cells of the previous generation and were not a product of the organism.[7] Although he revised his model many times, brilliantly inferred that the germplasm was halved in the germ cell, and correctly concluded that fertilization restored the total quantity of germplasm to the fertilized egg, Weissmann consistently pushed this distinction to the extreme of claiming that the germplasm had no relation at all to the organism surrounding it: the hereditary material was parceled out in the process of development, and would gradually disintegrate as an entity during embryonic growth.

Weissmann's final scheme was beautifully elaborate. The mass of germplasm, the starting point of a new individual, consisted of several pieces named *ids,* each containing all the possibilities—generic, specific, individual—of the new organism. In a vision that would have pleased Paracelsus, Weissmann thought of each id as a veritable microcosm, possessing a historical architecture that had been slowly elaborated during the multitudinous series of generations stretching backward in time from every living individual. This microcosm was in turn composed of minor vital units called *determinants,* arranged according to an orderly plan—and, in this specific account of morphogenesis, he was supported by another brilliant embryologist, Wilhelm Roux. There was a determinant for every part of the adult organism, and, inside each determinant—for these are the joys of *emboitement*—even smaller particles existed, the *biophores,* capable of passing into the cytoplasm of the cells, where they directed their vital activities.

The similarity to preformation was further stressed by the way Weissmann assumed that the germplasm would transit from generation to generation. In its first division, the germplasm would split into two parts, each of equal mass. One of the two portions would subsequently increase in bulk again, and would be transmitted unaltered to the next two cells. Eventually, this unchanged part of the germplasm would be marshaled in those organs of the adult from which new organisms were to arise. Therefore, the germplasm was continuously handed from generation to generation, forming an unbroken chain through each individual from grandparent to grandchild: herein lies the secret of the *immortality of the germ cells,* undoubtedly the part of the theory that had such a strong hold on the popular imagination. The second initial portion of the germplasm would meet a totally distinct fate, since its microcosms did not double at each division: the whole structure would slowly disappear as the determinants were distributed to the different parts of the body. This differentiating process occurred in an order determined by the historical architecture of the microcosm, so that the proper determinants were liberated at the proper time to modulate the formation of the tissues and organs. So, once again, development would be a slow unfolding of little elements already present in the germplasm. As Weissmann summarized it, "I believe that I have established that ontogeny can be explained only by evolution [preformation] and not by epigenesis."[8]

Hertwig, on the other hand, held that it was the chromatin in the cell nucleus that contained the material of heredity, which was exactly the same in every cell of a given organism. According to this scenario, development was epigenetic since, during its course, the hereditary material (which, by definition, was uniform and equally distributed to all cells of the organism), was differentially active. This type of activity implied some kind of interaction between chromatin and cell, as well as between each cell and its neighboring cells; and here the motif of epigenesis becomes even stronger, since this model assumes the impressing of different characters on identical material by the action of different surrounding forms.[9] After stating that preformation and epigenesis had long been the core of the debate on reproduction, Hertwig did a sort of patronizing justice to his predecessors by adding,

> That most of the great biologists of the seventeenth and eighteenth centuries were decided upholders of preformation was the natural result of the contemporary

knowledge of the facts. [Since] they knew only the external signs of the process of development . . . their mental picture of the germ or beginning of an organism was an exceedingly reduced image of the organism, an image requiring for its development nothing but nutrition and growth." He then stressed the continuity between the preformation of old and Weissmann's new proposals, noting that "thus, in our own days, after the controversy has been at rest for long, biologists are assembled in opposing groups, one under the standard of epigenesis, another under that of preformation.

Hertwig's last sentence leaves us with no hope that the battle will end anytime soon. Epigenesis may be an extremely sensible position to take, but preformation just will not go away. In his 1977 introduction to *The Biological Problem of To-day,* Mazzeo sided mainly with Hertwig, but in the end had to combine preformation with epigenesis to produce a more accurate description of our present state of knowledge:

It is clear that cell theory greatly fostered an epigenetic theory of development . . . The progress of cytology gradually uncovered more and more structures within the cell itself, especially the chromosomes which became visible during cell division and which were eventually identified as the carriers of the material substrate of heredity . . . The "materialization" of the gene, its location on chromosomes, the mapping of genes rested on the correlation of experimental genetics with cytology . . . In our own time, the progress of molecular biology has finally elucidated that structure of the gene, a one-dimensional segment of DNA, which can both duplicate itself and serve as a "template" for intermediary substances, messenger RNA and transfer RNA, which build the three-dimensional structure of the protein, and the organism is understood as the "translation" of the information contained in the gene. The gene is, thus, a "message," and the truth of preformation is that what is "preformed" is the information for making an organism.

ALL THE HOT TOPICS

Thus Hertwig and Weissmann died, but preformation lived on, and the debate continued. And it continued with a subtle use of derisive terms (just as subtle as the use of "homunculus"), clearly showing how strong concepts can persist through continuous transmutations. Perhaps embryology texts treated preformationists so badly because this old theory was a precursor of the genetics school of developmental biology. The battle between embryologists and geneticists has often been downplayed, but it consumed much of the energy in the field of reproduction during a large part of the twentieth century.[10]

This battle included twists of faith by the leading authorities as spectacular as Haller's changes from spermist to epigeneticist and then to ovist. Before

1911, the American geneticist Thomas Hunt Morgan (1866–1949) argued that interacting cytoplasmic fluids were responsible for the sexual phenotype of the offspring.[11] He argued strongly against those who, like Theodor Boveri, Nettie M. Stevens, and Edmund Beecher Wilson, held that this phenotype was determined by the chromosomes—asserting, in Wilson's words, that "the nucleus cannot operate without a cytoplasmic field in which its peculiar powers may come into play, but this field is created and molded by itself. Both are necessary to *development;* the nucleus alone suffices for the *inheritance* of specific possibilities of development." [12]

Another faction, supported by such scientists as Huxley and De Beer, held that the phenotypic fate of the offspring was determined by *morphogenetic fields,* a concept that, at first, appeared to be directly opposed to the gene theory.[13] For years, "fields" and "genes" wrestled to become *the* unit in development—and, now reborn as a geneticist, Morgan constantly squashed the idea of morphogenetic fields, with the support of his *Drosophila* school.[14]

Meanwhile, other leading experimental biologists (including such figures as E. G. Conklin, J. W. Jenkinson, F. R. Lillie, Albert Brachet, and Jacques Loeb) went so far as to claim that Mendelian genetics governed only those characteristics that did not exceed the framework of the species. In effect, they asserted that chromosomal genes determined trivial characteristics, claiming that the cytoplasm determined the fundamental constitution of plants and animals.[15]

In the end, the embryologists were, by and large, the losers in the battle against genetics. With time, the genetic approach to embryology (and evolution, and, for that matter, everything else under the biological sun) became the triumphant version. And the final successor of preformation, molecular biology, is currently threatening to take over the entire field of developmental biology. In the face of these disarming facts, there was, from the beginning, very little that classic embryologists could do—except call the geneticists names. So they called them preformationists. After all, preformation had lost the previous round. At least one could find some comfort in that memory.

This label was never completely erased from genetics—but, in another example of poetic justice, it was now applied to those who seemed to be the victors. Reaching forward to our own days, the modern incarnations of preformation pop up in any aspect of developmental biology that somehow touches the core of all the hot topics where our deepest perplexities and anxieties are crystallized—not to mention our more or less explicit social or scientific agendas. Think Genome Project. Think Bell Curve. This is where preformation holds the stage today.

But again, the battle cannot be over. I for one, as fond as I am of my beloved preformation of old, would definitely not like to be thrown into the dubious plot of any bell curve adventure. And, regardless of our own preferences, we still navigate troubled waters where each answer ultimately raises even more questions—and also uncovers a host of social dilemmas. Negotiations are still taking place between genetics and embryology at the boundaries of development. Newly found systems such as the *wnt* and *TRK* pathways certainly show the "truth" of epigenetic interactions on a more than descriptive level. But no one doubts that these molecules are genetically determined and placed in the appropriate cells through interactions of enhancers and transcription factors encoded by the "preformed" genome. Kant and Blumenbach had already foreseen something like this, but we are still unpacking Wilson's 1925 notion that ontogenesis is a cytoplasmic epigenesis underlain by a nuclear preformationism. As the example of why real dinosaur eggs must be used to make *Jurassic Park* possible clearly illustrates, we now know that the cytoplasm has just as much to say as the nucleus in the developmental fate of the organism. Slowly, we are coming to terms with this combinatory model.

However, some authors emphasize the preformed aspects, while others put their emphasis on the epigenetic side of the coin. These varying emphases often express a loaded social agenda, as becomes painfully clear whenever debates expand past the strict boundaries of basic science and reach social themes such as education. Why did J. P. Rushton's *Race, Evolution and Behavior* and R. J. Herrnstein and C. M. Murray's *The Bell Curve* both appear in 1994? Because the old debate is rearing its problematic head again, riding a wave that never ceases to crest in one form or another: these authors side with those who defend the *genetic* view of ontogenesis—generally conservatives who emphasize that our mental limits are established at fertilization, and that no program such as Head Start or ABC can change things substantially. On the other hand, the proponents of the *epigenetic* view emphasize the plasticity of the human brain and the fact that learning actually can cause new neuronal connections to form. Nobody said that reproduction was an easy matter during the Scientific Revolution, but the subject certainly has not become any easier today—especially since society now seems to expect final answers from developmental biology, just as it earlier looked for solutions in natural philosophy.

A THANK YOU

Our story cannot be put to rest, but we can at least meditate on the main contributions that preformation offered to its time, and hence to our collective future. Even if we pass on all the details, one thing we cannot deny: if so many torments and demons had not been juggled during the debate on reproduction, the Scientific Revolution would certainly have taken much longer to shed

the axiom of no difference between living and nonliving things, thus ruling out Descartes's postulate of no difference between a dog and a watch. Reproduction became the ultimate proof of the ultimate impossibility of strictly mechanical thought, and the efforts of all of those who dedicated their lives to the mysteries of eggs and sperm were the wheel behind the wheel. We do not know where we are going next. But we owe our thanks to the preformationists for having gotten us this far, this fast.

NOTES

Preface

1. Because NO, we DO NOT speak Spanish—no Spanish word would have an *e* immediately followed by an *i,* as in Corre*i*a!

2. If, until now, you were surprised by so much talking about myself, rest assured that this is the end of it. I too initially felt uncomfortable when my reviewers, my editor, and my intellectual hero alike kept urging me to "tell the reader more about yourself in the preface." The first person, I had always assumed, is the ultimate avatar of the essay; and essays, according to the canons I had learned, are allowed only to those who have already·established their ground as solid intellectual players within their culture. This indeed happened to me in Portugal, but Portugal is eight hundred years old and does not surpass the size of Indiana—with ten million people living in it. It is easy to become *anything* in such a crowded, we-all-know-each-other-and-each-other's-great-grand-parents-too kind of place. But I came to realize that this book, in spite of my sincere academic intentions, is in fact nothing but a larger-than-life essay—a long night's journey into day, the mirror image of Eugene O'Neill's unbearably beautiful tale of human anxiety. So there you have it. It was just as unexpected for me as it might have been for you.

3. For a detailed discussion of this dichotomy, see Peter J. Bowler's "Preformation and Pre-existence in the Seventeenth Century: A Brief Analysis," *Journal of the History of Biology* 4 (1971): 221–45.

4. Even though I cannot possibly forgive his never having bothered to read Nicholas Mosley's *Hopeful Monsters,* the one novel I truly wish I had written.

Prologue

1. As quoted by Shirley Ann Roe in *Matter, Life, and Generation: 18th Century Embryology and the Haller-Wolff Debate* (Cambridge: Cambridge University Press, 1981).

Chapter 1

1. As A. Schrierbeek points out in his *Jan Swammerdam, 1637–1680, His Life and Works* (Amsterdam: Swets and Zeitlinger, 1967), this vivid story, though oft-recounted, may not be totally accurate. Nonetheless, it is undeniable that Swammerdam frequented, most often as a silent guest, the salons of his patron Thévenot, and that he

made a strong impression with his occasional displays of different stages of insect development.

2. See Pierre Ducassé, *Malebranche, sa vie, son oeuvre, avec un exposé de sa philosophie* (Paris: Presses Universitaires de France, 1942).

3. This anecdote has been contested by several scholars, but it is still a clear illustration of the effect produced by the encounter between the young priest and the ideas of Descartes. For a detailed account of the development of Malebranche's scientific interests, see Ducassé's *Malebranche.*

4. For a detailed account of the relationship between the two philosophers, see *Malebranche et Leibniz, relations personelles,* by André Robinet (Paris: Librairie Philosophique J. Vrin, 1955).

5. In *La Mechanique des animaux;* cited by Shirley Ann Roe in *Matter, Life, and Generation: 18th Century Embryology and the Haller-Wolff Debate* (Cambridge: Cambridge University Press, 1981).

6. Cited by J. Langman and D. Wilson, "Embryology and Congenital Malformations of the Female Genital Tract," in *Pathology of the Female Genital Tract,* 2d ed., ed. A. Blaustein (New York: Springer-Verlag, 1982).

7. Mentioned by Arthur William Meyer in *The Rise of Embryology* (Stanford, Calif.: Stanford University Press, 1939).

8. There is an interesting subplot in this last statement by Leibniz: Notice that the names of Malebranche and Swammerdam, whom we would now think of as strict defenders of ovism, are mentioned side by side with the names of Leeuwenhoek and Hartsoeker, whom we would now regard as intransigent spermists. This casual grouping suggests that what we now consider a radical division between two camps may have been much less heated when the initial ideas on preformation first started to emerge. As a matter of fact, things were not necessarily clear at all for the authors of the Scientific Revolution who wrote about reproduction. Consider the following sentence from the book *Essai sur la manière de perfectionner l'espèce humaine,* published in 1751 by the French doctor M. Vandermonde: "Fabricius d'Aquapendente, Aldrovandi, and finally William Harvey, pretended that all animals came from an egg, and that the first eggs created contained all those which existed and shall exist in the Universe." When we think of Fabricius and Harvey as the main pioneers of modern epigenesis, and assume that epigenesis struggled in a head-to-head confrontation with the preformationist idea that "the first eggs created contained all those which existed and shall exist in the Universe," we can image all these authors spinning in their graves from seeing their contending ideas amalgamated into a single model!

9. See William A. Locy's 1935 *Biology and Its Makers* (New York: Henry Holt, 1935) for an excellent digest of Malpighi's life and work. For more extensive information, see Howard Adelmann's works, especially *Marcello Malpighi and the Evolution of Embryology.* (Ithaca, N.Y.: Cornell University Press, 1966).

10. Cited by Meyer in *The Rise of Embryology.*

11. If Swammerdam's condemnation of epigenesis seems deeply heartfelt and meant to be lethal, this did not stop him from writing (in *The Book of Nature, or, the Natural*

History of Insects [London: n.p., 1738] that the growth of the limbs proceeds very slowly, inside the larva's skin, "by an epigenesis, or accretion of the parts"; or to sound plainly epigenetic in passages such as "insects proceed from this Nymph, as the flower from the husk, and are also rendered fit for generation and reposition of their sperm. And as propagation is performed in plants, by the union of their seed with the moisture of the earth's womb, insects perform the act of generation by the conjunction of the fruitful, and as it were invisible particles of the male's sperm, with the conspicuous, vivifick and sensitive seed in the female. This seed of the female continues and perfects the life, motion and sense which it enjoys, when the spermatic virtue of the male is thrown into it; and it is in this continuance of motion that the fruitful conception of the seed is properly said to consist." These apparent contradictions offer an excellent example of how loose and interchangeable terms and ideas were at the onset of the generational debate in the seventeenth century.

12. This was a concept widely endorsed in Swammerdam's time, judging from M. Eliade's study in *The Forge and the Crucible* (trans. Stephen Corrin [Chicago: University of Chicago Press, 1956]), in which we find numerous references to embryological analogies wherein metals develop into gold in the womb of *Tellus Mater.*

13. This book contains an extensive account of Swammerdam's observations prior to his death in 1680, but was published posthumously.

14. Swammerdam's constant preoccupation with drawing parallels between human life and his observations in other animals are beautifully expressed in the following passage from *The Book of Nature,* concerning the difference between the male and the female nocturnal butterflies:

> The male is always provided with wings, whereas the female never has any. So that the male can enjoy the sweet refreshments which the free air affords, and ramble at pleasure over the smiling fields and fragrant flowers, when, on the other hand, the care at home, and management of the fruits of wedlock, are committed to the female only; for which reason she is always found with the hinder part of her body thrust out, in order, as it were, to induce the male to do its duty; nor does the male seem indisposed to perpetuate his species. Nature, therefore, intended to afford us in these insects the most striking examples of an affectionate mother, and a careful father.

15. Swammerdam's choice of specimens was not without reason. Most of the insects included in his studies are *holometabolous,* meaning that they undergo a dramatic and sudden transformation from the larval to the adult stage. Their larvae are covered by hard cuticles, and they must shed these cuticles and produce new, larger ones in order to grow. The process of embryonic development is thus very clear and nicely synchronized, culminating in a metamorphosis in which the larva becomes a *pupa,* which is no longer able to eat and derives its energy from the food ingested during the larval stages. Of all the different kinds of insect metamorphosis, this was certainly the one most fit to make sense in the eyes of a seventeenth-century naturalist.

16. Again, note that although Swammerdam tended to consider the egg (and thus the female body) the source of all lives, in this passage he seems to believe that males

carry other males in their loins—yet another reminder of how tentative the explanations offered by preformation were at this point.

17. Jan Jacobzoon Swammerdam, the father, an apothecary in Amsterdam, was also extremely interested in the wonders of nature and spent fifty years building up a magnificent collection of animals, insects, plants, and fossils, carefully arranged by order and place, from all the quarters of the world. This collection, completed with fine porcelains from India and Japan, became a famous private museum and a well-known tourist attraction, which its owner refused to sell at any price.

18. According to Hermann Boerhaave's introduction to *Biblia mundi,* even this decision was strongly influenced by spiritual factors. "His father intended him for the church," wrote Boerhaave, "and with this in view engaged a tutor to ground him in to understand the Holy Scriptures better; however, our author, after deep and earnest consideration of his disposition and talents, reached the conclusion that he was unequal to so great a responsibility. Of this he succeeded in convincing his father, who then consented to his son applying himself to medicine."

19. Antoinette Bourignon (1616–1680) was venerated uncritically by a circle of friends who called her (as she called herself) "the light of the word." In August 1673, she wrote to Swammerdam claiming that his former pursuits were only *"amusements de Satan":* if he really wanted to see the light, he was to abandon all of them and follow only Jesus Christ. Swammerdam obliged, and in 1675, with her permission, published *Ephemeri vita,* which contained the natural history of the mayfly amid numerous religious meditations and poems, including an extension of Psalm 139. However, this immensely rich woman, who liked to proclaim that she would rather throw all of her riches into the sea than give them to the poor, since we should all live free of materialistic appetites, did not seem to be able to fulfill Swammerdam's spiritual needs, and he left her ranks even sadder and more somber than before.

20. For a complete account of Swammerdam's life, see Schrierbeek's *Jan Swammerdam;* see also Locy's *Biology and Its Makers.*

21. A comprehensive and accessible description of imaginal disk molecular biology can be found in Scott F. Gilbert's *Developmental Biology,* 4th ed. (Sunderland, Mass.: Sinauer Associates, 1994,) For more details, both historical and scientific, see also Charles Knight, *Insect Transformations* (London: The Library of Entertaining Knowledge, 1830); H. Ursprung and R. Nothiger's *The Biology of Imaginal Disks* (Berlin: Springer-Verlag, 1972); Frederick W. Stehr's *Immature Insects* (Dubuque, Ia.: Kendall/Hunt Publishing, 1987); and Thomas Eisner and Edward O. Wilson's *The Insects* (San Francisco: W. H. Freeman, 1977).

22. In India this incubation was said to be performed by a goose called Hamsa, which represents the breath of the Divine Spirit.

23. This dichotomy is visible in the Hindu Brahmanda, which forms two spheres, one of silver and another of gold; in the egg shed by Leda after her intercourse with the swan, which gives birth to two separate hemispheres; and in the familiar black-and-white symbol of the polarization of primordial unity expressed in the Chinese yin-yang that has been the motif of so many earrings, necklaces, and other modern Western

ornaments. The Arabian philosopher Ibn al-Walid came even closer to the Stoic model borrowed by Paracelsus in describing the earth as the dense, coagulated yolk of the egg, whereas the white becomes the sky. So did the Chinese philosopher Chang Heng (78–139 A.D.) when he wrote: "Heaven is like an egg, and the earth is like the yolk of the egg."

24. Cited by Meyer in *The Rise of Embryology.*

25. As discussed by Joseph Needham in *A History of Embryology,* 1934.

26. The "heretic king" Akhnaton, who dared to replace the old worship of Amon-Ra with the cult of Aton, the sun-god, wrote a hymn to the sun ("Creator of the germ in woman/Maker of the seed in man/Giving life to the son in the body of the mother") with verses concerning the animation of the embryo inside the egg: "When the fledgling in the egg chirps in the shell/Thou givest him breath therein to preserve him alive/When thou hast brought him together/To the point of bursting out of the egg/He cometh forth from the egg/To chirp with all his might." (Cited by Will Durant in *Our Oriental Heritage,* vol. 1 of *The Story of Civilization* [New York: Simon & Schuster, 1963].)

27. Cited by Needham in *A History of Embryology.*

28. Needham, *A History of Embryology.*

29. Needham, *A History of Embryology.*

30. Aristotle's description of the observed development was largely plagiarized four centuries later in Pliny's *Natural History,* in the following manner:

> All eggs have within them in the midst of the yolk, a certain drop, as it were of blood, which some think to be the heart of the chicken, imagining that, to be the first that in every body is formed and made; and certainly a man shall see it within the very egg to pant and leap. As for the chick, it takes the corporal substance, and the body of it is made of the white waterish liquor in the egg, the yellow yolk serves for nourishment; while the chick is unhatched and within the egg, the head is bigger than all of the body besides . . . As the chick within grows bigger, the white turns to the middle, and is inclosed within the yolk. By the 20 day (if the egg be stirred) you shall hear the chick to peep inside the very shell, and from that time forward it begins to plume and gather feathers; and in this manner it lies within the shell, the head resting on the right foot, and the same foot under the right wing, and so the yolk by little and little decreases and fails. (Pliny, *Natural History,* translated by H. Rackham, edited by G. P. Goold. Cambridge, Mass.: Harvard University Press, 1983.)

31. Aristotle called infertile eggs "wind-eggs," a term retained by William Harvey in his treatise on epigenesis. In our days, farmers sometimes still use this term for unfertilized eggs.

32. Cited by Needham in *A History of Embryology.*

33. Mentioned by F. J. Cole in *Early Theories of Sexual Generation* (Oxford: Clarendon Press, 1930).

34. Cole, *Early Theories of Sexual Generation.*

35. Cole, *Early Theories of Sexual Generation.*

36. Schrierbeek, *Jan Swammerdam.* See this work for more details on the fight with de Graaf.

37. Swammerdam, *The Book of Nature.*

38. In the introduction to Haller's compilation of poems translated to French under the title *Poésies de Mr. Haller* (Bern, Switzerland: Chez Abr. Wagner Fils, 1760).

39. Otto Sontag calls it a "prickly personality" (in his introduction to *The Correspondence between Albrecht von Haller and Charles Bonnet* [Bern, Switzerland: Hans Huber Publishers, 1983]).

40. Cited in Margarete Hochdoerfer's *The Conflict between the Religious and the Scientific Views of Albrecht von Haller* (Lincoln, Nebr.: University of Nebraska Studies in Language, Literature, and Criticism, no. 12, 1932.) For further details on Haller's moral posture, see also Otto Sontag's "The Motivations of the Scientist: The Self-Image of Albrecht von Haller," *Isis* 65 (1974): 336–51.

41. Haller's book *Elementa physiologiae corporis humani,* the latest volume of which was published in 1765, was the first comprehensive textbook of physiology, bringing together endless scattered facts and presenting them as a self-contained whole, thus helping to make physiology an independent branch of science rather than a subset of medicine. Moreover, he made a crucial contribution to the shaping of modern physiology with his theory of Irritability and Sensibility, postulating that irritability was connected to the muscles and sensitivity to the nerves. For a stimulating discussion on the impact of Haller's contributions see Otto Sontag's "Albrecht von Haller and the Future of Science," *Journal of the History of Ideas* 35 (1974): 313–224; Otto Sontag's "The Idea of Natural Science in the Thought of Albrecht von Haller" (master's thesis, New York University, 1971; and *The Natural Philosophy of Albrecht von Haller,* edited by Shirley Ann Roe (New York: Arno Press, 1981).

42. The Portuguese poet Barbosa do Bocage wrote—in capricious, perfectly metric and carefully rhymed verses—a warm note of thanks to Haller for having shown us that mysterious art of the German language in which "beauty in all its glory can dispense with rime."

43. Consider, for instance, the following passage from "The Morning," which starts with a mellow description of the beauty in the fields at sunrise.

O Creator! All that I see is the work of your power! You are the Soul of Nature; the course and the light of the astral bodies, the mighty fire of the sun, are the work of your hand and carry your seal.

You light up the torch of the Moon to give us its glow: You give wings to the wind, and to the night You give the dew that it sheds on us: You rule over the course and the rest of the planets.

From clay and dust, you have made the mountains; from the sand you made the metals. You have expanded the Firmament and you have covered it with clouds like a tapestry. You have formed the veins of these fishes that jump in the rivers, and that excite somersaults beating their tails. From the mud you have built the Elephant, and you have animated its enormous mass, like a living hill.

You closed the shiny veils of the sky over the void, and by one single word You brought out of the nothingness this vast Universe, amazed at its own immensity.

Great God! The created spirits are too small to recount the glory of your works; they are endless, and to talk about them it is necessary to be infinite like you.

I am still bound by my limits, O Ununderstandable Being! Your glory blinds my weak eyes, and the One that created the very sky does not need the praise of a worm.

44. See Sontag, introduction to *The Correspondence between Albrecht von Haller and Charles Bonnet.*

45. *Letters from Baron Haller to His Daughter, on the Truths of the Christian Religion* (Albany, N.Y.: H. C. Southwick, 1816). These letters were published posthumously with Haller's permission, "for the benefit of the world at large."

46. See the introduction to *The Correspondence between Albrecht von Haller and Charles Bonnet.*

47. Cited by Roe in *Matter, Life, and Generation.*

48. Roe, *Matter, Life, and Generation.*

49. Roe, *Matter, Life, and Generation.*

50. Cited by Lorin Anderson in *Charles Bonnet and the Order of the Known* (Boston: D. Reidel, 1982).

51. Cited by Anderson in *Charles Bonnet and the Order of the Known.*

52. In a letter to Spallanzani dated November 1780, Bonnet expressed a clear pre-occupation with making the distinction between preformed bodies and pre-existing parts, insisting that "much has been said of the *emboîtement* of germs; the term is improper; germs are not little boxes enclosed one within another; they must have been integral parts of the first organized bodies that came from the hand of the Creator. I have insisted on this point . . . It is of consequence to fix the meaning of terms precisely."

53. For an excellent description of Bonnet's theory of the Chain of Being, see Stephen Jay Gould's *Ontogeny and Phylogeny* (Cambridge, Mass.: Belknap Press of Harvard University Press, 1977); see also Anderson's *Charles Bonnet and the Order of the Known* for a detailed account of Bonnet's main sources of inspiration.

54. In *Nouvelles recherches sur les découvertes microscopiques, et la génération des corps organisés* (Paris: Lacombe, 1769).

55. See the chapter dedicated to Spallanzani in James J. Walsh's *Catholic Churchmen in Science* (New York: Books for Libraries Press, 1917). These two women rarely got from the secondary sources the attention and recognition that they undoubtedly deserve. Laura Bassi does not even rate an entry in the monumental *Dictionary of Scientific Biography,* edited by Charles Coulston Gilliespy, first published in 1970 and reprinted through eight more editions after that! Londa Schiebinger, however, does her justice in *The Mind Has No Sex?: Women in the Origins of Modern Science* (Cambridge, Mass.: Harvard University Press, 1989): here, Laura Bassi is granted three full pages and a portrait. Of course, this book is a piece of intellectual archaeology on all the women scientists we never heard about—and *why* we never heard about them. It is hard to claim that the fact that Laura Bassi is fully described only here is nothing but a coincidence. For an illuminating account of Laura Bassi's career, see also Paula Findlen's "Science as a Carrier in Enlightenment Italy: The Strategies of Laura Bassi," *Isis* 84 (1993): 441–69.

Similarly, Marianna Spallanzani, the sister who helped the "Magnifico" so much, is consistently ignored by the literature. She is one of those *Little-Known Sisters of Well-Known Men,* as runs the title of an American book by Sarah Gertrude Pomeroy (Boston: D. Estes, 1912), which, incidentally, does not contain her name; nor, for that matter, the name of Carolina Herschel, who a little later was to be of so much assistance in his work to her brother, the famous astronomer Herschel. As mentioned in a footnote in *Catholic Churchmen in Science*:

> Even Dr. Mozans, in his book *Women in Science,* always so thorough, and usually so exhaustive, has missed Marianna Spallanzani's story. The arrangement of the cabinet of Natural History which came to be the focus of the scientific attention of Europe in Pavia was largely in her hands. Spallanzani often confessed that she knew more about it than he did. During his absence, distinguished visitors were taken through the cabinet by Marianna, and, as one of Spallanzani's biographers (Sénébier) says naively enough, "she knew the properties of the specimens contained in it, and was capable of reasoning upon them." He adds, moreover, the secret of her successful cultivation of natural science; for her mind was molded upon that very illustrious brother, whom it was pleasure to her to study and imitate.

56. Mentioned in Paul De Kruif's *Microbe Hunters* (New York: Harcourt, Brace, 1926).

57. For a detailed discussion of Spallanzani's contribution to the establishment of modern research methods, see Jean Rostand's *Les origines de la biologie experimentale et l'abbé Spallanzani* (Paris: Fasquelle, 1951).

58. Later it would be discovered that what was then called the "head" was actually a defective body part, lacking the essential components of the central nervous system. Nevertheless, these regenerated "heads" had eyes, mouths, tongues, teeth, and most of the sensory organs of the animal.

59. For a detailed—and highly entertaining, even if not rendered in orthodox academic terms—account of the Needham/Spallanzani controversy, see De Kruif's *Microbe Hunters.*

Chapter 2

1. In Nicolas Hartsoeker's *Cours de Physique* (La Haye: Jean Swart, 1730).

2. In "Spermatozoan Biology from Leeuwenhoek to Spallanzani," *Journal of the History of Biology* 6, no. 1 (1973): 37–68.

3. In *The Collected Letters of Antoni van Leeuwenhoek* (Amsterdam: Swets & Zeitlinger, 1952)

4. King Charles II, who was greatly interested in the work of the Royal Society.

5. The notes to *The Collected Letters of Antoni van Leeuwenhoek* (Amsterdam: Swets & Zeitlinger, 1952) suggest a cabinet of curiosities or an anatomical theater.

6. According to the notes appended to the English and Dutch translations of his letters, the person in question must have been Dionisius van der Sterre, a convinced ovist.

7. Leeuwenhoek later abandoned this idea of separate genders in the spermatozoa.

8. In *On the Origin of Species,* first mentioned in the 6th edition (1872).

9. This argument must have been very bitter indeed, judging from the sour tone used in most references to it in our days. Consider, for example, the following passage from Castellani's "Spermatozoan Biology from Leeuwenhoek to Spallanzani," written in 1973:

> Methodical research performed through such repeated observations as those of Leeuwenhoek was completely nonexistent throughout this period . . . In no way could the term "scientific observations" apply to those made by Hartsoeker, who went so far as to question the priority of Leeuwenhoek's discovery and who did serious harm to the animalculist cause with his fancy tales about having seen a miniature embryo inside a spermatozoon.

Or this passage from Arthur William Meyer's *The Rise of Embryology* (Stanford, Calif.: Stanford University Press, 1939):

> Although Hartsoeker's calculations helped to end the doctrine of preformation, I have not been able to find adequate evidence that his writings brought other than ridicule to the animalculists whose cause he had espoused. This probably was because he had made false claims regarding the discovery of spermatozoa and had actually represented a miniature homunculus with a disproportionately large head inside a spermatozoon.

10. There is one further chapter in the book, but it is not an exposition on the different kinds of worms: it is an interesting collection of "Aphorisms concerning Worms in Human Bodies."

11. For more details on this problem, see Castellani's "Spermatozoan Biology from Leeuwenhoek to Spallanzani."

12. Mentioned in Paul Johnson's *A History of Christianity,* 1976.

13. Mentioned in William R. Newman's "The Homunculus and His Forebears: Wonders of Art and Nature" (unpublished manuscript).

14. Leeuwenhoek's assumption, fully endorsed by all the other spermists, was strictly intuitive, since empirical knowledge of fertilization was nonexistent at that time, and remained so until Oscar Hertwig described it in the sea urchin in 1875. This firm belief, held in the absence of tangible proof and before a background not bound to be sympathetic to the waste of lives implied by the model, is a good indication of how dominant the preformationist's views had become at this time: if epigenesis had been more widely accepted, it could have provided for a number of scenarios absolutely free of any belief in putative massacres of millions of potential lives.

15. Calvin's hermeneutics, for instance, led him clearly to favor the second possibility. "To retreat on purpose from the woman so that the seed falls on the ground is a double monstrosity," wrote the Swiss reformist in 1554, in his "Commentary on the First Book of Moses, called the Genesis," "because it reduces the hope for further lineage and it kills the infant . . . before he is born" (*Calvin's Old Testament Commentaries: The Rutherford House Translation,* Grand Rapids, Mich.: Rutherford House, 1993).

Modern theologians also consider the possibility of Onan's sin being the violation of levirate marriage (and the failure to propagate his brother's line).

16. "Make no mistake: no fornicator or idolater, none who are guilty of adultery or homosexual perversion, no thieves or grabbers or drunkards or slanderers or swindlers, will possess the kingdom of God" (VI:9–10).

17. Mentioned in *L'histoire d'une grande peur: La masturbation,* by Jean Stengers and Anne Van Neck (Brussels, Belgium: Éditions de l'Université de Bruxelles, 1984).

18. Details on Medieval penances for masturbation can be found in *Sexual Variance in Society and History* by Vern L. Bullough (New York: Wiley, 1976); and in *Le Sexe et l'Occident. Évolution des attitudes et des comportements,* by Jean Louis Flandrin (Paris: Seuil, 1981).

19. Mentioned by Stengers and Van Neck in *L'histoire d'une grande peur.*

20. Mentioned by Stengers and Van Neck in *L'histoire d'une grande peur.*

21. In *Dissertatio in sextum Decalogi praeceptum,* by J. B. Bouvier (8th ed., Le Mans, France, 1836).

22. Benedicti's endorsement of Thomas Aquinas's idea of such nocturnal episodes being caused by the Devil is, however, the last claim of this sort found in the theological literature concerning masturbation. After him, nobody seems to have returned to this explanatory model.

23. Cited by André Robinet in *Malebranche et Leibniz, relations personelles—pre-sentées avec les textes complets des auteurs et de leurs correspondants revus, corrigés et inédits* (Paris: Librairie Philosophique J. Vrin, 1955).

24. According to F. J. Cole's *Early Theories of Sexual Generation* (Oxford: Claren-don Press, 1930), the eight-volume treatise *Dissertation sur le génération de l'homme, ou l'on rapporte des diverses opinions des modernes sur ce sujet, avec des réflections nou-velles et plusieirs faits singuliers,* published in Paris by the ovist P. Dionis in 1698, dis-missed the claims of "the Panspermatistes" that the seeds of life had floated in the air or lain scattered over the ground since the beginning of the world, waiting, *according to Hartsoeker,* to be taken up by living organisms through their food or their breath.

25. In *Practijke Criminele; naar het eenig bekend handschrift,* by Philips Wielant (Gent, Flanders, 1872).

26. According to the *Dictionnaire de Théologie catolique,* the original "De instruc-tione sacerdotume et peccatis capitalibus" was first printed in 1599, three years after Toledo's death.

27. According to the *Dictionnaire de Théologie catolique,* the original "Ad expellen-dum semen corruptum ac noxium, sanitatis causa" was first published in 1608, the year of Rebellus's death.

28. Caramuel was an extremely interesting and colorful character, more interested in Plato than in Aristotle and a fervent admirer of Ramón Llull (ca. 1235–1315), whose ideals of encyclopedic knowledge and the unity of science captured his interest by the time he was seventeen. After studying in Alcalá and Salamaca, he spent some time in Portugal, following his fascination with the developments of mathematics and oriental languages available in this country, then annexed by Spain; he was also among the first

Spaniards to foresee the upcoming Portuguese re-emancipation. Like Galileo and Van Langren, he was determined to find a sound way of determining longitude through the observation of the moon or some other celestial body. The son of an officer, in 1635 he put the knowledge on firearms acquired from his father to use as the chief engineer in the constructions destined to protect Louvain against attacks by the French and the Dutch. His biographers describe him as the first Scholastic to discover Descartes and exercise the Methodic Doubt in his theological pursuits. His texts provide for very stimulating reading. More details on his life and his writings can be found in Antonio Cestaro's *Juan Caramuel Vescovo di Satriano e Campagna, 1657–1673: Cultura e vita religiosa nella seconda meta del Seicento* (Salerno, Italy: Edisud, 1992); in Julian Velarde Lombrana's *Juan Caramuel: Vida y obra* (Oviedo, Spain: Pentalfa Ediciones, 1989); and in Dino Pastine's *Juan Caramuel: Probabilismo ed enciclopedia* (Firenze, Italy: La Nuova Italia, 1975).

29. "Diogenes," wrote Galen in book VI of *De locis affectis,* "is said to have been the most firm of all men in those works that demanded continence and constancy; however, he used venereal pleasures, in order to rid himself of the discomfort produced by the retained sperm, and not to seek the pleasure brought about by its emission. They tell of him that once he asked a courtesan to come meet him, and, since she was late, he gave free course to the seed using his hand; when the woman arrived he sent her away saying 'my hand was faster than you celebrating Hymen.' It is absolutely evident that chaste men do not use venereal pleasures for their enjoyment, but solely to cure an ailment, as if in reality they experience no enjoyment at all." This passage has been interpreted by many as barely concealed praise of masturbation.

30. Mentioned in *The Evolution of Childhood,* by Lloyd DeMause (New York: Psychohistory Press, 1974).

31. Mentioned by Stengers and Van Neck in *L'histoire d'une grande peur.*

32. In *The Diary of Samuel Pepys,* vol. 8, ed. by Latham and Matthews (London: Bell & Hyman, 1667).

33. Lyonet was suspicious of preformation altogether, but he chose not to debate ovism, "which I leave to its own fate."

34. Lyonet was also perplexed by the phenomenal growth rate that this system would entail, noting that, in the bitch, the animalcules would have to attain five hundred million times their original size in ten days, and wondering why, if this really happened in the female, it would not rather take place within the semen itself, which was the natural habitat of the "vermiculi"—and most likely, we assume, cause the poor male to burst.

35. Mentioned by Cole in *Early Theories of Sexual Generation.*

36. Cole, *Early Theories of Sexual Generation.*

37. Astruc's description of reproduction also included an interesting description of the passage of the fertilizing sperm cell through different layers of the ovary, until it reached the egg; and the clear assertion that the embryo is indeed of paternal descent, but that the egg is crucial for pregnancy in forming the placenta.

38. In his *Early Theories of Sexual Generation,* Cole says that Nicolas Andry had also

offered an explanatory model for the combination of characters from both parents in the progeny, using the mule as an example: if you mate a male horse with a female ass, the juices that the animalcule of the horse encounters in the uterus of the ass, being adapted to produce a greater development of the ears, stimulate the animalcule to grow ears longer than those of a horse—but in any event smaller than those of an ass, since the horse's animalcule cannot undergo all the ear growth that would occur in a donkey. This description is ascribed to Andry's *An account of the Breeding of Worms in Human Bodies.* However, I was unable to find it in the original I consulted.

39. Mentioned by Cole in *Early Theories of Sexual Generation.*

40. English translation with notes by E. Kegel-Bringgreve and A. M. Luyendijk-Elshout in the volume *Boerhaave's Orations* (Leiden: E. J. Brill/Leiden University Press, 1983).

41. Cited by Cole in *Early Theories of Sexual Generation.*

42. For a developed account of Wolff's rationales, see Shirley Ann Roe's *Matter, Life, and Generation: Eighteenth-Century Embryology and the Haller-Wolff Debate* (Cambridge: Cambridge University Press, 1981).

43. See Cole, *Early Theories of Sexual Generation;* Joseph Needham, *A History of Embryology* (Cambridge: Cambridge University Press, 1934); and Castellani, "Spermatozoan Biology from Leeuwenhoek to Spallanzani."

44. "Onanism" was a neologism created by the public right after the publication of *Onania.* The author of the book, however, did not make use of the term.

45. The author most likely invented several of his "letters from readers" so that he could discourse further upon the health-damaging consequences of the vice; he also seized the occasion to suggest several useless remedies.

46. In Vernon A. Rosario II, "Phantastical Pollutions: The Public Threat of Private Vice in France," in *Solitary Pleasures: The Historical, Literary and Artistic Discourse of Auto-Eroticism,* ed. Paula Bennett and Vernon A. Rosario II (New York: Routledge, 1995).

47. The Cabinet of Dr. Bertrand had been famous in Paris since the beginning of the nineteenth century, and included an entire room dedicated to onanism. Here, the visitors could see "a young man reduced to agony and in the last degree of emaciation because of masturbation"; "a young man with an interesting figure, enjoying splendid health"; "the same, having become hideous due to masturbation"; "a very beautiful maiden, enjoying splendid health"; "the same, six months later, now very ugly and exhausted for having delivered herself to the solitary vices, of which she happily got rid thanks to marriage"; and many others. In their work, Stengers and Van Neck quote Doussin-Dubreuil, a commentator of the period, who judged that "this very curious cabinet produced over the masturbators who visited it a much stronger effect than anything one can write about masturbation."

48. Mentioned in *Leopold II of the Belgians: King of Colonialism,* by Barbara Emerson (London: Weinfeld and Nicolson, 1979).

49. All the passages of Rousseau's work used in this text are Vernon Rosario's translations of the French originals, as presented in "Phantastical Pollutions."

50. Tissot's *L'Onanisme* was banned in Paris; Rousseau's *Émile* was equally condemned by the Sorbonne, and by both the Parliament and the Archbishop of Paris as an immoral and dangerous text. It eventually ended up being burned in the streets of the city.

51. Mentioned by Rosario in "Phantastical Pollutions."

52. In the writings of the Enlightenment, women appear not only as the main carriers of onanistic guilt, but also as the main victims of the masturbation furor. Like many of his medical followers, including Dr. J. T. D. de Bienville, Tissot gave masturbation an even more sinister status by describing its specific effects on females, whose nervous systems he considered to be "weaker . . . and naturally more predisposed to spasms," leading to violent fits. The resulting symptoms included hysteria, vapors, incurable jaundice, clitoral scabbing, and a uterine fury that "depriving them of their modesty and their reason, reduces them to the level of the most lascivious brutes," ultimately subjugating their minds to their genitals. This reasoning came to be of great usefulness during the campaign to discredit the French queen Marie-Antoinette as a new sex- and power-starved Messalina, with pamphlets circulating such as "The Uterine Furies of Marie-Antoinette, Wife of Louis XVI," or "The Private, Libertine, and Scandalous Life of Marie-Antoinette," the latter featuring a drawing of the queen being masturbated by one of her ladies-in-waiting.

53. Mentioned by Rosario in "Phantastical Pollutions."

54. In this entry of the *Encyclopédie* (cited by Stengers and Van Neck in *L'histoire d'une grande peur*), Ménuret de Chambaud states that "The mind continually absorbed in voluptuous thoughts, constantly directs the animal spirits to the generative organs, which by repeated handling, become more mobile, more obedient to the unruliness of the imagination; the result is almost continual erections, frequent pollutions, and the excessive evacuation of seed."

55. Haller's letters to Tissot, printed in French with notes in German, including bits of Tissot's responses, were collected in the volume *Albrecht Hallers Briefe an Auguste Tissot* (Bern, Switzerland: Huber, 1977).

56. The English author of *The New Theory of Generation,* published in 1764, in yet one last attempt to resurrect the credibility of spermism.

57. Caspar Wolff, the man who rescued epigenesis in the second half of the eighteenth century and thus became Haller's main opponent in the generational debate.

58. Haller was very worried by the damage to human physiology caused by a life too full of leisure, as is clear in this passage from his poem "The Alps":

Away from vain and miserable occupations, away from the smoke of the cities, the tranquility of the soul lives in this place. The active life of its peoples increases the strength of their robust bodies; they do not fatten themselves with a void idleness, work makes them happy and content; the pleasure and the health soothe their sorrows. A pure blood runs in their veins; no poison, inherited from a vicious father, has sneaked inside; this blood is neither corrupted by sorrow, nor inflamed by foreign wines, nor spoiled by a lascivious venom, nor embittered by spicy foods.

59. Cited in Cole, *Early Theories of Sexual Generation.*

60. *Sur une découverte particuliere concernant la génération des grenouilles,* 1752; cited by Cole in *Early Theories of Sexual Generation.*

61. Mentioned by Cole in *Early Theories of Sexual Generation.*

Chapter 3

1. In *The Collected Letters of Antoni van Leeuwenhoek* (Amsterdam: Swets & Zeitlinger, 1952).

2. For a more complete discussion of Ham's real history, see F. J. Cole, *Early Theories of Sexual Generation* (Oxford: Clarendon Press, 1930); and Joseph Needham, *A History of Embryology* (Cambridge: Cambridge University Press, 1934). In *Investigations into Generation, 1651–1828* (Baltimore, Md.: John Hopkins University Press, 1967), Elizabeth B. Gasking gives yet another account of the role played by Ham:

> Leeuwenhoek recounted how a certain Mr. Ham had visited him, bringing a tube containing human semen. According to Mr. Ham, the specimen came from a man who had been with a woman suffering from gonorrhoea, and it contained living animals which were killed if the patient took turpentine. Leeuwenhoek, however, went on to state that he had himself seen thousands of such animalcules in the semen of healthy men.

3. Two contradictory accounts of Hartsoeker's claim can be found in Fontenelle's preface to Hartsoeker's *Cours de Physique* (La Haye: Jean Swart, 1730) and in Clifford Dobell's *Antoni van Leeuwenhoek and His Little Animals* (New York: Dover Publications, 1960).

4. Huygens's participation in the publication of Hartsoeker's findings is discussed in Cole's *Early Theories of Sexual Generation.*

5. These three different dates have been offered in both textbooks and specialized literature by several authors.

6. For precise names and dates, see Cole's *Early Theories of Sexual Generation* and Needham's *A History of Embryology.*

7. Reported by Thomas L. Hankins in *Science and the Enlightenment* (New York: Cambridge University Press, 1985). Swammerdam's book proposing this term, published after the author's death and without his revision, was the Latin version of *Historia Generalis.*

8. In *The embryological treatises of Hieronymus Fabricius of Aquapendente,* facsimile edition with introduction, translation and commentary by Howard B. Adelmann (Ithaca, N.Y.: Cornell University Press, 1942).

9. The idea that all Aristotelian postulates had to be questioned and eventually overturned by more mechanistic approaches—one of the main obsessions of the Scientific Revolution—had not yet come into fashion when Fabricius performed his examinations. Although Fabricius differs from Aristotle in many respects, appearing much closer to Galen in the theory of generation in viviparous animals, he totally endorsed the Aristotelian theory of generation for oviparous organisms. It is possible that a more complete and better balanced explanation of his general views on this subject was contained in his lost manuscript *De instrumentis seminis.*

10. Fabricius quotes Aristotle in his own footnotes to *De formatione ovi et pulli.* In his theories of generation, Aristotle had already introduced the idea of a "spiritual" agency presiding over the acquisition of shape by unformed matter destined to become the new organism. By reviewing these writings in his work, Fabricius organized Aristotle's rather confusing notes into a clear concept—the action of the "irradiant faculty" of the sperm over the entire uterus.

11. In Cole's *Early Theories of Sexual Generation.*

12. In Kenneth Keele, *William Harvey, the Man, the Physician, and the Scientist* (London: Thomas Nelson and Sons, 1965).

13. In Keele, *William Harvey.*

14. This option applies mainly to the ovists. Spermists, for obvious reasons, were highly suspicious of the real existence of any kind of *aura seminalis.* In his letters to the Royal Society, Leeuwenhoek railed vehemently against this idea.

Our Harvey absolutely denies having found male sperm in the matrix, cut up immediately after copulation; and your Dr. de Graaf . . . has boldly . . . established as a fact that the testicles of a woman are two egg-nests or ovaries . . . Now when the egg is nearly ripe and has reached the matrix through the tuba Fallopiana, the foetus is formed, the male sperm being nothing but the vehicle of a certain extremely volatile animal spirit, impressing on the conception, i.e. the ovum of the woman, the perception of life . . . I consider (the egg of the hen) to have no other purpose but to serve as food for the semen of the cock and to make a chicken of it. And because we know that anxiety, fear, fright and uneasiness will cause abortion, this will happen even sooner in the case of animals when they are tied and cut in the sensitive parts, for this will cause such a fright, that the sperm which has been conceived will not only be deprived of further nutrient but that nature in its pain and anxiety will try to ease itself and to expel the seed.

15. These models ended up being associated mainly with the explanation of monstrous births in the secondary literature, thus heaping an added layer of disrespect on the concept of "spermatic worms."

16. Rachel Fink's *A Dozen Eggs* (Sunderland, Mass.: Sinauer Associates, 1991).

17. Cited by William Norton Wheeler in his preface to Réaumur's *The Natural History of Ants* (New York, London: A. A. Knopf, 1926).

18. This problem is extensively discussed in Wheeler's introduction to *The Natural History of Ants.*

19. This work became the principal rival of Diderot's thirty-two-volume *Encyclopédie* (the most impressive such venture of the age).

20. In Buffon (Georges-Louis Leclerc, Comte de), "Discours sur les animaux," in the beginning of vol. 4 (1753) of *Histoire naturelle, générale et particuliere, avec la description du Cabinet du Roi* (Paris: Imprimerie Royale, puis Plassans, 1749–1804).

21. In Buffon, "Discours sur les animaux."

22. In Francesco Redi, *Experiments on the generation of insects,* 1688. Redi's observations and experiments were varied and numerous, often very simple and highly revealing. He showed that putrefying meat generated "worms" only if flies had been able to lay their eggs on it, and that those so-called worms were nothing but the larvae of

future flies, which, upon reaching their adult state, could be shown to have inner ducts containing hundreds of eggs. These results allowed Redi to postulate that all kinds of plants and animals arise solely from seeds of plants or animals of the same kind, thus preserving the differences among species.

23. In Linnaeus, "Tvakonad alstring (generatio ambigena)," 1759. I thank Prof. Lisbet Koerner for the information and translation from Swedish.

24. Cited by Cole in *Early Theories of Sexual Generation.*

25. A detailed account of these events is given by Wheeler in his introduction to *The Natural History of Ants.* In *Buffon* (New York: Twayne Publishers, 1972), Otis E. Fellows and Stephen F. Milliken also give an account of Buffon's frequent visits to the salons of Madame de Pompadour, pointing out that, even to this powerful patron, the count's ideas seemed at times so outrageous that she once gallantly slapped him in the face with her fan.

26. In Buffon, "La théorie de la Terre," vol. 1 of *Histoire naturelle.* According to Fellows and Milliken, the data sources used for this work came largely from England and northern Europe, predominantly Protestant states in which science was regarded chiefly as a useful adjunct to the old theology. Most of the few systematic observations involved had been made by men whose primary concern was "documenting" certain biblical accounts, like Noah's Flood. Buffon's interest in this subject was not laid to rest with the publication of his first effort. By the time "Époques de la Nature" came out, in 1778, the hypothetical primitive ocean from "La théorie de la Terre" had become "that universal sea": "The waters covered the entire surface of the globe to a height of two thousand fathoms above the level of our present seas; the earth was then under the empire of the sea." However, while reviving and modernizing the debate on cosmogony and earth sciences, Buffon was so aware of the risks he faced by challenging the credos of the establishment that he carefully concealed the evolution of his own thought throughout the period, almost half a century, that he devoted to writing the *Histoire naturelle,* constantly adding supplements and notes to the original, rather than altering the text—and often protesting that nothing fundamental had been changed. His need to steer clear of any charge of heresy is quite transparent in pious and contradictory passages such as

> Our authors [Whiston, Woodward, Burnet] have made vain efforts to account for the Deluge; their errors in physics with regard to the secondary causes that they employ prove the truth of the fact exactly as it is related in the Holy Scriptures, and demonstrate that it could have been effected only by the primary cause, the Will of God. (*Histoire naturelle,* vol. 1, p. 169)

> To say that the sea formerly covered the globe itself quite entirely, and that it is for that reason that one finds the shells of the sea everywhere, this is to fail to take note of a very essential point, which is the unity in point of time of the Creation. (*Histoire naturelle,* vol. 1, p. 199)

> All truth coming equally from God, there is no difference between the truths that He has revealed to us and those that He has permitted us to discover by our observations and our researches (*Histoire naturelle,* vol. 5, p. 35)

27. Mentioned in Celestino da Costa's *Elementos de Embriologia* (Coimbra, Portugal: Livraria J. Rodrigues e Companhia, 1933).

28. Mentioned in A. Ruppert Hall, *The Scientific Revolution, 1500–1800: The Formation of the Modern Scientific Attitude* (Boston: Beacon Press, 1966).

29. In Schalom Ben-Chorin's 1977 *Mutter Mirjam* (cited in Eugen Drewermann, *Dying we live: Meditations for Lent and Easter,* translated from the German by Linda M. Maloney and John Drury [Maryknoll, N.Y.: Orbis Books, 1994]), the author notes that this combination is immediately evident in the Gospel's accounts of how Jesus was conceived. When the Holy Spirit comes over Mary, the Scriptures say that it will "overshadow" her (Luke 1:35). "Overshadowing" seems to be derived straight from the morning of Creation (Gen. 1:2), when the Spirit of God "broods" over the waters of primeval chaos like a bird on its nest—the world egg. For Ben-Chorin, "that God is accustomed to have intercourse in the shape of a bird with a daughter of a man is well known from the myth of Leda and the Swan . . . When the text speaks of overshadowing, this suggests the image of outspread wings beneath whose shadow the chosen virgin takes refuge. It also ties in with the Hebrew expression, *bezel kenaphecha* (in the shadow of your wings), a phrase that recurs in Psalms 17, 36, 57 and 63 and in the liturgy of the synagogue." For our own convenience and instruction, the tale of this specific virgin birth seems to have been carefully arranged. Even the angel Gabriel, who brings Mary the message of the divine birth of the Savior, bears a name that spells out his mission: "My husband is *(gabri)* God *(el).*" According to Ben-Chorin, "it corresponds . . . to an old Hebrew tradition that claims the messenger (the angel) and the message are identical. For this reason, in the Talmudic way of thinking, each angel can deliver only *one* message, can complete only *one* mission. The messenger *is* the message, even and especially in this case, because Mary becomes pregnant by the Holy Spirit through the annunciation itself."

30. Drewermann taught theology and history of religions at the University of Paderborn until the Vatican called him a "new Luther" and expelled him from its ranks after the publication of his *Kleriker: Psychogramm eines Ideals* in 1989.

31. In Drewermann, *Dying we live.*

32. "His face shone like the sun, and His clothes became white as the light. And they saw Moses and Elijah appear, conversing with him."

33. This hypothesis gained added likelihood when Bonnet, Trembley, and Lyonet were able to demonstrate the existence of both male and female aphids, with Bonnet even noting that the smaller male was "perhaps one of the most ardent that there are in Nature. It appears to me that it does nothing except have intercourse as soon as the day arrives." However, Bonnet was also prompt in dismissing the idea of coupling in utero, writing to Trembley that "this conjecture does not agree with the state of the Aphids enclosed in the womb, where they are not only washed away in a fluid that does not permit them to unite, but also where they are enclosed by a membrane that keeps all their parts better bound up then those of chrysalides. (Cited in Virginia Dawson's *Nature's Enigma: The Problem of the Polyp in the Letters of Bonnet, Trembley, and Réaumur* [Philadelphia, Pa.: American Philosophical Society, 1987].)

34. This question was mainly raised by Abraham Trembley, who by then was studying the regeneration of the freshwater polyp, but in the fall of 1740, at Réaumur's demand, took some time to observe the generation of aphids himself—and, albeit witnessing the birth of several consecutive generations, still wondered if some original mating could have been the cause of them all. Writing about how Trembley's doubts made him push his experiments even further, Bonnet would afterward say, "if this excellent friend had been able to foresee all the evil that this 'who knows' did to my eyes, I am very sure that his tender friendship for me would not have permitted him to express it. It was, however, in this simple 'who knows' that I undertook a new study that was much more laborious than the preceding one. I was young and full of ardor: it seemed that these two words reduced to nothing all my previous work." (Cited by Dawson in *Nature's Enigma*.)

35. Excellent accounts of Bonnet's work on the reproduction of lice can be found in Jean Rostand's *Un grand biologiste, Charles Bonnet, expérimentateur et théoricien* (Paris: Université de Paris, 1966); in Lorin Anderson's *Charles Bonnet and the Order of the Known* (Boston: D. Reidel, 1982); and in Dawson's *Nature's Enigma.*

36. Mentioned by Rostand in *Un grand biologiste.*

37. Abbé Pierquin, *Dissertation physico-théologique* (Amsterdam, 1742); mentioned by Rostand in *Un grand biologiste.*

38. Denis Diderot (1713–1784) was educated by the Jesuits and later married the devout Catholic Anne Toinette Champion, but neither his upbringing nor his tormented, short-lived marriage had a positive effect on his respect for the teachings of Christianity, leading him instead to a militant life of rationalistic objections to supernatural revelations and caustic denunciations of the hypocrisy of the clerical authorities. He became famous as a philosopher, novelist, and essayist, and also as the creator (in collaboration with d'Alembert) of the monumental *Encyclopédie,* intended to collect under one roof all the active writers, all the new ideas, all the new knowledge, that were then stirring the cultivated class to its depths, but were still comparatively ineffectual by reason of their dispersion. The *Encyclopédie* is arguably Diderot's foremost legacy; however, his contemporaries knew him best for his vagabond lifestyle and his thundering appearances in the philosophical salons, where his tirades against clerics and religion delighted the audiences. Some of his writings were so incendiary (as in the case of *Les Bijoux indiscrets*) that the author himself eventually repented their publication.

39. In *Discours d'un philosophe sur un roi,* vol. 4 of *Oeuvres completes;* cited by Rostand in *Un grand biologiste.*

40. The version consulted here was the 1885 edition in *Collectanea Adamantea.*

41. Mentioned by Shirley Ann Roe in *Matter, Life, and Generation: Eighteenth-Century Embryology and the Haller-Wolff Debate* (Cambridge: Cambridge University Press, 1981).

42. Mentioned by Hall in *The Scientific Revolution.* In *Investigations into Generation,* Elizabeth B. Gasking gives the following account of Harvey's words on the subject: "on these points we shall say more when we show that many animals, especially insects, arise and propagate from elements and seeds so small as to be invisible (like atoms flying

in the air), scattered and dispersed here and there by the winds; yet these animals are supposed to have arisen spontaneously, or from decomposition, because their ova are nowhere to be found."

43. The original quote is from Pliny's *Natural History* VIII, translated by H. Rackham, edited by G. P. Goold (Cambridge, Mass.: Harvard University Press, 1983), p. 166.

44. In Jorge Luis Borges, *The Book of Imaginary Beasts* (New York: Dutton, 1969). Virgil's quote mentioned here by Borges, in the translation presented in *Lucina sine concubitu* and attributed by John Hill to his friend "Mr. Dryden," is as follows:

"The mares to cliffs of rugged rocks repair,

And, with wide nostrils, snuff the Western air:

When (wondrous to relate) the Parent Wind,

Without the stallion, propagates the kind."

45. Solveig Thorsteinsdottir, personal communication.

46. Another method of birth control, although more rarely acknowledged, was, obviously, infanticide. In this case, the decision was often dictated by men. However, a strange profession, assigned to women alone, flourished in Portugal for centuries, and is said to persist to this day in very remote areas. These women were called "Weavers of Angels." Their "official" job consisted of being paid by mothers to "take care" of their newborn children. Such "care," however, meant killing the babies—generally by suffocation or neck traction—and afterward reporting to the community that the baby had died of a "high fever," therefore sparing everyone from any share of guilt. At least according to oral tradition, the father was not involved in these transactions—unless he happened to be the priest of the village.

47. Cited by Will Durant in *Caesar and Christ,* vol. 3 of *The Story of Civilization* (New York: Simon & Schuster, 1971).

48. In Amato Lusitano, *Centúrias de Curas Medicinais,* Book I, translated by Firmino Crespo (Lisbon: Universidade Nova de Lisboa, 1980).

49. In Caroline A. F. Rhys Davids, *Buddhist Psychology: An Inquiry into the Analysis and the Theory of Mind in Pali Literature* (London: G. Bell and Sons, 1914) and Paul Deussen, *Fundamental Philosophy of Upanishads,* authorized English translation by A. S. Geden (Delhi: Kanti Publications, 1989).

50. It is noteworthy that our tales of spiritual penetration are not confined to the genital ducts. The alimentary canal is another universally functional site of entrance. When we take the Consecrated Host at Communion, aren't we, literally, supposed to be eating flesh and blood penetrated by a spirit? According to Robert Briffault (*The Mothers,* vol. 2 [1927; London: Allen & Unwin, 1959]), we are repeating what the Gallas of Abyssinia did long before us, every time they ate in solemn ceremony the fish that they worshiped—and said, as seriously as we say "amen," "we feel the spirit moving within us as we eat."

51. Detailed accounts of this particular mythology are given in W. G. Sumner and A. G. Keller, *The Science of Society,* vol. 3 (New Haven: Yale University Press; London: H. Milford, Oxford, 1927) and Julius Lippert, *The Evolution of Culture,* trans. and ed.

by George Peter Murdoch (New York: Macmillan, 1931). See also the opening chapter of Arthur William Meyer's *The Rise of Embryology,* "Aboriginal Ideas of Reproduction" (Stanford, Calif.: Stanford University Press, 1939), for an account of the different expressions of the same belief among several peoples.

52. In Will Durant, *Our Oriental Heritage,* vol. 1 of *The Story of Civilization* [New York: Simon & Schuster, 1963].)

53. In Durant, *Our Oriental Heritage.*

54. In E. Westermarck, *The Origin and Development of Moral Ideas* (London: Macmillan, 1924–1926).

55. In G. de Tarde, *The Laws of Imitation,* trans. from the second French edition by Elsie Clews Parson, with an introduction by Franklin H. Giddings (New York: H. Holt, 1903).

56. In Durant, *Our Oriental Heritage.*

57. In Briffault, *The Mothers,* vol. 3.

58. In Clifford Howard, *Sex Worship: An Exposition of the Phallic Origin of Religion* (Chicago: Chicago Medical Book Co., 1909).

59. In Havelock Ellis, *Man and Woman: A Study of Human Secondary Sexual Characters* (London: Walter Scott; New York: Scribner, 1900).

60. Cited by Georges Cattaui in "Paracelse et sa postérité," in *Paracelse: l'homme, le médecin, l'alchimiste,* by Béatrice Whiteside (Paris: Table Ronde, 1966).

61. In Eugen Drewermann, *Discovering the God Child Within: A Spiritual Psychology of the Infancy of Jesus,* translated from the German by Peter Heinegg (New York: Crossroad, 1994.

62. Cited in Drewermann, *Discovering the God Child Within.*

63. In A. Erman, "Die Religion der Agypter: Ihr Werden und Vergehen in vier Jahrtausenden," 1934; cited in Drewermann, *Discovering the God Child Within.*

64. Drewermann, again, makes a good point about this everlasting vision in *Discovering the God Child Within,* by simply quoting what he calls "three unfinished conversations."

In the first one, set many decades ago, a young Lebanese woman in charge of an anthropology museum in Beirut expresses her loss of Catholic faith as follows:

"Everything Christianity teaches is thousands of years older than itself. Have you seen the mother goddesses—Inanna, Cybelle, Isis? They all have a child die on them, or a mate, or the god they love. And the world holds its breath; they go down in the underworld and awaken the dead one. These are myths, images, dreams. Can we believe in dreams?"

In the second one, he recalls the words of an auto dealer from Merzin, in Turkey:

"The Old Testament was really new. It declared that there was only one God. All Muhammad had to do was to pick up on that idea. God doesn't beget children with a woman, says the Qur'an, and he doesn't make himself the son of a woman. Islam is the only religion that no longer needs myths. From the standpoint of cultural history, Christianity is still part of Asia Minor."

The third and final dialogue takes place with an Indian tourist, at the Indian Museum of Calcutta:

"I saw how in the morning in Benares the people went down to the Ganges to carry out the holy ablutions . . . Sometimes it seems to me as if even in Christianity, especially in Catholicism, we're still drinking from these ancient sources of religion, but as if they were flowing only in tiny trickles. When we enter a church, we sign ourselves with the water of rebirth and purification. But our so-called holy water stoups and baptismal fonts look like dried-up pools from the torrents that pour from the head of Shiva."

And then, pointing to a picture from the second century B.C.:

"Here you see the stages of the Buddha's life: Here's the scene where his mother Mamaya, wife of the Sakya prince Suddhodana, virginally conceives the Buddha after years of infertility . . . A white elephant approaches her, a symbol of the clouds raining down and fertilizing Mother Earth. And here is Mamaya . . . bringing the future redeemer into the world, delivering him from her side—in other words, virginally, even during birth."

65. In Cristina Castel-Branco, "O Lugar e o Significado—Os Jardins dos Vice-Reis," Ph.D. thesis, Lisbon, Portugal, 1993.

Chapter 4

1. This version is quoted in Cole's *Early Theories of Sexual Generation* (Oxford: Clarendon Press, 1930). The author offers K. F. Burdach's *Die Physiologie als Erfahrungswissenchschaft,* 1835–1837, as an example of how early this fairy tale was taken for granted as a historical reality.

2. In *An account of the Breeding of Worms in Human Bodies* (London: H. Rhodes and A. Bell, 1701).

3. The latest piece of evolutionary news reported on the Sperm-Net was the discovery by Scott Pitnick of the world's longest sperm cell, published in an issue of *Nature* from May 1995: the 5.8 cm long spermatozoa of the 3 mm long *Drosophila bifurca,* in which the testes account for 11% of the total body mass. Since we are no longer supposed to believe in nature's practical jokes, the implications of these findings promise a lot of cyberdebate. There is also a sperm home page available on the World Wide Web, but its contents are somewhat less inspiring. For more serious spermatology reports see A. F. Dixson's "Sexual Selection, Sperm Competition and the Evolution of Sperm length," *Folia Primatologica* (Basel) 61 (1993): 221–27; J. T. Manning and A. T. Chamberlain's "Sib Competition and Sperm Competitiveness: An Answer to 'Why So Many Sperms?' and the Recombination/Sperm Number Correlation," *Proceedings of the Royal Society London B. Biological Sciences* 256 (1994): 177–82; G. A. Parker and M. E. Begon's "Sperm Competition Games: Sperm Size and Number under Gametic Control," *Proceedings of the Royal Society London B. Biological Sciences* 253 (1993): 255–62; and M. Gomendio and E. R. Roldan's "Sperm Competition Influences Sperm Size in Mammals," *Proceedings of the Royal Society London B. Biological Sciences,* 243 (1991): 181–85.

4. The Cyclops was notorious for having only one eye in the middle of his forehead, and tales of such one-eyed creatures *(monoculi),* generally allocated to the East, abound in Western mythology. Interestingly enough, according to Claude Kappler in his *Mon-*

stres, démons et merveilles à la fin du Moyen Age (Paris: Payot, 1980), peoples in the East held the same kind of beliefs concerning the monstrous creatures that inhabited the West.

5. The horns of the Unicorn per se, as displays of the jokes of Nature, became an important collector's item in the *cabinets de curiosités* of the early modern period.

6. In *Monsters and Demons in the Ancient and Medieval Worlds: Papers Presented in Honor of Edith Porada* (Mainz on Rhine, Germany: Verlag Philipp von Zarben, 1987).

7. In *Monsters and Demons in the Ancient and Medieval Worlds.* For a detailed discussion of the same theme, see also Arnold Davidson's "The Horror of Monsters," in *The Boundaries of Humanity,* edited by J. J. Sheenan and M. Sisna (Berkeley: University of California Press, 1991).

8. In Kappler's *Monstres, démons et merveilles.*

9. Translated into English under the title *Certaine Secrete Wonders of Nature* (London: H. Bynnemann, 1569).

10. The strange fate of the Babylonian emperor, "having all, understanding nothing, fearing everything" before the nonexistent hope for salvation offered by his own religion, is beautifully discussed in Will Durant's *Our Oriental Heritage,* vol. 1 of *The Story of Civilization* (New York: Simon & Schuster, 1963).

11. In Francis Bacon, *Novum organon,* as quoted by Park and Daston in "Unnatural Conceptions: The Study of Monsters in Sixteenth- and Seventeenth-Century Germany and France," *Past and Present* 92 (1981): 20–54.

12. Paré had some problems with the prudish atmosphere of his time, though. Reportedly, he was forced to eliminate from his work on monsters an entire section on lesbianism, including detailed descriptions of female genitals, which seems to indicate that many influential readers saw these proto-medical approaches as disguised forms of pornography.

13. Mentioned by William R. Newman in "The Homunculus and His Forebears: Wonders of Art and Nature" (unpublished manuscript).

14. Newman, "The Homunculus."

15. For more details on the meaning of the Centaurs, see Peter H. von Blanckenhagen's chapter "Easy Monsters" in *Monsters and Demons in the Ancient and Medieval Worlds.*

16. Cited by Javier Moscoso in "Experimentos de Regeneración Animal: 1686–1765. Cómo Defender la Pre-Existencia," *Dynamis, Acta Hispanica ad Medicinae Scientiarumque Historiam Illustrandam* 15 (1995): 341–73.

17. In *The Historie of Foure-Footed Beastes,* (London: William Iaggard, 1607), Edward Topsell states that this animal is "sometimes called Gyrapha and Zirafa, from the Arabian translation Sarapha."

18. In *De la génération de l'homme ou le tableau de l'amour conjugal* (Cologne: Chez Claude Joly, 1702).

19. Mentioned in Needham's *A History of Embryology* (Cambridge: Cambridge University Press, 1934).

20. In Needham's *A History of Embryology.*

21. For a detailed account of Réaumur's experiments in this field, see William Norton Wheeler's introduction to Réaumur's *The Natural History of Ants* (New York, London: A. A. Knopf, 1926).

22. In Needham's *A History of Embryology*.

23. In Needham's *A History of Embryology*.

24. In "Jokes of Nature and Jokes of Knowledge: The Playfulness of Scientific Discourse in Early Modern Europe," *Renaissance Quarterly* 43, no. 2 (1990): 292–331.

25. Cited by Paula Findlen in "Jokes of Nature and Jokes of Knowledge." Note that the concept of Nature's sense of humor and of learned men imitating that same sense of humor is extremely widespread: it could have been the main reason why François de Plantade drew his imaginative "little men inside the sperm," as discussed in chapters 2 and 6. Linnaeus made some of his reputation by exposing a fake chimera; and, to show how enduring this concept could be, the British Association first refused to recognize the existence of the platypus, because its members assumed that it had to be some crude Aussie hoax: they sent one of their own to investigate the veracity of this unbelievable duck-billed, web-footed, oviparous animal.

26. Cited by Park and Daston in "Unnatural Conceptions."

27. Reported by Park and Daston in "Unnatural Conceptions."

28. For a comprehensive discussion of this important topic, see Carolyn Merchant's *The Death of Nature: Women, Ecology and the Scientific Revolution* (San Francisco: Harper San Francisco, 1980).

29. Right after Trembley's discovery, several naturalists speculated that this organism was a plant, or even an intermediate form between the plant and the animal kingdoms. Trembley himself took a long time to reach a definitive conclusion, and this was most likely one of the reasons why he took so long to publish his work. In his correspondence with Réaumur, we see him hesitating and agonizing again and again over this issue: "There is much resemblance between what happens to this animal, and plants which grow from cuttings. Perhaps the threads [in a figure included in the letter] are sorts of roots and that what I call an animal is perhaps an ambulant and sensitive plant, but which has characteristics very different from all those which one knows"; or "It is currently rather difficult to mark the difference between plants and animals. Mr. Boerhaave indicates principally the distinction between the location of the parts through which they draw their nourishment. According to him, plants have external roots, and animals, interior ones. It is not possible to say yet how our little bodies nourish themselves. I have some possibilities which lead me to believe that these organized bodies are a particular species of a new class of organized body until now unknown and intermediate between plants and animals." For a detailed account of this debate, see Virginia P. Dawson's *Nature's Enigma: The Problem of the Polyp and the Letters of Bonnet, Trembley, and Réaumur* (Philadelphia, Pa.: American Philosophical Society, 1987).

30. This experiment was so difficult to perform that many nineteenth-century scientists failed when attempting to reproduce it. Trembley himself was plainly aware of the unlikeliness of what he had done, and often took the precaution of repeating the experiment before reliable witnesses, "in order to be able to cite the testimony of

others . . . to prove the truth of a Fact as strange as this one." These witnesses included Pierre Lyonet and Bernard Albinus, Boerhaave's successor and the most highly regarded anatomist of the period. A good summary of Trembley's experiments is given by Thomas L. Hankins in *Science and the Enlightenment* (New York: Cambridge University Press, 1985). For a detailed account of the regeneration studies, see Virginia P. Dawson's *Nature's Enigma*. See also Javier Moscoso's "Experimentos de regeneration animal" for a stimulating debate on regeneration's implications for the evolution of natural sciences during the Enlightenment.

31. *Mémoires pour servir à l'histoire d'un genre de polypes d'eau douce à bras en forme de cornes* (Leiden: Jean & Herman Verbeek, 1744).

32. This is not to say that we now know everything. The case of the hydra is relatively clear. The whole organism is in fact a colony of individual cells, albeit an extremely complex one: it has different cell types as well as polarity, together with an integrated nerve net. This colony undergoes two alternate reproductive cycles: one is sexual, with production and fusion of germ cells (which Réaumur was the first to observe, and to describe as "eggs"), and the other is asexual, with vegetative propagation through budding. (For modern research on propagation in the hydra, see H. Meinhardt's 1993 "A Model for Pattern Formation of Hypostome, Tentacles, and Foot in Hydra: How to Form Structures Close to Each Other, How to Form Them at a Distance," *Developmental Biology* 157 (1993): 321–33; and W. A. Muller's "Competition for Factors and Cellular Resources as a Principle of Pattern Formation in *Hydra:* I. Increase of the Potentials for Head and Bud Formation and Rescue of the Regeneration-Deficient Mutant *reg-16* by Treatment with Diacylglycerol and Arachidonic Acid," *Developmental Biology* 167 (1995): 159–74.

The regeneration of flatworms is still shrouded in mystery. The problem now referred to as "pattern formation" is a complex area of developmental biology, and is under the scrutiny of hundreds of laboratories worldwide, with lengthy chapters dedicated to it in textbooks, submitted to constant updatings. The immensity of the puzzle is so overwhelming that it continues to mesmerize the brightest embryologists of our era. His inability to understand his own results in the regeneration of the sea urchin led Hans Driesch, in 1894, to conclude that development could not be explained by mechanical forces, and to postulate the existence of a vital force, which he called *entelechy.* This internal goal-directed force, reminiscent of both Caspar Wolff's *vis essentialis* and Immanuel Kant's *Bildungstrieb,* endowed the embryo with an internal psyche and wisdom that allowed it to accomplish goals despite all the obstacles that the embryologists could put in its path. Unable to deal with his own conclusions in biological terms, Driesch renounced the study of developmental physiology and became a professor of philosophy, defending vitalism until his death in 1941.

33. Mentioned and translated from the French in Dawson's *Nature's Enigma.*

34. Mentioned by Moscoso in "Experimentos de regeneration animal."

35. The fact that they kept moving after they had been separated from the body, though, had already posed a philosophical dilemma that in a certain sense anticipated the case of the polyp; and the same problem was posed by the growth of hair and nails observed in cadavers, which was widely debated by the Académie des Sciences in 1698.

See Moscoso's "Experimentos de regeneration animal" for a lengthier discussion of this issue. The same kind of problem is addressed in one of Bonnet's letters to Réaumur, dated 1741. He had cut off the heads and tails of worms, and these parts promptly died. However, the trunk continued to live and move, and even dug a hole in the mud with the extremity where the head used to be, trying to hide itself. As Bonnet wrote,

> Where then does the principle of life reside in such worms, if after having cut their heads, they still demonstrate the same movements, I say! Why do they make the bending motions? . . . Are these worms only simple machines? Or are they Composites in which the soul makes their springs move? And if they have in them such a principle, how can this principle find itself in each portion? Will one admit that there are as many souls in these worms as there are portions of these worms which can themselves become complete worms?

Bonnet used to poke fun at his cousin Trembley's philosophical enquiries, but the properties of worms led him toward an increasingly metaphysical posture. He was particularly perplexed by the fact that worms with severed tails and heads regenerated two tails (he verified that a perfectly normal tail grew where the head used to be), and anxiously asked Réaumur in 1742,

> where does the soul reside, *the self,* in this portion which instead of a head has regained a tail? It would be very strange that such a noble principle would be found lodged in a part which is so insignificant. But this is not among the questions to which one can ever promise to have enlightenment.

36. Cited in Elizabeth B. Gasking's *Investigations into Generation, 1651–1828* (Baltimore, Md.: John Hopkins University Press, 1967).

37. Bonnet, *Considérations sur les corps organisés* (Amsterdam: Chez Marc-Michel Rey, 1762).

38. Cited by Stephen Jay Gould in *The Flamingo's Smile: Reflections on Natural History* (New York: Norton, 1985).

39. Cited by Moscoso in "Experimentos de regeneration animal."

40. Reported by P. M. Rattansi in "Voltaire and the Enlightenment Image of Newton," in *History and Imagination: Essays in Honor of H. R. Trevor-Roper,* ed. Hugh Lloyd-Jones, Valerie Pearl, and Blair Worden (New York: Holmes & Meier, 1982).

41. See the excellent subchapters "Mechanics and Enlightenment Philosophy" and "Three Tests of Universal Gravitation" in Thomas L. Hankins' *Science and the Enlightenment* for a succinct account of Maupertuis's and Voltaire's combined efforts on behalf of Newtonian concepts.

42. This term, used here for the sake of simplicity, was first coined by Blumenbach in 1784.

43. Initially proposed by Hippocrates, the theory of the double semen held that both parents contributed their own semen at reproduction (the maternal semen being the vaginal secretions); and that the embryo resulted from the mingling of these two substances.

44. Quoted in Gasking's *Investigations into Generation.*

45. Cited by Gould in *The Flamingo's Smile.*

46. See Cole's *Early Theories of Sexual Generation.*

47. In an experiment reported to the Royal Society in 1743, Needham referred to some microscopic filaments that he had observed as "eels," since they were "an Aquatic animal not unlike the Fresh-Water Eel, within this Difference, that in them both extremities are alike without any Appearance of Mouth or Head," an interpretation clearly influenced by the notion of microscopic-macroscopic correspondence.

48. For a detailed account of the Voltaire-Needham controversy, see the chapter "The Life Force" in Otis E. Fellows and Stephen F. Milliken's *Buffon* (New York: Twayne Publishers, 1972).

49. Cited by Cole in *Early Theories of Sexual Generation.*

50. Wolff was credited by several nineteenth-century authors as having laid the foundations of "rational teratology," although his work obtained its due recognition only after it was translated into German in 1812. For first-hand acknowledgment of Wolff's importance in teratology, see L. J. Moreau de la Sarthe's *Déscription des principales monstruosités dans l'homme et dans les animaux* (Paris: Fournier Freres, 1808); and Barton Cooke Hirts and George A. Piersol's *Human Monstrosities* (Philadelphia, Pa.: Lea Brothers, 1891).

51. For a complete description of Wolff's unfinished work on monsters, see Shirley A. Roe's *Matter, Life, and Generation: Eighteenth-Century Embryology and the Haller-Wolff Debate* (Cambridge: Cambridge University Press, 1981).

52. This fetus, born in 1752, apparently survived for four days; among its main deformities, Haller noted that "the eyes were placed irregularly . . . Instead of the left ear . . . there was a Carbuncle resembling a wart . . . The nostrils were wide spread, without any Septum . . . From the skin of the right side of the face, there was an open canal leading to the nostrils. But there was no palate to be found."

53. To make matters worse, the single eye was positioned in the place of the nose, was "large and protuberant" and had a "remarkably broad" cornea and lacked the lower eyelid.

54. For a complete discussion of estimates made through the ages of how many Christians could actually reach Heaven—a number that, at some points, dropped to a chilling one in one thousand or even one in ten thousand—see Paul Johnson's *A History of Christianity* (New York: Atheneum, 1976.)

55. Saint Stephanus Granditonensis actually argued that the pains of Hell were so unspeakable that if a human conceived of them, he would immediately die of terror.

56. This idea was one of the main agreements between Catholics and orthodox Calvinists.

57. Several bright minds of the Scientific Revolution, and the recipients of their legacy in the times that followed, agonized over the issue of God allowing bad things to happen. Nicolas Malebranche, the man who told us that "We see everything in God," believed that monsters were produced by maternal imagination during pregnancy, arguing in *De la recherche de la verité* (Paris: Chez Christophe David, 1721) that the ability of the mother's brain to imprint all kinds of information in the brain of the fetus was necessary for survival, "since it is very convenient for young lambs to know early on

that they have to run away from wolves"; and that it was through this cerebral communication that species continuity was possible, since "it teaches the parts of the fetus to arrange themselves in the fashion of the mother's body, so that the fetus comes to resemble the mother and thus belong to the mother's species." In Malebranche's reasoning, although "God did not desire to produce monsters, because it seems evident to me that if God would only produce one animal He would not make it a monster," the sporadic occurrence of monstrous births resulted from the fact that God had "the most admirable goal of producing a vast work by simple means and to connect all of His creatures with each other." Therefore, "He programmed certain effects that would follow necessarily the order and nature of things, and this [the birth of monsters] did not deter Him from His final plan. Because, after all, although a monster in itself is an imperfect thing, when considered together with all the other creatures it does not make the world imperfect."

Interestingly, two centuries later, Charles Darwin ended up debating similar aspects of this moral dilemma in the context of natural selection. The American botanist Asa Gray, who was otherwise supportive of Darwin's ideas, pointed out that it would be inconceivable that God would allow poorly adapted, but not poorly behaved, species to simply disappear, without care and without any form of divine intervention. In a letter dated 1860, Darwin replied that "the theological view of the question" was "always painful" to him. "I am bewildered," Darwin wrote.

> I had no intention to write atheistically . . . I cannot persuade myself that a beneficent and omnipotent God would have designedly created the Ichneumidae with the express intention of their feeding within the bodies of Caterpillars, or that a cat should play with mice . . . I cannot anyhow be contented to view this wonderful universe, and especially the nature of man, and to conclude that everything is the result of brute force. I am inclined to look at everything as resulting from designed laws, with the details, either good or bad, left to the working out of what we may call chance. Not that this notion *at all* satisfies me. I feel most deeply that the whole subject is too profound for the human intellect. The lightning kills a man, whether a good or a bad one, owing to the excessively complex action of natural laws. A child (who may turn out an idiot) is born by the action of even more complex laws, and I can see no reason why a man, or other animal, may not have been aboriginally produced by other laws, and that all these laws may have been expressly designed by an omniscient Creator, who foresaw every future event and consequence. But the more I think the more bewildered I become.

58. In Jean Palfyn's *Description anatomique des parties de la femme qui servent à la génération avec un traité des monstres et leurs causes, de leur Nature & de leurs differences* (Leiden: Chez da Veve de Bastiaan Schouten, 1708).

59. Mentioned in Roe's *Matter, Life, and Generation.*

Chapter 5

1. For an excellent account of the debates on spontaneous generation during the Enlightenment, see John Farley's *The Spontaneous Generation Controversy From Descartes to Oparin* (Baltimore, Md.: Johns Hopkins University Press, 1974).

2. Note that, in his preoccupation with linking Buffon's conclusions with his superior vision as a microscopist, Vandermonde carefully avoids the issue of Buffon's repeated statements concerning the irrelevance of the microscope, including pronouncements such as "the discoveries that one can make with the microscope amount to very little, for one sees with the mind's eye and without the microscope the real existence of all these little beings."

3. See Needham's *A History of Embryology.*

4. As mentioned by Paul De Kruif in *Microbe Hunters* (New York: Harcourt, Brace, 1926).

5. De Kruif, *Microbe Hunters.* In the same passage, De Kruif also recalls that some of Buffon's contemporaries did not hesitate to extend this kind of propagation to organisms as complex as the mouse—and, "if anybody doubted this, let him go to Egypt, and there he would find the fields literally swarming with mice, begot of the muds of the river Nile, to the great calamity of the inhabitants!"

6. As quoted by Farley in *The Spontaneous Generation Controversy.*

7. Note that De Kruif wrote his book in 1926. His comparison to "present day talk about relativity" bears yet another testimony to how colorful, easily transmittable ideas about obscure and exciting facts can find their way into the general consciousness and remain solidly entrenched there, hardly clear but readily mentioned. Only a few years ago, a respected Portuguese dramatist called me on the phone to ask me to write the introductory text for the program of a new play he was staging. It was something called "$E = MC^2$." "You know," he said, "the formula for the atomic bomb."

8. For a detailed account of all these experiments rendered by Spallanzani himself, see the first volume of *Opuscules de physique, animale et végétale* (Geneva, Switzerland: Chez Barthelemi Chirol, 1777). See also *Nouvelles recherches sur les découvertes microscopiques, et la génération des corps organisés* (Paris: Lacombe, 1769).

9. The squid, according to Needham, was the macroscopic version of the hydra, in the context of the "little-big" theory that we examined in chapter 4.

10. Needham himself had never done observations of this kind.

11. As quoted by Castellani in "Spermatozoan Biology from Leeuwenhoek to Spallanzani," *Journal of the History of Biology* 6, no. 1 (1973): 37–68.

12. Castellani, "Spermatozoan Biology from Leeuwenhoek to Spallanzani."

13. In other words, Buffon and Needham could be talking about different aspects of what we now call spermatogenesis, or the maturation of sperm cells. But this process is complex, and testes are a very delicate tissue to handle; if they really noticed the existence of distinct stages of spermatogenesis, Needham had to be a very fine microscopist indeed.

14. Again, although it seems rather unlikely considering the crude microscopy of the time and the fragility of the tissues of the testes, we can admit some scenarios that would not result from imagination alone. Spermatocytes and spermatids (both tailless) attach themselves in a regular fashion to the outside of the much larger, supporting Sertoli cells as they make their way to the lumen of the organ. Alternatively, spermatocytes remain attached to each other by small cytoplasmic bridges, thus giving the impression

of organized ranks. But none of these phenomena involves *brisk* movements. And once more, if Buffon and Needham really ever saw any of this, three cheers for Needham's microscopy!

15. Mentioned by Elizabeth B. Gasking in *Investigations into Generation, 1651–1828* (Baltimore, Md.: John Hopkins University Press, 1967). Note that in this comment, Spallanzani is being highly complimentary toward Leeuwenhoek—a position that reveals, once again, that the gap between ovists and spermists was not as insurmountable as we now tend to present it.

16. This claim was most likely due to Bonnet's influence, since, in the light of Bonnet's theory of pre-existing germs, it made perfect sense to assume that spermatozoa "germs" were brought to the egg at fertilization. These germs would be born only later, enlarging themselves inside the mature semen of their own, preassigned generation.

17. See Spallanzani's *Experiences pour servir à l'histoire de la génération des animaux et des plantes* (Geneva, Switzerland: Chez Barthelemi Chirol, 1785).

18. Included in *Dissertations Relative to the Natural History of Animals and Vegetables* (London: J. Murray, 1789).

19. Spallanzani does not give the name of the species, but he insists it is not the *Rana viridis aquatica* used by others in their experiments.

20. In a later passage Spallanzani explains that, when the weather is warmer, the copulating embraces last for four or five days, but that they can continue for up to ten days in cooler temperatures.

21. Spallanzani correctly noted that the eggs retained in the ovary during copulation were smaller than those passed on to the uterus through the oviducts, and he also correctly guessed that the smaller eggs left behind in the ovary were at earlier, immature stages of development.

22. Spallanzani later described having also found this substance in green tree-frogs and in different species of toads, carefully describing the particulars of each mating, each system of laying eggs, and the anatomy of each "penis."

23. Pricking eggs with needles was a common method of parthenogenetic activation from the second half of the nineteenth century to the first half of the twentieth century; those eggs, however, were *not* amphibian eggs, like those used by Spallanzani. In experiments of this type involving amphibians, the main objectives were the study of deformities induced in tadpoles.

24. The centrosome was discovered by Boveri only in 1900, and its functioning in somatic and sexual cells is still a hot topic of controversy. Centrosomic behavior at fertilization, when two cells merge, but only one centrosome can remain active to ensure correct division, is one of the most difficult problems in this controversy.

Chapter 6

1. According to Daniel C. Dennett in *Consciousness Explained* (Boston: Little Brown, 1991), "homuncularism" as a vice in methods of thinking can be noted ever since Descartes: understanding that the eye functions as a "darkroom" where the per-

ceptive image is projected, the philosopher wonders what entity "sees" this virtual image, invoking, probably for the first time, a rough notion of "recursiveness" in mental processes. The cognitive homunculus is therefore an ideal entity endowed with a psychological "ability" by pure analogy with a human subject. In contemporary AI language, homunculi are referred to as "modules" (Fodor) or "agencies" (Minsky). A more modern version of AI ("connectionism" or "determinism") criticizes these attempts and proposes a version simultaneously more "atomistic" and more physiological, based on synaptic potentials aggregated in neuronal networks. This model would be, a priori, free from homunculi. But both concepts are debatable—and highly debated, namely on the Internet newsgroup comp.ai.philosophy, where one can find such sharp sound bites as "if you are DEFINING 'sensorimotor transduction' as whatever is required in order for a system to be grouped you make the whole discussion VOID!" or "a brain with ALL of its sensorimotor transducer ablated would be DEAD (or just a hapless hunk of cytoplasm) rather than a computer running a program that implements a sophisticated mind." It is interesting to note that, as often happens, these notions are starting to permeate the realm of fiction. The short story "Influenza," by Daniel Menaker, published in *The New Yorker* of 16 January 1995, contains a paragraph in which the narrator refers to his analyst in the following terms: "It's true that life improved for me as I went to him, but whether if I could do it all over again I would actually choose to have the homunculus of an insane, bodybuilding, black-bearded Cuban Catholic Freudian shouting at me from inside my own head I am not sure."

2. In V. Gordon Childe, *Most Ancient East: The Oriental Prelude to European Prehistory* (New York: Knopf, 1929).

3. In Leonard Wooley, *The Sumerians* (Oxford: Clarendon Press, 1928).

4. From Pierre Lory's introduction to Jâbir ibn Hayân's *Dix traités d'alchimie* (Paris: Sindbad, 1983).

5. In Paul Kraus, *Jâbir Ibn Hayân,* vol. 1 of *Contribution à l'histoire des idées scientifiques dans l'Islam* (Cairo, Egypt: Imprimerie de l'Institut Français d'Archéologie Orientale, 1943).

6. In his list of the "Seven Arts," Jâbir mentions as the last one the "science of forms," or "science of generation," as the conclusion of all the preceding disciplines. This culminating science aims at the artificial production of beings belonging to the Three Kingdoms, and, most of all, living beings. In its full scope, alchemy is not only meant to transmute certain bodies into others, but also to try to form new bodies from their constitutive elements. So the artificial generation of stones would necessarily correspond to the generation of plants, and of animals—and, most importantly, of men. In Jâbir's own words, "there are two sorts of creation, a first (by God) and a second (by Man); and the second, represented by the Art, is similar to the first."

7. In Kraus, *Jâbir Ibn Hayân.*

8. This type of conception survived until our days. Serge Hutin quotes a certain Papus, a.k.a. Gerard Encausse, one of the many *esotheristes* of the French *belle époque,* as explaining in his "Traité élémentaire de science occulte" that "the philosophers' stone is simply an energetic condensation of life in a small amount of matter and it acts

as a ferment over the bodies exposed to it. A little ferment is all that is needed to grow a big mass of bread; similarly, a little philosophers stone is all that is needed to develop the life contained in any matter, mineral, vegetal or animal." (From "Les doctrines secretes," in *Paracelse: l'homme, le médecin, l'alchimiste,* by Béatrice Whiteside [Paris: Table Ronde, 1966].)

9. Literally, the "unformed mass."

10. Other experts in the same field, most notably Moshe Idel, later refuted Scholem's connection between the Golem and the homunculus. It is nevertheless remarkable how promptly the latter term surfaces whenever the issue is the artificial creation of human beings.

11. In Gershom Scholem, *Major Trends in Jewish Mysticism* (New York: Schocken Books, 1941). For a recent account of the Golem legend, see also Marge Piercy's *He, She and It* (New York: Knopf/Random House, 1991).

12. Baer, in *Zion,* vol. 3 (1938), p. 1–50; cited by Scholem in *Major Trends in Jewish Mysticism.*

13. From Chaim Potok's foreword to Martin Buber's *Tales of the Hasidim* (New York: Schocken Books, 1991).

14. In Scholem, *Major Trends in Jewish Mysticism.*

15. As Byron Sherwin puts it in *The Golem Legend: Origins and Implications* (New York: University Press of America, 1985), "magical mastery of these names is not a perverted form of knowledge but rather a pure and sacred knowledge that belongs to God's image. The use of these names and words for magical purposes—including the creation of a Golem—is viewed in the literature of German Hasidism as an act of *imitatio dei* par excellence." Thus the creation of a Golem eventually came to be viewed as a mystical rite of initiation.

16. This "Book of Creation," less than two thousand words in length, has been ascribed to Abraham the Patriarch, to Rabbi Akika, and to an anonymous Palestinian scholar writing sometime between the third and the sixth century.

17. In Scholem, *Major Trends in Jewish Mysticism.*

18. According to Sherwin in *The Golem Legend,* in the most pervasive of these stories the Golems could grow to gigantic proportions. In the sixteenth century, it was said that Rabbi Elijah of Helm had seen the Golem he created grow to such an enormous size that he feared the creature could become dangerous and caused it to revert to dirt. But, in its collapse, the Golem injured the rabbi. In later variants of the legend, the Golem actually went on a rampage before it was destroyed.

19. The silent movie *The Golem,* directed by Paul Wegener in 1920, starts with the statement "the learned Rabbi Loew reads in the stars misfortune to come for the Jewish people," and portrays the scene of the Golem gone bad, setting houses on fire and tearing down gates.

20. In Harry Collins and Trevor Pinch, *The Golem: What Everybody Should Know About Science* (Cambridge: Cambridge University Press, 1993).

21. According to Sherwin in *The Golem Legend,* the animation of the Golem triggered by the inscription of the name Emeth in its forehead, or on a parchment attached

to it, is part of the legendary material and was not really practiced by the Hasids. The same applies to the tale that the Golem could be deactivated by erasing the first aleph (the letter in the Hebrew alphabet transliterated as *e*): *Emeth* would thus become *Meth* (death) and the Golem would die.

22. According to Lynn Thorndike in *A History of Magic and Experimental Science* (New York: Macmillan, 1929), the *Liber vaccae* is ascribed to Galen, who in turn says he revised and abbreviated it from Plato, and is possibly related to Albertus Magnus's *De mirabilius mundi.* It has also appeared under the titles *Liber angenis, The Book of Active Institutes,* and *The Book of Aggregations of Divers Philosophers,* as well as being cited as *Liber de prophetiis* by Pedro Alfonso in his *Disciplina clericalis,* written at the close of the eleventh century, possibly from an Arabic or Hebrew translation of the original.

23. As Thorndike advises in the introduction to the chapter on the *Liber vaccae,* this work is just about impossible to read for scholars not trained in deciphering Hermetic texts. I owe thanks for help in understanding the recipe for "rational beings" to Prof. William Newman.

24. Quoted in Mircea Eliade, *The Forge and the Crucible,* trans. Stephen Corrin (Chicago: University of Chicago Press, 1956).

25. Quoted by Serge Hutin in "Les doctrines secretes."

26. As with everything related to Paracelsus, the exact origin of the recipe and the true authorship of the book are hard to establish. Even the title, *The nature of things,* is but a reprise of Lucretius's work published in the first century B.C., which was in turn a simple translation of the *Peri physeos,* which the pre-Socratics had used as a common name for their treatises. To add further confusion to its real origins, the recipe was repeated with slight variations in a number of other books. Actually, it is not even clear that Paracelsus was the real author of the book containing the first version. William Newman argues that it is a pseudo-Paracelsian tract, most likely the work of one of his disciples. The quotes given here are from *Hermetic Chemistry,* vol. 1 of *The hermetic and alchemical writings of Aureolus Philippus Theophrastus Bombast, of Hohenheim, called Paracelsus the Great,* edited by Arthur Edward Waite (Berkeley, Calif.: Shambhala, 1976).

27. In Hutin, "Les doctrines secretes."

28. This text, mentioned by Thorndike in *A History of Magic and Experimental Science,* vol. 8, is available in a German version from 1855, *Drei Bucher der magnetischen Heilkunde* (Stuttgart: Scheible, 1855).

29. In Thorndike, *A History of Magic and Experimental Science,* vol. 8. According to Thorndike, Castelan was particularly skeptical about the belief that "witches and rustics" transform the root of the plant into a human form. As we have seen, Samuel Beckett also mentions the "homunculus" in the roof of the mandrake in *Waiting for Godot,* but here it is the semen of the hanged man, rather than his urine, that causes the growth of the plant—yet another oft-reprised variant of the legend.

30. In Thorndike, *A History of Magic and Experimental Science,* vol 8.

31. In Thorndike, *A History of Magic and Experimental Science,* vol. 8. This book

had six different editions, both by Rolfink and under the form of "disquisitiones" by some of his students. The same claims can be found in Zacharias Brendel's *Zachariae Brendelii Chimia in artis formam redacta* (Lugduni Batavorum: Ex officina Arnoldi Doude, 1671).

32. This claim, however, is highly debatable. Although Rolfink insists that procedures such as the transmutation of metals are "chemical non-entities," we have to bear in mind that metal transmutation is only one of the several interests of alchemy. Besides, he seems to think of chemistry solely in the apothecary sense, thus reducing the discipline (rather than elevating it) to a mere subsidiary of medicine.

33. In Thorndike, *A History of Magic and Experimental Science,* vol. 7. Another version of the same document, attributed to Johann Sigismund Elsholtz, is available in English under the title *The curious distillatory, or, The art of distilling coloured liquors, spirits, oils, etc.* (London: Printed by J. D. for Robert Boulter, 1677).

34. In Thorndike, *A History of Magic and Experimental Science,* vol. 8. Thorndike states here that although the original manuscript is supposed to be from 1676, he could only find a copy from 1681.

35. In Thorndike, *A History of Magic and Experimental Science,* vol. 8.

36. In *Athanasii Kircheri & Soc. Jesu mundus subterraneus, in XII libros digestus* (Amsterdam: Joannem Janssonium a Waesberge & fillios, 1678).

37. In Thorndike, *A History of Magic and Experimental Science,* vol. 7.

38. In Scholem's *Major Trends in Jewish Mysticism.*

39. In Thorndike, *A History of Magic and Experimental Science,* vol. 8.

40. Mentioned in M. V. DePorte's foreword to More's *Enthusiasmus triumphatus* (Los Angeles: William Andrews Clark Memorial Library, University of California at Los Angeles, 1966). For more information on the Cambridge Platonists, More's theosophical doctrines, and the Enthusiasts, see also Flora Isabel Mackinnon's *Philosophical Writings of Henry More* (New York: AMC Press, 1969) and Serge Hutin's *Henry More: Éssai sur les doctrines théosophiques chez les Platoniciens de Cambridge* (Hildesheim, Germany: Georg Olms Verlagsbuchhandlung, 1966).

41. In F. J. Cole, *Early Theories of Sexual Generation* (Oxford: Clarendon Press, 1930).

42. In Cattaui, "Paracelse et sa postérité," in *Paracelse: l'homme, le médecin, l'alchimiste,* by Béatrice Whiteside (Paris: Table Ronde, 1966).

43. In Hutin's "Les doctrines secretes."

44. In *Investigations into Generation, 1651–1828* (Baltimore, Md.: John Hopkins University Press, 1967), Elizabeth B. Gasking quotes the following sentence from Leeuwenhoek, commenting on Dalenpatius's drawings: "the Royal Society would have none of the seminal homunculi." However, I was unable to find this term in either the Latin or the Dutch-English versions of Leeuwenhoek's complete writings that I have consulted.

45. Mentioned by Cole in *Early Theories of Sexual Generation.*

46. From Marcel Barral's introduction to Plantade's *Le conte des Fées du Mont des Pucelles* (Montpellier, France: Publications de l'Entente Bibliophile, 1988).

47. In Plantade, "Extrait d'une Lettre de M. Dalenpatius à l'auteur de ces Nouvelles" (Amsterdam: Nouvelles de la République des Lettres, art. V, 1699).

48. In Joseph Needham's *A History of Embryology* (Cambridge: Cambridge University Press, 1934) and A. Ruppert Hall's *The Scientific Revolution, 1500–1800: The Formation of the Modern Scientific Attitude* (Boston: Beacon Press, 1966), respectively. In *Investigations into Generation,* Gasking also says that "Indeed, it was soon suggested that the whole letter was the work of a man called Plantade, and that it was intended as an hoax."

49. In Cole's *Early Theories of Sexual Generation.*

50. Cole, *Early Theories of Sexual Generation.*

51. In "Epistolae ad Societatem Regiam Anglicam," 1719. The spelling is a reproduction *ipsis verbis* from the original.

52. *The Collected Letters of Antoni van Leeuwenhoek* (Amsterdam: Swets & Zeitlinger, 1952).

53. In *Anthropogeniae ichnographia,* 1671. The English translation used here is by Needham, in *A History of Embryology.*

54. "It seems to me that these 'homunculi,' acting in such a way, are certainly nothing but fables like many others"; in Martin Lister, *Dissertatio de humoribus,* as cited by Cole in *Early Theories of Sexual Generation.*

55. Thanks to its ironic brilliance, this work became vastly popular and had several reprints. The version consulted here is *Lucina sine concubitu: A letter humbly addressed to the Royal Society* (Edinburgh: Collectanea adamantea, 7, 1885).

56. In J. T. Needham, *Observations on the generation, composition and decomposition of animal and vegetable substances* (London, 1749).

57. In Maupertuis, "Lettres de M. de Maupertuis" (Dresden, 1752).

58. Mentioned by Cole in *Early Theories of Sexual Generation.*

Chapter 7

1. In Boris G. Kouznetzov's "Le soleil comme centre du monde, et l'homogénéité de l'espace chez Galilée," in *Le Soleil à la Renaissance: Sciences et mythes* (Paris: Presses Universitaires de France, 1965).

2. According to this author, it was the legacy of the Pythagoreans that paved the way for the shift from the geocentric to the heliocentric system, and inspired Kepler to produce his final 1619 masterpiece, *Harmonia Mundi:*

> the Renaissance was saturated in the lore of Pythagoras. His doctrine . . . was aimed at purification of the mind, achieved through successive stages of knowledge about the universe until an ultimate knowledge of final principles was acquired. At this culmination, cleansed of corporeal distraction, possessed of infinite wisdom, the mind could re-unite with, and participate in, deity. This spiritual perfection was the aim of man's study, but yet the goal could be reached only by the observation of nature. For one of the rare moments in history, science and ethics were incorporated into a single philosophical system . . . For this reason . . . the Pythagorean philosophy appealed so strongly to the Renaissance. Without diminishing the central importance of man nor the possibility of his perfection, it urged the study of physics. It provided the humanists with a scientific orientation that Platonism

lacked, absorbed as it was with mysticism. Moreover, it provided a mathematical tradition of number, weight, and measure, a quantitative approach that academic Aristotelianism lacked, absorbed as it was with qualitative analysis and logic.

3. Cited in Heninger's "Pythagorean Cosmology and the Triumph of Heliocentrism," in *Le Soleil à la Renaissance.*

4. This concept was bound to become a serious philosophical problem for the Aristotelians of the Middle Ages and the Renaissance, since it was obvious that nature did not totally conform to it: dry land extended above the waters, and fire was visible on the surface of the earth. In *In Defense of the Earth's Centrality and Immobility: Scholastic Reaction to Copernicanism in the Seventeenth Century* (Philadelphia, Pa.: The American Philosophical Society, 1984), Edward Grant describes some of the elaborate scenarios invoked to overcome this problem. In the fourteenth century, John Buridan, convinced that the spheres of earth and water were concentric with respect to the center of the world,

> assumed that the water did not completely surround the earth because some part of it flowed naturally downward and filled the bowels of the earth, while other parts of it mixed with air after evaporation. The quantity of water was thus insufficient to cover the entire earth and so, inevitably, part of the earth was left exposed above the waters.

Buridan also addressed another inescapable objection: over long periods of time, geologic processes, especially those wherein the waters flowing to the seas carry matter down from the mountains, would wear down the ridges and other elevations, leaving the earth everywhere submerged below the waters.

> Buridan explained how this potentially drastic consequence was avoided. The earthy matter continually deposited in the seas by the waters flowing down from the mountains and elevations makes the submerged portions of the earth heavier, which, in turn, causes the earth's center of gravity to shift continually. With each such shift, the earth moves rectilinearly until its center of gravity coincides with the center of the world. These minute, but incessant, rectilinear shifts of the earth's center of gravity will cause previously submerged parts of the earth to rise above the surface of the seas and oceans. Because this geologic process is cyclic and continuous, part of the earth will always remain elevated above the waters.

5. For a detailed account of the evolution of astronomical thought in Greece, see D. R. Dicks's *Early Greek Astronomy to Aristotle* (Ithaca, N.Y.: Cornell University Press, 1970).

6. Originally titled *The Mathematical Composition* and then rebaptized *The Great Astronomer.* In their translation of the work, the Arabs titled the book *The Greatest,* prefixing the article *al* to the Greek *megiste,* leading to the final title *Almagest,* used ever since until our times.

7. As explained by Angus Armitage in *Copernicus and the Reformation of Astronomy* (London: Published for the Historical Association by George Philip & Son, 1950).

8. As quoted by Grant in *In Defense of the Earth's Centrality and Immobility.*

9. Copernicus was stricken with apoplexy in 1542, and was half-paralyzed after that.

Reportedly, he died right after receiving the first copy of his newly printed book. He was buried in the Frawenburg Cathedral, in whose tower he had performed most of his astronomical observations.

10. *De Revolutionibus Orbium Coeslestium,* Libri VI.

11. Tycho Brahe was a flamboyant and controversial character, who wore a tin nose after the original was cut off during a duel. He was extremely wealthy and founded a state-of-the-art astronomical observatory on an isolated island, but he feared that his government would confiscate all his data as a means of retaliation against his rebellious social and religious positions.

12. "After the emanation of the Superior Sphere, the emanation continues through the conjoint production of an Intellect and a Sphere. Thus, from the second Intellect, is produced a third Intellect and the Sphere of the fixed stars; from the Third Intellect comes a fourth one, a fifth one and the Sphere of Jupiter; from the fifth Intellect comes a sixth one and the Sphere of Mars; from the sixth Intellect, a seventh one and the Sphere of the Sun; from the seventh Intellect, an eighth one and the Sphere of Venus; from the eighth Intellect, a ninth one and the Sphere of Mercury; from the ninth Intellect, a tenth one and the Sphere of the Moon."

13. In Jean Chevalier and Alain Gheerbrant's *Dictionnaire des symboles: Mythes, rêves, coutumes, gestes, formes, figures, couleurs, nombres* (Paris: Éditions Robert Laffont and Éditions Jupiter, 1982.)

14. When the Europeans reached Polynesia for the first time, the Samoans thought that the newcomers had broken through one of the solid cupolas of the heavens, and called them *papalangi,* or "heaven-busters."

15. In Kjell Akerblom's *Astronomy and Navigation in Polynesia and Micronesia: A Survey* (The Ethnographical Museum Monograph Series, Stockholm, Sweden, 1968).

16. "Of all the nations," wrote the Portuguese Jesuit Alvaro Semedo in 1669, "China is the one that most appreciates Mathematics and Astronomy, with such excess that it seems to make depend on these sciences the conservation of the monarchy and the good government of the State." (As quoted by Francisco Rodrigues in *Jesuítas Portugueses, Astrónomos na China, 1583–1805* [Macau, Portugal: Instituto Cultural de Macau, 1990].)

17. As quoted by Rodrigues in *Jesuítas Portugueses.*

18. Rodrigues, *Jesuítas Portugueses.*

19. See, for instance, among countless other possible examples, Robert M. Sadowski's "Stone Rings of Northern Poland," and R. P. Norris and P. N. Appleton's "A Survey of the Barbrook Stone Circles and Their Claimed Astronomical Alignments," both in *Archaeoastronomy in the Old World,* ed. D. C. Heggie (London: Cambridge University Press, 1982).

20. As discussed by Heninger in "Pythagorean Cosmology and The Triumph of Heliocentrism."

21. As claimed by de Vries in *Dictionary of Symbols and Imagery* (Amsterdam: North-Holland Publishing Company, 1974).

22. In M. W. Feldman, *Rabbinical Mathematics and Astronomy* (New York: Hermon Press, 1978).

23. One of the most widespread stories about how Pythagoras discovered the numerical ratios between the notes of the musical scale recounts that, one day, he passed in front of a blacksmith's shop, were several men where at work, and noticed that the sounds they beat out were in perfect harmony. Upon investigation, he determined that the blacksmiths were using hammers with weights in the ratio of whole numbers. Inspired by this observation, Pythagoras constructed a full set of hammers with appropriate weights, and thereby reproduced the musical scale. He likewise set up lengths of gut in similar ratios, and again contrived the diapason.

24. In *A History of Astronomy from Thales to Kepler* (New York: Dover Publications, 1953).

25. In 1979, Yale University professors Willie Ruff (player of bass and French horn and associate professor of music) and John Rodgers (an "enthusiastic pianist" and Sillian Professor of Geology) produced a record for "The Kepler Label," entitled *The Harmony of the World—A Realization for the Ear of Johannes Kepler's Astronomical Data from Harmonia Mundi, 1619.* As the literature included in the album announces, "three and a half centuries after their conception, Kepler's data plotting the harmonic movement of the planets have been realized in sound with the help of modern astronomical knowledge and a computer-sound synthesizer." Their project was triggered by an earlier variation on the musical connotations of Kepler's work, Paul Hindemith's opera *Die Harmonie der Welt.* The computer program by which the astronomical data were converted into sounds was written by Mark Rosenberg, using the Music 4bF software developed at Bell Laboratories and the IBM 360/91 computer at the Princeton University Computer Center. According to the composers' notes, the process was rather simple: "If we assume with Kepler that the planets obey his laws and that the notes they sing are proportional to their angular velocities around the sun, then all we need to reconstruct Kepler's harmony of the world is, for each planet p, the period of revolution around the sun; e, the eccentricity of its orbit; and t', some measure of how much of its revolution the planet, at some particular moment in time, has covered since it was last at its perihelion (when it was singing its higher note), so that its song can be correlated to real time and to the songs of other planets." Of course, since Kepler's death, three new planets have been introduced into the picture, but for this, too, there was a computer-generated solution: "The periods of the newly discovered planets are all much longer than Saturn's, which is already 2½ minutes in our realization. Pluto, whose period is 248 years, would take over 20 minutes, and, despite its large eccentricity, its changes of pitch would be too slow to hear. It seemed best therefore to remain with Kepler's choices, in which case the new planets would vibrate only two to ten times per second; in other words, they are not part of the vocal polyphony but of the *rhythm* section."

26. See "Mathematics and the Exact Sciences," in Thomas L. Hankins's *Science and the Enlightenment* (New York: Cambridge University Press, 1985) for a complete description of the reaction to Newton's proposal and of the bold experiments realized to prove its veracity.

27. See Meyer's *The Rise of Embryology* (Stanford, Calif.: Stanford University Press, 1939) for a complete discussion of Aristotle's conflicting views on sex determination.

28. As mentioned by Jacques Languirand in *La magie des nombres: de McLuhan à Pythagore* (France: Editions de Mortagne, 1995).

29. See Robert Briffault's *The Mothers,* vol. 2 (1927; London: Allen & Unwin, 1959).

30. See W. G. Sumner's *Folkways* (1906; New York: New American Library, 1940).

31. As quoted by Scott F. Gilbert in *Developmental Biology,* 4th ed. (Sunderland, Mass.: Sinauer Associates, 1994). For an excellent and comprehensive discussion of the subject of "Woman-as-man," see Thomas Laqueur's *Making Sex: Body and Gender from the Greeks to Freud* (Cambridge, Mass.: Harvard University Press, 1990). See also Catherine Gallagher and Thomas Laqueur, eds., *The Making of the Modern Body: Sexuality and Society in the Nineteenth Century* (Berkeley: University of California Press, 1987), for an account of modern ramifications of the same issue. For detailed information on nineteenth-century Western scientists' and philosophers' efforts to prove women's inferiority to men, see Cynthia Eagle Russett's *Sexual Science: The Victorian Construction of Womanhood* (Cambridge, Mass.: Harvard University Press, 1989).

32. In "Nouvelles recherches physiques et metaphysiques sur la nature et la réligion, avec une nouvelle théorie de la Terre," bound with the second volume of Spallanzani's *Nouvelles recherches sur les découvertes microscopiques, et la génération des corps organisés* (Paris: Lacombe, 1769).

33. This is a unique case within organogenesis. Generally, the embryonic rudiments of one organ can become only that specific organ: primordial lungs can produce only lungs, just as primordial liver can become only liver. Confirming the classic belief that in the beginning there is only one sex, primordial gonads are the only embryonic organs able to develop in more than one way, either as ovaries or as testes, depending on the chromosomal composition of the embryonic cells.

34. The literature produced on this subject over the last decade is abundant. For a comprehensive review of the subject, see Ann McLaren's "Sex Determination in Mammals," *Trends in Genetics* 4, no. 6 (1988): 153–57, and "Sex Determination in Mammals," *Oxford Review of Reproductive Biology* 13 (1991): 1–33; see also "Sex Determination: What Makes a Man a Man," *Nature* 346, no. 6281 (1990): 240–44, and "Sex Determination: The Making of Male Mice," *Nature* 351, no. 6322 (1991): 117–21, both by the same author.

35. The mechanisms through which estrogens feminize the body are still poorly understood, but a good description of these effects can be found in Langman and Wilson's "Embryology and Congenital Malformations of the Female Genital Tract," in *Pathology of the Female Genital Tract,* 2d ed., ed. A. Blaustein (New York: Springer-Verlag, 1982).

36. Reported by Ann McLaren in "Development of the Mammalian Gonad: The Fate of the Supporting Cell Lineage," *Bioessays* 13, no. 4 (1991): 151–56.

37. More than 90% of these animals observed in scientific experiments display ectopic primordial germ cells.

38. See K. Yoshinaga et al., "Germinal Cell Ectopism in the Strepsirhine Prosimian *Galago crassicaudatus crassicaudatus,*" *American Journal of Anatomy* 187, no. 3 (1990): 213–31.

39. See Ann McLaren, "Somatic and Germ-Cells in Mammals," *Philosophical Trans-*

actions of the Royal Society of London, Biological Sciences 322, no. 1208 (1988): 3–9; H. Hogg and A. McLaren, "Absence of a Sex Vesicle in Meiotic Foetal Germ Cells Is Consistent with an XY Sex Chromosome Constitution," *Journal of Embryology and Experimental Morphology* 88 (1985): 327–32; A. McLaren, "Meiosis and Differentiation of Mouse Germ Cells," *Symposia of the Society of Experimental Biology* 38 (1984): 7–23; A. McLaren, "Studies on Mouse Germ Cells Inside and Outside the Gonad," *Journal of Experimental Zoology* 228, no. 2 (1983): 167–71; S. Francavilla and L. Zamboni, "Differentiation of Mouse Ectopic Germinal Cells in Intra- and Perigonadal Locations," *Journal of Experimental Zoology* 233, no. 1 (1985): 101–9; and L. Zamboni and S. Upadhyay, "Germ Cell Differentiation in Mouse Adrenal Glands," *Journal of Experimental Zoology* 228, no. 2 (1983): 173–93.

40. Although the functioning of the section of the Y chromosome responsible for sex determination, and the cascade of genes involved in this process, are still poorly understood, it is undeniable that this segment is extremely powerful. A person with five X chromosomes and only one Y chromosome (XXXXXY, a rare but occasionally found genotype) will still develop as a male.

41. For a detailed exploration of the symbolic masculine-feminine complementarity, see Chevalier and Gheerbrant, *Dictionnaire des symboles.*

42. As quoted by Will Durant in *Our Oriental Heritage,* vol. 1 of *The Story of Civilization* (New York: Simon & Schuster, 1963).

43. Durant, *Our Oriental Heritage.*

44. Durant, *Our Oriental Heritage.*

45. In Ronaldo Rogério de Freitas Mourão, *Astronomia do Macunaíma* (Bahia, Brazil: Francisco Alves, 1984).

46. As described by Newman in "The Homunculus and His Forebears: Wonders of Art and Nature" (unpublished manuscript).

47. See Count Hermann Keyserling's *Travel Diary of a Philosopher* (New York: Harcourt, Brace, 1925).

48. See Will Durant's *Caesar and Christ,* vol. 3 of *The Story of Civilization* (New York: Simon & Schuster, 1971).

49. From my personal database, circa 1994: "so, you *really* mean that your husband lets you travel all the time by yourself?" This was a university professor. In Massachusetts.

50. The symbol ♂ was also sometimes used in chemistry to represent zinc, although this metal was more often associated with one of the three signs used for Jupiter.

Chapter 8

1. Some modern scholars refer to this trend as "Quantophrenia."

2. For the benefit of the reader whom Hartsoeker's numbers may strike as unlikely measures of immensity, it is worth pointing out that the current unit of largeness in the world of "large numbers" is 10^{100} (or unity followed by one hundred zeros), the number called "googol"—the largest named number, at least in the West. It was so named,

urban legend has it, by a nine-year-old nephew of mathematician Edward Kasner, upon being asked by his uncle how many drops of rain would fall in New York on a rainy day. As far as current estimations go, there is nothing of which there is a googol: the number of electrons in the universe, for instance, is on the order of only 10^{89}. For a detailed discussion of "large numbers," see Eli Maor's 1986 *To Infinity and Beyond* (Boston: Birkhauser, 1986).

3. Cited by Gould in *The Flamingo's Smile: Reflections on Natural History* (New York: Norton, 1985). The same theme is developed by Abbé Regley in his preface to Spallanzani's *Nouvelles recherches sur les découvertes microscopiques, et la génération des corps organisés* (Paris: Lacombe, 1769) in the following terms:

> There is a calculation by Leeuwenhoek whose enormity renders it frightening, and which thus loses credibility. I'm talking about his sort of frog . . . He said the female frog had 9,334,000 eggs, and since he needs 10,000 animalcules in the male for each egg in the female, it follows that one single male frog has inside many more living frogs than there are men all over the surface of earth. [Leeuwenhoek] even produced a chart with the space occupied by the waters on one side and the space occupied by the dry land on the other. Afterward, considering that the entire earth was as populated as Holland, the place he knew the best, he found that the dry land had a number of living humans rather inferior to the number of presumed animalcules within his male frog.

4. There are some rare cases of immobile sperm, namely in marine organisms, but they are irrelevant to this point. Not only are they rare, but I doubt that any spermist would have been using these animals for his observations.

5. Mentioned in James J. Walsh's *Catholic Churchmen in Science* (New York: Books for Libraries Press, 1917).

6. "Rational Individuals versus Laws of Society: From Probability to Statistics," in *Probability Since 1800: Interdisciplinary Studies of Scientific Development,* ed. Michael Heidelberger, Lorenz Kruger, and Rosemarie Rheinwald (Germany: B. Kleine-Verlag, 1983).

7. "The Invention of Ethical Calculus," in *The Seventeenth Century: Studies in the History of English Thought and Literature from Bacon to Pope by Richard Foster Jones and Others Writing in His Honor* (Stanford, Calif.: Stanford University Press, 1951).

8. As cited by Louis Bredvold in "The Invention of Ethical Calculus."

9. Weigel conceived a new science called *Pantometria,* which would subordinate moral philosophy and jurisprudence as well as metaphysics.

10. Leibniz's earlier attempts were directed toward the solution of political questions by geometric methods. He presented his propositions to the new king of Poland, and two years later to Louis XIV, insisting before the French monarch that geometry recommended the dispatch of his armies from the Lower Countries to Egypt in order to break the power of the Turks. Neither political demonstration had any influence on the subsequent course of events, and, as he matured, Leibniz moved away from geometry and toward calculus. In the remaining fragments of Leibniz's unfinished encyclopedic treatise, the author sought a "general science" that would include not only geometry, in

a transcendent form, and mechanics, but also a "civilian logic," or "logic of life," that would provide a calculation of probabilities applicable to all practical questions, and in particular, to legal problems. For this purpose, he devised a sort of mathematical logic called "the universal characteristic," by which he hoped to reduce all reasoning to a combination of signs, hence to a process of calculation.

11. Mentioned by Bredvold in "The Invention of Ethical Calculus."

12. Bredvold, "The Invention of Ethical Calculus." Bredvold quotes Laurence Sterne mocking Francis Hutcheson's equations included in the section dedicated to "Introducing Mathematical Calculations in Subjects of Morality" in Hutcheson's 1725 *Inquiry into the Original Ideas of Beauty and Virtue* on the grounds that "Hutcheson plus or minus's you to heaven or hell, by algebraic equations—so that no one but an expert mathematician can ever be able to settle his accounts with S. Peter—and perhaps S. Matthew, who had been an officer in the customs, must be called in to audit them."

13. *Opuscules de physique, animale et vegetale* (Geneva, Switzerland: Chez Barthelemi Chirol, 1777).

14. Notice that fifteen years passed between Galileo's first use of the awkward microscope and the date of the first publication in the field; and that early microscopy produced almost no illustrations until 1650, leading during this time to fewer than ten publications, most of minor value.

15. "Chromatic aberration" is a well-known phenomenon in optics: unless lenses are correctly constructed, light of different wavelengths bends to different degrees when passing through the glass, thus giving colored fringes to the objects under examination. The problem was so complex that even Newton wrongly proclaimed the correction of this aberration impossible. Later, scientists found that a combination of components of different types of glass, varying in refractive index at different rates with the wavelength of the light transmitted, could circumvent the difficulty. For a detailed account of this technical process, see the chapter "Development of Microscopical Observation" in Arthur Hughes's *A History of Cytology* (London: Abelard-Schuman, 1959).

16. The early microscopists certainly used the term "cell" quite generously, but its real meaning remained rather loose. In Hooke's illustrations of *Micrographia,* the cellular structure of the cork is very clear, but the author was by no means aware of what these "pores or cells" really represented. He even used the word "cell" for a number of technical artifacts, such as honeycombs. It was only in the nineteenth century that several botanists formed the idea that cells were not just the spaces between a network of fibers, but, rather, separate and separable units. A similar approach was soon taken toward the corpuscles in animal body fluids, such as the red blood cells described two centuries earlier by Jan Swammerdam and Marcello Malpighi, and also, in his characteristically disorganized fashion, by Leeuwenhoek—three fine microscopists who, like Hooke and his fellow botanists, had the cell before their eyes, but did not have the mental tools to understand what they were seeing.

The next important step toward the formulation of the cell theory was the description of the nucleus in an adult animal cell. On this front, as often happens, Nature played a mean trick on microscopists. Mature mammalian erythrocytes are enucleated, whereas those of all other animals have a nucleus; and, sure enough, the nucleus in the red blood cell was the first to be described, creating great headaches for those who were trying to shed some light on the nuclear puzzle. As it also often happens, the ground-breaking discovery was part of a totally random process. In 1781, Felix Fontana published a book dedicated mainly to the viper and its venom. At the end of the book, the author included a section on his miscellaneous observations with a microscope, including one on the slime from the skin of an eel. Within this substance there were several globules (epithelial cells), and within each globule there were consistent look-alike oviform bodies. There was the nucleus, in its first formal introduction to the world. Each oviform body even had inside it a little spot, which may well have been the nucleolus.

All these findings finally emerged as a synthesis in the 1830s, when both botanists and zoologists came to recognize the existence of a basic correspondence in minute structure between plants and animals. This era is generally regarded as the time of the foundation of the cell theory, with the ensuing recognition of the primary living substance, initially called "protoplasm" and then more correctly divided into the modern terms of "nucleus" and "cytoplasm." Two names are usually most prominently associated with the establishment of the cell theory: Theodor Schwann and M. J. Schleiden.

17. For an excellent discussion of Bonnet's ideas on this subject, see Gould's *Ontogeny and Phylogeny* (Cambridge, Mass.: Belknap Press of Harvard University Press, 1977). See also Anderson's "Charles Bonnet's Taxonomy and Chain of Being," *Journal of the History of Ideas* 37 (1976): 45–58.

18. From the preface to Spallanzani's *Opuscules de physique.*

19. As cited by Dobell in *Antoni van Leeuwenhoek and His Little Animals* (New York: Dover Publications, 1960). According to Elizabeth B. Gasking's *Investigations into Generation, 1651–1828* (Baltimore, Md.: John Hopkins University Press, 1967), Leeuwenhoek referred to these calculations in a letter addressed to Huygens.

20. In Abbé Regley's preface to Spallanzani's *Nouvelles recherches sur les découvertes microscopiques.*

21. The founder of the Academy of the Lynx, Duke Frederico Cesi, was particularly interested in these insects.

22. It may also have discouraged others from pursuing these types of observations, since nobody could really keep up with his prolixity and accuracy. As early as 1691 we find Hooke lamenting that "[microscopists] are now reduced to a single votary which is Mr. Leeuwenhoek, besides whom I hear of none that make any other use of this instrument."

23. In "Super octo libros de physico auditus subtillissimae quaestiones." Cited by Norma E. Emerton in *The Scientific Reinterpretation of Form* (Ithaca, N.Y.: Cornell University Press, 1984).

24. In the preface to Spallanzani's *Opuscules de physique.*

25. In "On Generation and Corruption." Cited by Emerton in *The Scientific Reinterpretation of Form.*

26. As quoted in G. S. Kirk, J. E. Raven, and M. Schofield's *The Presocratic Philosophers,* 2d ed. (Cambridge: Cambridge University Press, 1983).

27. Kirk, Raven, and Schofield, *The Presocratic Philosophers.*

28. Again, this type of perception is echoed in eighteenth-century commentaries on the microscope, as in the case of Jean Sénébier pointing out, in his preface to Spallanzani's *Opuscules de physique,* that blood loses its color when it reaches the degree of dilution required to pass through the capillaries.

29. In his section titled "Gas aquae," included in the treatise *Ortus medicinae,* Helmont asserted that water could not be turned into air, but that it could be attenuated to the point of becoming "vapor," or if still more rare, "gas." These products were only "extenuated water," brought into that state by "local division" and "extraversion of parts." Hence water was vaporized by mere attenuation or attrition of its particles into "atoms," but gas was produced when these were further divided and literally turned inside out by the process of "extraversion." These particles or "atoms" were forced to descend by the exhalations of the stars, eventually coming into contact with the tepid air of the lower atmospheric regions, which would break the sulfurous covering of the corpuscles. This interpretation may sound rather baroque, but—as William Newman brilliantly points out in his 1993 "The Corpuscular Theory of J. B. van Helmont and its Medieval Sources," *Vivarium* 31, no. 1 (1993): 161–91—it is definitely to Helmont that we owe the notion of gas as distinct from vapor, and of different types of gases.

30. For a detailed account of Helmont's corpuscular theory, see William Newman's "The Corpuscular Theory of J. B. van Helmont." For a more extended analysis of the notion of "ferments of transmutation" in the seventeenth century—namely, the corpus of work produced by the Harvard-trained American alchemist George Starkey—see William Newman's "The Corpuscular Transmutational Theory of Eirenaeus Philalethes," in *Alchemy and Chemistry in the XVI and XVII Centuries,* edited by P. Rattansi and A. Clericuzio (Dordrecht: Kluwer Academic Publishers, 1994).

31. According to the legend, Pythagoras's father was a gem-cutter, and it was on a crystallographic basis, through observing his father's work, that Pythagoras developed his concept that "all things are numbers."

32. According to the translation of an Aristotelian fragment offered in Kirk, Raven, and Schofield's *The Presocratic Philosophers.*

33. Several Hebrew sects produced different systems for measuring the body of God. This often entailed a fall from grace with the dominant Hebrew powers.

34. Both Arabs and Jews started from a representation of numbers directly associated with the letters of their respective alphabets, but the Arabs evolved toward an independent notation, whereas the Jews kept the old correlation between the 22 letters and their numerical equivalents. It is possible that the Jewish *Sephirot* represents an aspect of the transition between the archaic system based on 22 (the branches of the *arbor vitalis,* which connect the sephirots) and the decimal systems (represented by the sephirots proper, each an expression of the ten names of God).

35. This question was only settled by the First Vatican Council, in 1870.

36. Mentioned by Margaret Hochdoerfer in *The Conflict between the Religious and the Scientific Views of Albrecht von Haller (1708–1777)* (Lincoln, Nebr.: University of Nebraska Studies in Language, Literature, and Criticism, no. 12, 1932).

37. As quoted by Jonathan Spence in *The Memory Palace of Matteo Ricci* (New York: Viking Penguin, 1984).

38. Associations between numbers and musical scales had been common practice ever since the Pythagorean school demonstrated that music could be expressed through fractions of numbers.

39. In *Centúrias de Curas Medicinais,* trans. Firmino Crespo (Lisbon: Universidade Nova de Lisboa, 1980).

40. As for the Chinese themselves, one of Ricci's first pupils in the art of mnemonics, the eldest son of Governor Lu Wangai, did brilliantly in his exams, but prepared himself mostly through the traditional Chinese methods of repetition and recitation. After the exams, he told a confidant that he had read Ricci's memory book with great care and realized that "though the precepts are the true rules of memory, one has to have a remarkably fine memory to make any use of them."

41. For a further development of Llull's system, see Frances Amelia Yates, *Assaigs sobre Ramon Llull* (Barcelona, Spain: Editorial Empurias, 1985).

42. For a detailed description of Giordano Bruno's philosophical system, see Frances Amelia Yates, *Giordano Bruno and the Hermetic Tradition* (Chicago: University of Chicago Press, 1964).

43. Here are some symbolic properties of a chosen few numbers:

Zero: As the cosmic egg, it symbolizes "all potentialities." It also symbolizes the object that, without value by itself, changes the qualities of others through its position (this "personality" is employed, for instance, in the Tarot).

One: Symbolizes the man standing up, and verticality is sometimes considered prevalent over reason in distinguishing man from all other beings. Its phallic "personality" represents man associated with the work of reproduction. It also symbolizes the Beginning, independent of all manifestation; through this characteristic, it comes to represent the Revelation and the mystical center from which the Spirit shines like the Sun. Ultimately, it is the encompassing figure of Unity, the core of all monotheistic religions.

Two: As the number of all ambivalences and splittings, it represents the opposition and the conflict as well as all achieved balances and latent threats. As the first and most radical of all divisions, it is the emblem of all dialectics, all combats, all movements, and all progresses; and, since division is the basis of multiplication and synthesis, two ultimately stands for the latent antagonism that becomes manifest.

Three: The powerful symbol of the Trinity, which reaches far beyond Father, Son, and Holy Ghost. A good example of the ramifications springing from the Trinitarian organization is the work of Ramón Llull. Llull developed his entire art of memory based on the Trinitarian structure, through which it became a reflection of the Trinity, which Llull intended to be used by all those three powers of the soul that Augustine defined as the reflection of the Trinity in man. As *intellectus,* it was an art of knowing or finding the truth; as *voluntas,* it was an art of training the will toward loving truth; as *memoria,* it was an art of memory for remembering the truth. Llull's system was not only continuous with the Augustinian description of how Trinity was embodied in man, but was also a spinoff of the Scholastic formulations concerning Prudence, itself considered a Trinitarian structure, encompassing *memoria* (the location of artificial memory), *intelligentia,* and *providentia.*

Four: As the implicit principle of the square and the cross, it represents the main divisions of the universe: the four cardinal points (the colors of the four horses in the Book of Revelation); the four humors (the four plagues brought about by the four horsemen in the Book of Revelation); the four winds, the four pillars of the universe, the four seasons, the four rivers of Paradise, the four letters of God's name (YHVH) and of the name of the first man (ADAM), the four writers of the Gospels, etc. In this wholeness it comes to symbolize the totality of all that was created and revealed.

Five: As the sum of the first even and the first odd (2 + 3) and the center of the first nine numbers, it symbolizes the union and the center. Due to its "personality" as the number of the five senses and of the five sensible forms of matter, it represents the totality of the perceivable world. In *The Works and the Days,* Hesiodus postulated five successive humanities, ours being the latest: men of gold, men of silver, men of bronze, and demigods (exterminated during the Trojan War) had inhabited the earth before us. Plutarch considered five to be the number for the succession of species, and the same idea surfaces in Genesis, where the fifth day of Creation brings about all the creatures of the waters and all the creatures of the air. Saint Hildegard von Bingen developed an entire theory of the number five as the number of man. The Pentahedron as the symbol of perfection has been reprised in all kinds of variations ever since the Vespusian drawings.

Six: Since the world was created in six days, this number symbolizes mediation between the beginning and the manifestation; but it also marks the end of times, since the age of the earth was supposed to be six millennia. Thus, in the Book of Revelation, six is the number of sin, and the number of the Antichrist: "the Antichrist shall be marked by the name of the Beast or by the number of its name . . . this number is 666" (13, 17–18). This "personality" fits well with six's identity as the number of Nero, the sixth Roman emperor and the embodiment of the State divinized.

Seven: The theme of the seven days of the week of Creation infiltrates all the Christian basic texts with a special emphasis on the number seven, as exemplified by such minute details as the fact that the two disciples are seven miles outside Emmaus when the resurrected Jesus first appears to them. According to the *Dictionnaire des symboles,* the number seven corresponds to the seven days of the week, the seven planets, the seven notes of the musical scale, the seven degrees of perfection, the seven celestial spheres, the seven petals of the rose, in a pattern recurrent in numerous world mythologies. Several septets are symbols of other septets: the rose of seven petals represents the seven skies, the seven angelic hierarchies, all the perfect groups. Thus seven represents the totality of the planetary orders, the totality of the heavenly homes, the totality of moral orders, and the totality of the energies, especially at the spiritual levels. Ultimately, seven stands in diverse cultures as the representation of eternal life.

Eight: Combining the four cardinal points with their intermediary degrees, eight symbolizes the cosmic balance. Since it comes after the six days of creation and the seventh day of rest, it stands for resurrection and transfiguration, announcing a future eternal era: according to Augustine, eight is number of the day of the just and of the condemnation of the impious.

Nine: As 3 × 3, it embodies the trinity of trinities. Since human gestation takes nine months, nine appears as a ritual value from the writings of Homer onward. Demeter travels through the world in nine days, looking for her daughter Persephone. Leto suffers the pains of childbirth during nine days and nine nights. The nine Muses are born from Zeus after nine nights of love. Thus nine symbolizes all gestations, all fruitful researches, the success of any effort, and the accomplishment of any creation. Also, since each world is symbolized by a triangle, nine corresponds to the unity of Earth, Heaven, and Hell: it can symbolize a celestial sphere as well as the circles of Hell.

Ten: According to the Pythagorean tradition, ten is the sum of the first four numbers (1 + 2 + 3 + 4), and therefore it ultimately represents the universal creation and the totality of movement. Since it is the first binary number, it symbolizes dualism and therefore life and death, their alternation, and their coexistence. This numerical position also endows ten with the "personality" of the ultimate totalizer, as expressed in the Decalogue, which represents the whole of the law in ten commandments that constitute one unity in themselves.

The list could go on forever, since the combination of different "personalities" through the combination of more than one digit provides for all kinds of symbolic values. Recurrent magical numbers are, for instance, ten thousand, seventeen or seventy-two, forty, thirty-six, twenty-one, seventy, one hundred, and one thousand.

Epilogue

1. As this genre is not exactly my cup of tea, I never read the book. Consequently, I will be talking strictly about the film.

2. There are exceptions to this rule, such as the sea urchin sperm. However, the rule is overwhelmingly widespread, and we have no indication that it did not apply to dinosaurs.

3. The mouse is an exception to this rule, but no other exceptions have been recorded so far, in studies ranging from sea urchins to humans.

4. Make no mistake: those little cells preserved in liquid cryogen in in vitro fertilization clinics all over the world, which are presently causing so many headaches for governments, legislators, and ethics committees, are *fertilized embryos,* not *unfertilized eggs.*

5. The mouse, again, is an exception to this rule: this so-called *maternal-zygotic transition* occurs right after the first cell cycle, in the two-cell embryo. This is why mice are even harder to clone than other mammals. After so many decades of genetic engineering, what have we done to our mice, anyway?

6. He was also opposed to the Darwinian theory (which previously included the notion of inheritance of acquired characteristics, but was at Hertwig's time in the process of being identified solely with natural selection) on both philosophical and moral grounds, and came to adopt some moderate Lamarckian views. This serves as yet an-

other example of the immense complexities underlying the philosophical problem of preformation versus epigenesis. Regardless of the social dangers lurking behind the concept of natural selection, Darwin, after all, had written that "Inheritance must be looked at as merely a form of growth"—a statement not bound to please any convinced epigeneticist.

7. Weissmann defined the germplasm as a presumed part of the cell's nucleus, separated from the stock of the parent, endowed with peculiar reproductive powers, and destined to transmit the genetic information from one generation to the next. He proposed this theory as part of his lifelong campaign to rescue Darwin's concept of evolution, ascribing to himself the mission of purging evolutionary theory of any recourse to the idea of inheritance of acquired characteristics.

8. In making this claim, Weissmann cited the observations of contemporary surgeons and physicians, who noted that the growths of disease always conformed strictly, in their cellular nature, to the tissues from which they arose; and that in the healing of wounds, like only grew from cellular like. But this type of empirical knowledge was by no means the only source of inspiration behind Weissmann's model. After Bonnet's "germs" had faded away, the biologists of the nineteenth century searched desperately for a satisfactory substitute for that concept, under more refined and updated guises. As Jan Sapp summarizes the issue in his *Beyond the Gene: Cytoplasmic Inheritance and the Struggle for Authority in Genetics* (New York: Oxford University Press, 1987), "during the second half of the nineteenth century, biology was dominated by particulate theories of heredity which postulated some sort of material corpuscle as the ultimate basis of life. Recognized as the 'bearers of heredity,' a large number of such corpuscles were thought to, somehow, build up the individual organism. The 'gemmules' of Charles Darwin, the 'pangenes' of Hugo de Vries, the 'physiological units' of Herbert Spencer, the 'granules' of Richard Altmann, and other hypothetical entities were highly considered by biologists for almost half a century."

9. Hertwig's conception of the developmental process gained strong supporting evidence with the experiments first performed by Hans Driesch in 1891, in which he separated the first two cells of a sea urchin embryo and showed that both cells grew to form perfect larvae. If, as Weissmann had proposed, changes in the hereditary content of the nucleus had taken place, this normal development of isolated cells from an already formed embryo could never have occurred. This discovery led to a frenzy of cutting, destroying, transplanting, and centrifuging of isolated cells from early embryos. These experiments generally reproduced Driesch's results, so that by the 1930s, an important generalization had emerged: in practically all early embryos, if necessary, an isolated part could give rise to the whole organism (a phenomenon now called *"totipotency of the early blastomeres"*). This, in turn, led to another generalized truism: the principle that "the organism as a whole controls the formative process going on in each part." The end result of all these discoveries, however, was not destined to please Hertwig: the regulative qualities of the egg and the functional equivalence of the nuclei in the early embryonic cells led many embryologists between 1891 and 1910 to localize the primary seat of differentiation in the cell's cytoplasm.

10. See Gilbert's "Cellular Politics: Just, Goldsmith, and the Attempts to Reconcile

Embryology and Genetics," in *The American Development of Biology,* ed. R. Rainger, K. Benson, and J. Maienschein (Philadelphia, Pa.: University of Pennsylvania Press, 1988).

11. See Gilbert's "The Embryological Origin of the Gene Theory," *Journal of the History of Biology* 11 (1978): 307–51.

12. As quoted by Sapp in *Beyond the Gene.*

13. According to Gilbert in *Developmental Biology,* 4th ed. (Sunderland, Mass.: Sinauer Associates, 1994), a morphogenetic field can be described as a group of cells whose position and fate are specified with respect to the same set of boundaries. Thus a particular field of cells will give rise to its particular, predestined organ (a forelimb, an eye, or a tail), even when transplanted to a different part of the embryo; and the cells of the field can regulate their fate to make up for missing cells.

14. For an interesting account of this struggle, see Gregg Mitman and Anne Fausto-Sterling's "Whatever Happened to Planaria?," in *The Right Tools for the Job: At Work in Twentieth-Century Life Sciences,* edited by A. E. Clarke and J. H. Fujimura (Princeton, N.J.: Princeton University Press, 1991).

15. For a detailed account of this specific debate, see Sapp's *Beyond the Gene.*

BIBLIOGRAPHY

Adelmann, Howard B. *Marcello Malpighi and the Evolution of Embryology.* Ithaca, N.Y.: Cornell University Press, 1966.

Akerblom, Kjell. *Astronomy and Navigation in Polynesia and Micronesia: A Survey.* Stockholm, Sweden: The Ethnographical Museum Monograph Series, 1968.

Anderson, Lorin. "Charles Bonnet's Taxonomy and Chain of Being." *Journal of the History of Ideas* 37 (1976): 45–58.

———. *Charles Bonnet and the Order of the Known.* Boston: D. Reidel, 1982.

Andry du Bois-Regard, Nicolas. *An account of the Breeding of Worms in Human Bodies; Their Nature, and Several Sorts; Their Effects, Symptoms and Prognostics. With the true Means of to avoid them, and Med'cines to cure them.* London: H. Rhodes and A. Bell, 1701.

———. *Orthopaedia, or the Art of Correcting Deformities in Children.* Facsimile reproduction of the first edition in English, London, 1743. Philadelphia: Lippincott, 1961.

Aristotle. *Aristotle's History of Animals.* Translated by Richard Cresswell. London: H. G. Bohn, 1869.

———. *The Generation of Animals.* Translated by A. L. Peck and edited by G. P. Goold. Cambridge, Mass.: Harvard University Press, 1990.

Armitage, Angus. *Copernicus and the Reformation of Astronomy.* London: Published for the Historical Association by George Philip & Son, 1950.

Astruc, Jean. *A Treatise on All the Diseases Incident to Women.* New York: Garland, 1985.

———. *A Treatise of the Venereal Diseases.* New York: Garland, 1986.

Barth, Lester George. *Embryology.* Rev. and enl. ed. Dryden Press Publications in the Biological Sciences. New York: Holt, Rinehart & Winston, 1953.

Bilsky, Emily D., Moshe Idel, and Elfriede Ledig. *Golem! Danger, Deliverance and Art.* New York: Jewish Museum, 1988.

Blanckenhagen, Peter H. "Easy Monsters." In *Monsters and Demons in the Ancient and Medieval Worlds: Papers Presented in Honor of Edith Porada.* Mainz on Rhine, Germany: Verlag Philipp von Zarben, 1987.

Boaistuau, Pierre. *Certaine Secrete Wonders of Nature, Containing a Description of Sundry Strange Things, Seming Monstrous to our Eyes and Judgement, bicause we are not Priuie to the Reasons of Them. Gathered out of Divers Learned Authors as Well as Greek as Latine, Sacred as Prophane.* London: H. Bynnemann, 1569.

Boerhaave, Hermann. *Sermo academicus, de Comparando certo in Physicis; quem habuit in Academia Lugduno-Batava.* Lugduni Batavorum, apud. Leiden: Petrum Vander Aa, Bibliopolam, 1715.

———. *An essay on the virtue and efficient cause of magnetical cures; To which is added a new method of curing wounds without pains, and without the application of remedies. Hitherto kept as a secret in private families.* London, 1743.

———. *Boerhaave's Orations.* Translation and notes by E. Kegel-Bringgreve and A. M. Luyendijk-Elshout. Leiden: E. J. Brill/Leiden University Press, 1983.

Bonnet, Charles. *Considérations sur les corps organisés; Ou l'on traite de leur Origine, de leur Dévelopment, de leur Réproduction, etc; & ou l'on a rassemblé en abrégé tout ce que l'Histoire Naturelle offre de plus certain & de plus intéressant sur se sujet.* Amsterdam: Chez Marc-Michel Rey, 1762.

Borges, Jorge Luis. *The Book of Imaginary Beasts.* With Margarita Guerrero. Reviewed, enlarged, and translated by Norman Thomas di Giovanni, in collaboration with the author. New York: Dutton, 1969.

Bouvier, J. B. *Dissertatio in sextum Decalogi praeceptum.* 8th ed. Le Mans, France: n.p., 1836.

Bowler, Peter. "Preformation and Pre-existence in the Seventeenth Century: A Brief Analysis." *Journal of the History of Biology* 4 (1971): 221–45.

Bredvold, Louis I. "The Invention of Ethical Calculus." In *The Seventeenth Century: Studies in the History of English Thought and Literature from Bacon to Pope by Richard Foster Jones and Others Writing in His Honor.* Stanford, Calif.: Stanford University Press, 1951.

Brendel, Zacharias. *Zachariae Brendelli Chimia in artis formam redacta . . . ; disquisitio curata de famosissima praeparatione, auri potabilis instituitur . . .* Lugduni Batavorum: Ex officina Arnoldi Doude, 1671.

Briffault, Robert. *The Mothers.* London: Allen & Unwin, 1959.

Buber, Martin. *Tales of the Hasidim.* New York: Schocken Books, 1991.

Buffon, Georges-Louis Leclerc, comte de. *Histoire naturelle, générale et particuliere, avec la description du Cabinet du Roi.* 44 vols. Paris: Imprimerie Royale, puis Plassans, 1749–1804.

Bullough, Vern L. *Sexual Variance in Society and History.* New York: Wiley, 1976.

Burnham, Sophy. *Le livre des anges.* Alleur, Belgium: Marabout, 1994.

Calvin, Jean. *Calvin's Old Testament Commentaries: The Rutherford House Translation.* Grand Rapids, Mich.: Rutherford House, 1993.

Cantimpré, Thomas de. *Incipit liber qui dicit[ur] Bonu[m] vniu[er]sale de p[ro]p[ri]etatibus apum.* Cologne: Pr. of Augustine, De fide, 1473.

Carlson, Bruce M. *Patten's Foundations of Embryology.* New York: McGraw-Hill, 1981.

Cassirer, Ernst. *The Philosophy of the Enlightenment.* Princeton, N.J: Princeton University Press, 1951.

Castel-Branco, Cristina. "O Lugar e o Significado—Os Jardins dos Vice-Reis." Ph.D. thesis, Lisbon, Portugal, 1993.

Castellani, Carlo. "Spermatozoan Biology from Leeuwenhoek to Spallanzani." *Journal of the History of Biology* 6, no. 1 (1973): 37–68.

Castelli, Bartholomaei. *Lexicon medicum graeco-latinum.* Leipzig: apud. Thoman Fritsch, 1713.

Cattaui, Georges. "Paracelse et sa postérité." In *Paracelse I: l'homme, le médecin, l'alchimiste,* by Béatrice Whiteside. Paris: Table Ronde, 1966.

Cestaro, Antonio. *Juan Caramuel Vescovo di Satriano e Campagna, 1657–1673: Cultura e vita religiosa nella seconda meta del Seicento.* Salerno, Italy: Edisud, 1992.

Chevalier, Jean, and Alain Gheerbrant. *Dictionnaire des symboles: Mythes, rêves, coutumes, gestes, formes, figures, couleurs, nombres.* Paris: Éditions Robert Laffont and Éditions Jupiter, 1982.

Childe, V. Gordon. *Most Ancient East: The Oriental Prelude to European Prehistory.* New York: Knopf, 1929.

Cole, F. J. *Early Theories of Sexual Generation.* Oxford: Clarendon Press, 1930.

Collins, Harry, and Trevor Pinch. *The Golem: What Everybody Should Know about Science.* Cambridge: Cambridge University Press, 1993.

The Complete Master-Piece: Displaying the Secrets of Nature in the Generation of Man. Vol. 1 of *Aristotle's Works Completed.* 22d ed. Printed, and sold by the Booksellers. London, 1741.

Correia, Maximino, J. O. Miller Guerra, L. Leibowitz, M. Pina, M. Meneses, José Dias, L. Glesinger, Caria Mendes, and José Boléo. *Amato Lusitano: Miscelânea de Cartas e Documentos.* Castelo Branco, Portugal: Estudos de Castelo Branco, 1968.

Costa, Celestino da. *Elementos de Embriologia.* Coimbra, Portugal: Livraria J. Rodrigues e Companhia, 1933.

Coulter, Harris L. *Divided Legacy: A History of the Schism in Medical Thought.* Washington: Wehawken Book Co., 1973–1977.

d'Agoty, Gautier. *Anatomie des parties de la generation de l'homme et de la femme, representées avec leurs couleurs naturelles, selon le nouvel art. Jonite à L'angéologie de tout le corps humain, et à ce qui concerne la grosesse et les acouchemens.* Paris: chez J. B. Brunet, 1773.

Darwin, Charles. *On the Origin of Species, by Means of Natural Selection; Or the Preservation of Favoured Races in the Struggle for Life.* London: G. Richards, 1902.

Daston, Lorraine. "Rational Individuals versus Laws of Society: From Probability to Statistics." In *Probability since 1800: Interdisciplinary Studies of Scientific Development,* edited by Michael Heidelberger, Lorenz Kruger, and Rosemarie Rheinwald. Germany: B. Kleine-Verlag, 1983.

Davidson, Arnold. "The Horror of Monsters." In *The Boundaries of Humanity,* edited by J. J. Sheenan and M. Sisna. Berkeley: University of California Press, 1991.

Davidson, Gustav. *A Dictionary of Angels, Including the Fallen Angels.* New York: The Free Press, 1967.

Dawson, Virginia P. *Nature's Enigma: The Problem of the Polyp in the Letters of Bonnet, Trembley and Réaumur.* Philadelphia, Pa.: American Philosophical Society, 1987.

De Kruif, Paul. *Microbe hunters.* New York: Harcourt, Brace, 1926.

DeMause, Lloyd. *The History of Childhood.* New York: Psychohistory Press, 1974.

Dennett, Daniel Clement. *Consciousness Explained.* Boston: Little Brown, 1991.

Deussen, Paul. *Fundamental Philosophy of Upanishads.* Authorized English translation by A. S. Geden. Delhi: Kanti Publications, 1989.

Dias, José Lopes. Dr. *João Rodrigues de Castelo Branco, Amato Lusitano, ensaio bio-bibliográfico.* Lisbon: Publicações do Congresso do Mundo Português, Vol. 13, Tomo 2°, 1940.

Dicks, D. R. *Early Greek Astronomy to Aristotle.* Ithaca, N.Y.: Cornell University Press, 1970.

Dictionary of the History of Science. Edited by William F. Bynum, E. Janet Browne, and Roy Porter. Princeton, N.J.: Princeton University Press, 1984.

Dixson, A. F. "Sexual Selection, Sperm Competition and the Evolution of Sperm Length." *Folia Primatologica* (Basel) 61 (1993): 221–27.

Dobell, Clifford. *Antoni van Leeuwenhoek and His Little Animals: Being Some Account of the Father of Protozoology and Bacteriology and his Multifarious Discoveries in These Disciplines.* New York: Dover Publications, 1960.

Drewermann, Eugen. *Discovering the God Child Within: A Spiritual Psychology of the Infancy of Jesus.* Translated from the German by Peter Heinegg. New York: Cross-road, 1994.

———. *Dying We Live: Meditations for Lent and Easter.* Translated from the German by Linda M. Maloney and John Drury. Maryknoll, N.Y.: Orbis Books, 1994.

Dreyer, J. L. E. *A History of Astronomy from Thales to Kepler.* New York: Dover Publications, 1953.

Ducassé, Pierre. *Malebranche, sa vie, son oeuvre, avec un exposé de sa philosophie.* Paris: Presses Universitaires de France, 1942.

Dujovich, Adolfo. *Amato Lusitano, medico y botanico sefardi, su epoca, vida y obra.* Coleccion Grandes Figuras del Judaismo, vol. 80. Buenos Aires, Argentina: Biblioteca Popular Judia, 1974.

Durant, Will. *Our Oriental Heritage: Being a History of Civilization in Egypt and the Near East to the Death of Alexander, and in India, China and Japan from the Beginning to Our Own Day; with an Introduction on the Nature and Foundations of Civilization.* Vol. 1 of *The Story of Civilization.* New York: Simon & Schuster, 1963.

———. *Caesar and Christ: A History of Roman Civilization and of Christianity from Their Beginnings to A.D. 325.* Vol. 3 of *The Story of Civilization.* New York: Simon & Schuster, 1971.

Edwards, John. *A demonstration of the Existence and Providence of God, from the contemplation of the visible structure of the greater and lesser world, in two parts, the first, shewing the Excellent Contrivance of the Heavens, Earth, Sea, etc; the second, the Wonderful Formation of the Body of Man.* London: Printed by J. D. for Jonathan Robinson at the Golden Lion, and John Wyat at the Rofe in St. Paul's Church-Yard, 1696.

Eisner, Thomas, and Edward O. Wilson. *The Insects.* San Francisco: W. H. Freeman, 1977.

Eliade, Mircea. *The Forge and the Crucible.* Translated by Stephen Corrin. Chicago: University of Chicago Press, 1956.

Ellis, Havelock. *Man and Woman: A Study of Human Secondary Sexual Characters.* London: Walter Scott; New York: Scribner, 1900.

Elsholtz, Johann Sigismund. *The curious distillatory, or, The art of distilling coloured liquors, spirits, oyls, etc, from vegetables, animals, minerals and metals . . . containing many experiments . . . relating to the production of colours, consistence and heat . . .: together with several experiments upon the blood (and its serum) of diseased persons, with divers other collateral experiments.* London: Printed by J. D. for Robert Boulter, 1677.

Emerson, Barbara. *Leopold II of the Belgians: King of Colonialism.* London: Weinfeld and Nicolson, 1979.

Emerton, Norma E. *The Scientific Reinterpretation of Form.* Ithaca, N.Y.: Cornell University Press, 1984.

Fabricius of Aquapendente, Hieronymus. *The embryological treatises of Hieronymus Fabricius of Aquapendente: "The formation of the egg and the chick" and "The formed fetus."* Facsimile edition with an introduction, a translation, and a commentary by Howard B. Adelmann. Ithaca, N.Y.: Cornell University Press, 1942.

Farley, John. *The Spontaneous Generation Controversy from Descartes to Oparin.* Baltimore, Md.: Johns Hopkins University Press, 1974.

———. *Gametes and Spores: Ideas about Sexual Reproduction, 1750–1914.* Baltimore, Md.: Johns Hopkins University Press, 1982.

Feldman, W. M. *Rabbinical Mathematics and Astronomy.* New York: Hermon Press, 1978.

Fellows, Otis E., and Stephen F. Milliken. *Buffon.* New York: Twayne Publishers, 1972.

Findlen, Paula. "Jokes of Nature and Jokes of Knowledge: The Playfulness of Scientific Discourse in Early Modern Europe." *Renaissance Quarterly* 43, no. 2 (1990): 292–331.

———. "Science as a Career in Enlightenment Italy: The Strategies of Laura Bassi." *Isis* 84 (1993): 441–69.

Fink, Rachel. *A Dozen Eggs.* VHS format. Sunderland, Mass.: Sinauer Associates, 1991.

Fischer, Henry G. "The Ancient Egyptian Attitude towards the Monstrous." In *Monsters and Demons in the Ancient and Medieval Worlds: Papers Presented in Honor of Edith Porada.* Mainz on Rhine, Germany: Verlag Philipp von Zarben, 1987.

Flandrin, Jean Louis. *Le sexe et l'Occident: Évolution des attitudes et des comportements.* Paris: Seuil, 1981.

Fodor, Jerry A. *Representations: Philosophical Essays on the Foundations of Cognitive Science.* Cambridge, Mass.: MIT Press, 1981.

Francavilla, A., and L. Zamboni. "Differentiation of Mouse Ectopic Germ Cells in Intra- and Perigonadal Locations." *Journal of Experimental Zoology* 233, no. 1 (1985): 101–9.

Freitas Mourão, Ronaldo Rogério de. *Astronomia do Macunaíma.* Bahia, Brazil: Francisco Alves, 1984.

Fristrom, J. W., D. Fristrom, E. Fekete, and A. H. Kuniyuki. "The Mechanism of Evagination of Imaginal Discs of *Drosophila melanogaster. American Zoologist* 17 (1977): 671–84.

Fristrom, J. W., R. Raikow, W. Petri, and D. Stewart. "In Vitro Evagination and RNA Synthesis in Imaginal Discs of *Drosophila melanogaster.*" In *Problems in Biology: RNA in Development.* Salt Lake City: University of Utah Press, 1969.

Galen. *Galen on the Affected Parts: Translation from the Greek Text with Explanatory Notes.* New York: S. Karger, 1976.

Gallagher, Catherine, and Thomas Laqueur (eds.). *The Making of the Modern Body: Sexuality and Society in the Nineteenth Century.* Berkeley: University of California Press, 1987.

Gandhi, Mahatma. *La jeune Inde.* Paris: Librarie Stock, 1924.

Garden, George. "A Discourse Concerning the Modern Theory of Generation." *Philosophical Transactions of the Royal Society of London* 17 (1691): 474–83.

Gasking, Elizabeth B. *Investigations into Generation, 1651–1828.* Baltimore, Md.: John Hopkins University Press, 1967.

Gilbert, Scott F. "The Embryological Origin of the Gene Theory." *Journal of the History of Biology* 11 (1978): 307–51.

———. "Cellular Politics: Just, Goldschmidt, and the Attempts to Reconcile Embryology and Genetics." In *The American Development of Biology,* edited by R. Rainger, K. Benson, and J. Maienschein. Philadelphia, Pa.: University of Pennsylvania Press, 1988.

———. "Commentary: Cytoplasmic Action in Development." *Quarterly Review of Biology* 66 (1991): 309–16.

———. *Developmental Biology.* 4th ed. Sunderland, Mass.: Sinauer Associates, 1994.

Gold, Judith Taylor. *Monsters and Madonnas: The Roots of Christian Anti-Semitism.* New York: New Amsterdam, 1988.

Gomendio, M., and E. R. Roldan. "Sperm Competition Influences Sperm Size in Mammals." *Proceedings of the Royal Society of London B. Biological Sciences* 243 (1991): 181–85.

Gould, Stephen Jay. *Ontogeny and Phylogeny.* Cambridge, Mass.: Belknap Press of Harvard University Press, 1977.

———. *The Flamingo's Smile: Reflections on Natural History.* New York: Norton, 1985.

Gouveia, A. J. Andrade. *Garcia d'Orta e Amato Lusitano na ciência do seu tempo.* Biblioteca Breve, vol. 102. Lisbon: Escolar Editora, 1985.

Graaf, Reinier de. *De mulierum organis generationi inservientibus, 1672.* Facsimile with an introduction by J. A. van Dongen. Nieuwkoop: B. de Graaf, 1965.

Grant, Edward. *In Defense of Earth's Centrality and Immobility: Scholastic Reaction to Copernicanism in the Seventeenth Century.* Transactions of the American Philosophical Society, vol. 74, pt. 4. Philadelphia, Pa.: The American Philosophical Society, 1984.

Hall, A. Ruppert. *The Scientific Revolution, 1500–1800: The Formation of the Modern Scientific Attitude.* Boston: Beacon Press, 1966.

Haller, Albrecht von. *De monstris dissertatio II. Qua trium monstrorum anatome ed ad contraria D. Lemeryi argumenta responsiones continentur.* Göttingen: n.p., n.d.

————. *Eléments de physiologie: ou, Traité de la structure et des usages des différentes parties du corps humain. Traduit du Latin.* Paris: Prault, 1752.

————. *Pathological Observations, Chiefly from Dissections of Morbid Bodies.* London: D. Wilson and T. Durham, 1756.

————. *Sur la formation du coeur dans le poulet: Sur l'oeil, sur la structure du jaune, etc: Premier mémoire. Expose des faits: Second mémoire.* Lausanne, Switzerland: Chez Marc.-Mich. Bousquet & Co., 1758.

————. *Poesies de Mr. Haller, traduites de l'Allemand.* Bern, Switzerland: Chez Abr. Wagner Fils, 1760.

————. *Albrecht von Hallers Briefe an August Tissot, 1754–1777.* Bern, Switzerland: Huber, 1777.

————. *Letters from Baron Haller to His Daughter, on the Truths of the Christian Religion.* Albany, N.Y.: H. C. Southwick, 1816.

————. *The Natural Philosophy of Albrecht von Haller.* Edited by Shirley Ann Roe. New York: Arno Press, 1981.

Hankins, Thomas L. *Science and the Enlightenment.* New York: Cambridge University Press, 1985.

Hartsoeker, Nicolas. *Éssai de dioptrique.* Paris: Jean Anisson, 1694.

————. *Cours de physique; accompagné de plusieures pièces concernant la physique qui ont déjà paru et d'un extrait critique des letters de m. Leeuwenhoek.* La Haye: Jean Swart, 1730.

Harvey, William. *Exercitationes de generatione animalium: quibus accedunt quaedum de partu, de membranis ac humoribus uteri, & de conceptione.* London: Typis Du-Gardinis: Impensis Octaviani Pulleyn, 1651.

Heninger, S. K. Jr. "Pythagorean Cosmology and the Triumph of Heliocentrism." In *Le Soleil à la Renaissance: Sciences et mythes.* Paris: Presses Universitaires de France, 1965.

Herrnstein, Richard J., and Charles M. Murray. *The Bell Curve: Intelligence and Class Structure in American Life.* New York: Free Press, 1994.

Hertwig, Oscar. *The Biological Problem of To-day: Preformation or Epigenesis? The Basis of a Theory of Organic Development.* Introduction by Joseph Anthony Mazzeo; authorized translation by P. Chalmers Mitchell. Oceanside, N.J.: Dabor Science Publications, 1977.

Hill, John. *Lucina sine concubitu: A letter humbly addressed to the Royal Society, in which it is proved, by most incontestable evidence, drawn from reason and practise, that a woman may conceive and be brought to bed, without any commerce with man.* Edinburgh: Collectanea adamantea, 7, 1885.

Hirts, Barton Cooke, and George A. Piersol. *Human Monstrosities.* Philadelphia, Pa.: Lea Brothers, 1891.

Hobart, Michael E. *Science and Religion in the Thought of Nicolas Malebranche.* Chapel Hill: University of North Carolina Press, 1982.

Hochdoerfer, Margarete. *The Conflict between the Religious and Scientific Views of Albrecht von Haller (1708–1777).* Lincoln, Nebr.: University of Nebraska Studies in Language, Literature, and Criticism, no. 12, 1932.

Hogg, H., and A. McLaren. "Absence of a Sex Vesicle in Meiotic Foetal Germ Cells Is Consistent with an XY Sex Chromosome Constitution." *Journal of Embryology and Experimental Morphology* 88 (1985): 327–32.

Howard, Clifford. *Sex Worship: An Exposition of the Phallic Origin of Religion.* Chicago: Chicago Medical Book Co., 1909.

Hughes, Arthur. *A History of Cytology.* London: Abelard-Schuman, 1959.

Hutin, Serge. "Les doctrines secretes." In *Paracelse: l'homme, le médecin, l'alchimiste,* by Béatrice Whiteside. Paris: Table Ronde, 1966.

———. *Henry More: Éssai sur les doctrines théosophiques chez les Platoniciens de Cambridge.* Hildesheim, Germany: Georg Olms Verlagsbuchhandlung, 1966.

Huxley, Julian S., and G. R. de Beer. *The Elements of Experimental Embryology.* London: Cambridge University Press, 1963.

Idel, Moshe. *Le Golem.* Paris: Cerf, 1992.

Jâbir ibn Hayân. *Dix traités d'alchimie.* Translated and presented by Pierre Lory. La Bibliothèque de l'Islam. Paris: Sindbad, 1983.

Johnson, Francis R. *Astronomical Thought in Renaissance England: A Study of English Scientific Writings from 1500 to 1645.* Baltimore, Md.: Johns Hopkins University Press, 1937.

Johnson, Paul. *A History of Christianity.* New York: Atheneum, 1976.

———. *A History of the Jews.* New York: Harper & Row, 1987.

Jorge, Ricardo. *Comentários à Vida, Época e Obra de Amato Lusitano.* Porto, Portugal: Tip. A Vapor da Enciclopédia Portuguesa, 1916.

Kappler, Claude. *Monstres, démons et merveilles à la fin du Moyen Age.* Paris: Payot, 1980.

Keele, Kenneth D. *William Harvey: The Man, the Physician, and the Scientist.* London: Thomas Nelson and Sons, 1965.

Kepler, Johannes. *The Harmonies of the World.* Translated by Charles Glenn Wallis. Great Books of the Western World, vol. 16. Chicago: Encyclopaedia Britannica, 1952.

Keyserling, Hermann, Graf von. *The Travel Diary of a Philosopher.* New York: Harcourt, Brace, 1925.

Kircher, Athanasius. *Athanasii Kircheri & Soc. Jesu mundus subterraneus, in XII libros digestus.* Amsterdam: Joannem Janssonium a Waesberge & fillios, 1678.

Kirk, G. S., J. E. Raven, and M. Schofield. *The Presocratic Philosophers.* 2d ed. Cambridge: Cambridge University Press, 1983.

Knight, Charles. *Insect Transformations.* London: The Library of Entertaining Knowledge, 1830.

Kouznetzov, Boris G. "Le Soleil comme le centre du monde, et l'homogénéité de l'espace chez Galilée." In *Le Soleil à la Renaissance: Sciences et mythes.* Paris: Presses Universitaires de France, 1965.

Kraus, Paul. *Jâbir Ibn Hayân.* Vol. 1 of *Contribution à l'histoire des idées scientifiques dans l'Islam.* Cairo, Egypt: Imprimerie de l'Institut Français d'Archéologie Orientale, 1943.

Langman, J., and D. Wilson. "Embryology and Congenital Malformations of the Female Genital Tract." In *Pathology of the Female Genital Tract,* 2d ed., edited by A. Blaustein. New York: Springer-Verlag, 1982.

Languirand, Jacques. *La magie des nombres: de McLuhan á Pythagore.* France: Editions de Mortagne, 1995.

Laqueur, Thomas. *Making Sex: Body and Gender from the Greeks to Freud.* Cambridge, Mass.: Harvard University Press, 1990.

Lécuyer, Bernard-Pierre. "Interconnections between Bio-medical and Sociological Trends of Development of Probabilistic Thinking under Napoleon, the Restorated Bourbon France and the July Monarchy." In *Probability since 1800,* edited by Michael Heidelberger, Lorenz Kruger, and Rosemarie Rheinwald. Germany: B. Kleine-Verlag, 1983.

Leeuwenhoek, Antonii A. *Epistolae ad Societatem Regiam Anglicam, et alios illuistres viros seu Continuatio mirandorum Arcanorum Naturae detectorum. Quadraginta Epistolis contentorum.* Lugduni Batavorum: Apud. Joh. Arnold. Langerak, 1719.

———. *The Collected Letters of Antoni van Leeuwenhoek.* Edited, illustrated, and annotated by a committee of Dutch scientists. Amsterdam, : Swets & Zeitlinger, 1952.

Leibniz, Gottfried Wilhelm von. *Monadology and Other Philosophical Essays.* Translated by Paul Schrecker and Anne Martin Schrecker. Introduction and notes by Paul Schrecker. New York: Bobbs-Merrill, 1951.

Lemos, Maximiano de. *Amato Lusitano, Sua Vida e Sua Obra.* Porto, Portugal: Martins, 1907.

Linnaeus, Carolus. *Systema Naturae.* Facsimile of the first edition (1735), with an introduction and a first English translation of the *Observationes* by M. S. J. Engel-Ledeboer and H. Engel. Nieuwkoop: n.p., 1964.

Lippert, Julius. *The Evolution of Culture.* Translated and edited by George Peter Murdoch. New York: Macmillan, 1931.

Lister, Martin. *Dissertatio de humoribus.* London, 1709.

Liungman, Carl G. *Dictionary of Symbols.* Santa Barbara, Calif.: ABC-Clio, 1991.

Locy, William A. *The Story of Biology.* Garden City, N.Y.: Garden City Publishing, 1925.

———. *Biology and Its Makers.* New York: Henry Holt, 1935.

Lusitano, Amato. *Centúrias de Curas Medicinais.* Translated by Firmino Crespo. Lisbon: Universidade Nova de Lisboa, 1980.

Mackinnon, Flora Isabel. *Philosophical Writings of Henry More.* New York: AMC Press, 1969.

Malebranche, Nicolas. *De la recherche de la verité ou l'on traite de la nature de l'esprit de l'homme, & de l'usage qu'il doit en faire pour éviter l'erreur dans les Sciences— Septieme édition, revue & augmenté de plusieurs Eclaircissements.* Paris: Chez Christophe David, 1721.

Manning, J. T., and A. T. Chamberlain. "Sib Competition and Sperm Competitiveness: An Answer to 'Why So Many Sperms?' and the Recombination/Sperm Number

Correlation." *Proceedings of the Royal Society London B. Biological Sciences* 256 (1994): 177–82.

Maor, Eli. *To Infinity and Beyond.* Boston: Birkhauser, 1986.

Maupertuis, P. L. M. de. "Lettres de M. de Maupertuis." Dresden, Germany, 1752.

Maxwell, William. *Drei Bucher der magnetischen Heilkunde; worin sowohl die Teorie als Praxis dieser Wissencschaft enthalten ist, viele geheime Naturwunder geoffenbart, die bisher unbekannten Wirkumgen des Lebensgeistes enthullt und die Fundamente dieser ganzen verborgenen Kunst mit der starksten, auf Erfahrung gestutzten Beweingrunden dargestellt werden.* Kleiner Wunder-Schauplatz, Th. 3. Stuttgart: Scheible, 1855.

McLaren, Ann. "Studies on Mouse Germ Cells Inside and Outside the Gonad." *Journal of Experimental Zoology* 228, no. 2 (1983): 167–71.

———. "Meiosis and Differentiation of Mouse Germ Cells." *Symposia of the Society of Experimental Biology* 38 (1984): 7–23.

———. "Sex Determination in Mammals." *Trends in Genetics* 4, no. 6 (1988): 153–57.

———. "Somatic and Germ-Cells in Mammals." *Philosophical Transactions of the Royal Society of London, Biological Sciences* 322, no. 1208 (1988): 3–9.

———. "Sex Determination: What Makes a Man a Man?" *Nature* 346, no. 6281 (1990): 240–44.

———. "Development of the Mammalian Gonad: The Fate of the Supporting Cell Lineage." *Bioessays* 13, no. 4 (1991): 151–56.

———. "Sex Determination: The Making of Male Mice." *Nature* 351, no. 6322 (1991): 117–21.

———. "Sex Determination in Mammals." *Oxford Review of Reproductive Biology* 13 (1991): 1–33.

Meinhardt, H. "A Model for Pattern Formation of Hypostome, Tentacles, and Foot in Hydra: How to Form Structures Close to Each Other, How to Form Them at a Distance." *Developmental Biology* 157 (1993): 321–33.

Merchant, Carolyn. *The Death of Nature: Women, Ecology and the Scientific Revolution.* San Francisco: Harper San Francisco, 1980.

Mettrie, Julien Offray de la. *L'homme machine, avec une introduction et des notes de J. Assezat.* Paris: Frederic Henry, 1865.

Meyer, Arthur William. *The Rise of Embryology.* Stanford, Calif.: Stanford University Press, 1939.

Minsky, Marvin Lee. *Computation: Finite and Infinite Machines.* Prentice-Hall Series on Automatic Computation. Englewood Cliffs, N.J.: Prentice-Hall, 1967.

Mitman, Gregg, and Anne Fausto-Sterling. "Whatever Happened to Planaria?" In *The Right Tools for the Job: At Work in Twentieth-Century Life Sciences,* edited by A. E. Clarke and J. H. Fujimura. Princeton, N.J.: Princeton University Press, 1991.

More, Henry. *Enthusiasmus Triumphatus.* The Augustan Reprint Society 118. Los Angeles: William Andrews Clark Memorial Library, University of California at Los Angeles, 1966.

Moscoso, Javier. "Experimentos de regeneración animal: 1686–1765. Cómo defender la pre-existencia." *Dynamis, Acta Hispanica ad Medicinae Scientiarumque Historiam Illustrandam* 15 (1995): 341–73.

Muller, W. A. "Competition for Factors and Cellular Resources as a Principle of Pattern Formation in *Hydra.* I. Increase of the Potentials for Head and Bud Formation and Rescue of the Regeneration-Deficient Mutant *reg-16* by Treatment with Diacylglycerol and Arachidonic Acid." *Developmental Biology* 167 (1995): 159–74.

Needham, J. T. *Observations on the generation, composition and decomposition, of animal and vegetable substances.* London, 1749. (Summary in *Philosophical Transactions of the Royal Society of London* 45 (1750): 615.

Needham, Joseph. *A History of Embryology.* Cambridge: Cambridge University Press, 1934.

Nelkin, Dorothy, and M. Susan Lindee. *DNA Mystique: The Gene as a Cultural Icon.* New York: W. H. Freeman, 1995.

Newman, William R. "The Corpuscular Theory of J. B. van Helmont and Its Medieval Sources." *Vivarium* 31, no. 1 (1993): 161–91.

———. "The Corpuscular Transmutational Theory of Eirenaeus Philalethes." In *Alchemy and Chemistry in the XVI and XVII Centuries,* edited by P. Rattansi and A. Clericuzio. Dordrecht: Kluwer Academic Publishers, 1994.

Nordenskiold, Erik. *The History of Biology: A Survey.* New York: Tudor Publishing, 1928.

Norris, R. P., and P. N. Appleton. "A Survey of the Barbrook Stone Circles and Their Claimed Astronomical Alignments." In *Archaeoastronomy in the Old World,* edited by D. C. Heggie. London: Cambridge University Press, 1982.

Onania; or The heinous sin of self-pollution, and all its frightful consequences, in both sexes, considered: With spiritual and physical advice to those, who have already injur'd themselves by this abominable practice: And seasonable admonition to the youth (of both sexes) and those whose tuition they are under, whether parents, guardians, masters, or mistresses.: To which is added, a letter from a lady (very curious) concerning the use and abuse of the marriage bed. With the author's answer thereto. Six lines from Genesis. London printed; re-printed at Boston, for John Phillips, 1724.

Oppenheimer, Steven B., and George Lefevre Jr. *Introduction to Embryonic Development.* Boston: Allyn and Bacon, 1984.

Palfyn, Jean. *Description Anatomique des Parties de la Femme qui servent à la Génération, avec un Traité des Monstres, de leurs causes, de leus Nature, & de leurs differences: Et une Description Anatomique de la disposition surprenante de quelques Parties Externes & Internes de Deux Enfants Nés dans la ville de Gand, Capitale de Flandres le 28 Avril 1703. Etc. Etc. Lesquels ouvrages on peut considérer comme une suite de L'ACCOUCHEMENT DES FEMMES par M. Mauriceau. Avec figures.* Leiden: Chez la Veve de Bastiaan Schouten, 1703.

Palmer, Tobias. *Un ange aupres de moi.* Translated by Laurence E. Fritsch. Paris: Table Ronde, 1995.

Paracelsus. *Hermetic Chemistry.* Vol. 1 of *The hermetic and alchemical writings of Au-*

reolus Philippus Theophrastus Bombast, of Hohenheim, called Paracelsus the Great, now for the first time faithfully translated into English, edited by Arthur Edward Waite. Berkeley, Calif.: Shambhala, 1976.

Paré, Ambroise. *On Monsters and Marvels.* Translation, introduction, and notes by Janis L. Pallister. Chicago: University of Chicago Press, 1982.

Park, Katharine, and Lorraine J. Daston. "Unnatural Conceptions: The Study of Monsters in Sixteenth- and Seventeenth-Century France and England." *Past and Present* 92 (1981): 20–54.

Parker, G. A., and M. E. Begon. "Sperm Competition Games: Sperm Size and Number under Gametic Control." *Proceedings of the Royal Society London B. Biological Sciences* 253 (1993): 255–62.

Pastine, Dino. *Juan Caramuel: Probabilismo ed enciclopedia.* Firenze, Italy: La Nuova Italia, 1975.

Patai, Raphael. *The Jewish Alchemists: A History and Source Book.* Princeton, N.J.: Princeton University Press, 1994.

Pepys, Samuel. *The Diary of Samuel Pepys.* Vol. 8, ed. by R. Latham and W. Matthews. London: Bell & Hyman, 1667.

Perrault, Claude. *Mémoires pour servir à l'histoire naturelle des animaux.* Amsterdam: P. Mortier, 1736.

Piercy, Marge. *He, She and It.* New York: Knopf/Random House, 1991.

Plantade, François de. *Extrait d'une lettre de M. Dalenpatius à l'auteur de ces Nouvelles, contenant une découverte curieuse, faite par le moyen du microscope.* Amsterdam: Nouvelles de la République des Lettres, art. V, 1699.

———. *Le conte des fées du Mont des Pucelles.* Introduction and notes by Marcel Barral. Montpellier, France: Publications de l'Entente Bibliophile, 1988.

Pliny. *Natural History.* Translated by H. Rackham, edited by G. P. Goold. Cambridge, Mass.: Harvard University Press, 1983.

Pomeroy, Sarah Gertrude. *Little-Known Sisters of Well-Known Men.* Boston: D. Estes, 1912.

Porta, Giambattista della. *Natural Magik.* The Collector's Series in Science. New York: Basic Books, 1957.

Poupard, Paul, ed. *Dictionnaire des Religions.* Paris: Presses Universitaires de France, 1984.

Ramnoux, Clémence. "Héliocentrisme et Christocentrisme (sur un texte du Cardinal de Bérulle)." In *Le Soleil à la Renaissance: Sciences et mythes.* Paris: Presses Universitaires de France, 1965.

Rattansi, P. M. "Voltaire and the Enlightenment Image of Newton." In *History and Imagination: Essays in Honor of H. R. Trevor-Roper,* ed. Hugh Lloyd-Jones, Valerie Pearl, and Blair Worden. New York: Holmes & Meier, 1982.

Réaumur, René Antoine Ferchault de. *The Natural History of Ants, from an unpublished manuscript in the archives of the Academy of Sciences in Paris.* Translated and annotated by William Norton Wheeler. New York, London: A. A. Knopf, 1926.

Redi, Francesco. *Experiments on the Generation of Insects.* Translated from the Italian edition by Mab Bigelow. Chicago: Open Court, 1909.

Rhys Davids, Caroline A. F. *Buddhist Psychology: An Inquiry into the Analysis and the Theory of Mind in Pali Literature.* London: G. Bell and Sons, 1914.

Richards, Aute. *Outline of Comparative Embryology.* New York: John Wiley & Sons, 1931.

Robinet, André. *Malebranche et Leibniz, relations personelles—presentées avec les textes complets des auteurs et de leurs correspondants revus, corrigés et inédits.* Paris: Librairie Philosophique J. Vrin, 1955.

Rodrigues, Francisco. *Jesuítas Portugueses, Astrónomos na China, 1583–1805.* Macau: Instituto Cultural de Macau, 1990.

Roe, Shirley A. *Matter, Life, and Generation: Eighteenth-Century Embryology and the Haller-Wolff Debate.* Cambridge: Cambridge University Press, 1981.

Rosario, Vernon A. II. "Phantastical Pollutions: The Public Threat of Private Vice in France." In *Solitary Pleasures: The Historical, Literary and Artistic Discourse of Auto-Eroticism,"* edited by Paula Bennett and Vernon A. Rosario II. New York: Routledge, 1995.

Rostand, Jean. *Les origines de la biologie experimentale et l'abbé Spallanzani.* Paris: Fasquelle, 1951.

———. *Un grand biologiste: Charles Bonnet, expérimentateur et théoricien.* Paris: Université de Paris, 1966.

Rousseau, Jean-Jacques. *Les conféssions.* Paris: Chez madame veuve Perroneau, 1819.

———. *Émile, ou, De l'éducation.* Paris: Garnier, 1860.

Rusconi, M. *Amours des salamandres aquatiques, et development du tetard de ces salamandres, depuis l'oeuf jusqu'a l'animal parfait.* Milan, Italy: Chez Paolo Emilio Giusti, 1821.

Rushton, J. Philippe. *Race, Evolution and Behavior: A Life History Perspective.* New Brunswick, N.J.: Transaction Publishers, 1994.

Russell, E. S. *The Interpretation of Development and Heredity: A Study in Biological Method.* Oxford: Clarendon Press, 1930.

Russett, Cynthia Eagle. *Sexual Science: The Victorian Construction of Womanhood.* Cambridge, Mass.: Harvard University Press, 1989.

Sadowski, Robert M. "Stone Rings in Northern Poland." In *Archaeoastronomy in the Old World,* edited by D. C. Heggie. London: Cambridge University Press, 1982.

Sapp, Jan. *Beyond the Gene: Cytoplasmic Inheritance and the Struggle for Authority in Genetics.* New York: Oxford University Press, 1987.

Sarthe, L. J. Moreau de la. *Déscription des principales monstruosités dans l'homme et dans les animaux.* Paris: Fournier Freres, 1808.

Schiebinger, Londa. *The Mind Has No Sex?: Women in the Origins of Modern Science.* Cambridge, Mass.: Harvard University Press, 1989.

Scholem, Gershom. *Major Trends in Jewish Mysticism.* New York: Schocken Books, 1941.

Schrierbeek, A. *Jan Swammerdam, 1637–1680, His Life and Works.* Amsterdam: Swets and Zeitlinger, 1967.

Schurigio, Martino. *Spermatologia historico-medica, h. e. Seminis Humani Consideratio, Physico-Mecido-Legalis, qua Ejus natura et. usus, infilmulque Opus generationes et

*varia de coitu aliaque huc pertinentia, v. g. De Castratione Herniotomia, Phimiosi,
Circumcisione, Rectutitione, & Infibulatione, item De Hermaphroditis & sexum mu-
tatibus, raris & selectis Observationibus, annexo Indice lucupletissimo.* Francofurti
ad Moenum: Johannis Beckii, 1720.

Scott, Gaspar. *Physica curiosa, sive Mirabilia natura et artis, libris XII. Comprehensa,
quibus pleraque, quae de angelis, daemonibus, hominibus, spectris, energumensis,
monstris, portentis, animalibus, meteoris, etc, rara, arcana, curiosa; circumferunter,
ad veritatis trutinam expenduntur, variis ex historia ac philosophia petitis disquisi-
tionibus excutiuntur & innumeris exemplis illustrantur.* Herbipoli: Simptibus
Johannis Andres Enoteri & Wolfgangi jun. haeredum, excudebat Jobus Hertz,
1667.

Secret, François. "Le soleil chez les kabbalistes chrétiens de la Renaissance." In *Le Soleil
à la Renaissance: Sciences et mythes.* Paris: Presses Universitaires de France, 1965.

*The secrets of Nature Revealed, or the Mystery of Procreation and Copulation considered
and Explained. Translated from the Original Latin of the Celebrated Michael Scotus and by him Written, for the use of the Emperor of Germany. In which many
vulgar and pernicious errors are corrected, and Many wholesome and Salutary Directions Given with Regard to the Conduct of the Marriage-Bed; Being a Work Necessary to Be read by all Married people, and one which May Produce the Greatest Good
to Society in General. To which are added Safe and Certain Methods for Curing
impotency in Men and Barrenness in Women.* Sold by the Booksellers of London
and Westminster, ca. 1730.

Sherwin, Byron L. *The Golem Legend: Origins and Implications.* New York: University
Press of America, 1985.

Small, Meredith F. "Sperm Wars." *Discover,* July (1991): 48–53.

Small, Robert. *An Account of the Astronomical Discoveries of Kepler.* Madison: University of Wisconsin Press, 1963.

Sontag, Otto. "The Idea of Natural Science in the Thought of Albrecht von Haller."
Master's thesis, New York University, 1971.

———. "Albrecht von Haller and the Future of Science." *Journal of the History of Ideas*
35 (1974): 313–22.

———. "The Motivations of the Scientist: The Self-Image of Albrecht von Haller." *Isis*
65 (1974): 336–51.

———. *The Correspondence between Albrecht von Haller and Charles Bonnet.* Bern,
Switzerland: Hans Huber Publishers, 1983.

Spallanzani, Lazzaro. *Nouvelles recherches sur les découvertes microscopiques, et la génération des corps organisés.* Paris: Lacombe, 1769.

———. *Opuscules de physique, animale et vegetale.* Geneva, Switzerland: Chez Barthelemi Chirol, 1777.

———. *Experiences pour servir à l'histoire de la géneration des animaux et des plantes.*
Geneva, Switzerland: Chez Barthelemi Chirol, 1785.

———. *Dissertations Relative to the Natural History of Animals and Vegetables.* London: J. Murray, 1789.

———. *Tracts on the Nature of Animals and Vegetables.* London: Printed for William Creech, and Ar. Constable, and T. Cadell & W. Davis, and J. White, 1799.

Speert, Harold. *Obstetrics and Gynecology: A History and Iconography.* San Francisco, Calif.: Norman Publishing, 1994.

Spence, Jonathan D. *The Memory Palace of Matteo Ricci.* Elisabeth Sifton Books. New York: Viking Penguin, 1984.

Stehr, Frederick W. *Immature Insects.* Dubuque, Ia.: Kendall/Hunt Publishing, 1987.

Stengers, Jean, and Anne van Neck. *L'histoire d'une grande peur: La masturbation.* Brussels, Belgium: Éditions de l'Université de Bruxelles, 1984.

Sterne, Laurence. *The Life and Opinions of Tristram Shandy, Gentleman.* The Temple Classics. London: J. M. Dent, 1899.

Sumner, William Graham. *Folkways.* New York: New American Library, 1940.

Sumner, William Graham, and Albert Galloway Keller. *The Science of Society.* New Haven: Yale University Press; London: H. Milford, Oxford, 1927.

Swammerdam, Jan. *Historia generalis.* Lugd. Batav.: Henninius; Utrecht, Holland: Ribbius, 1685.

———. *Historia insectorum generalis, in qua quaecunque ad insecta eorumque mutationes spectant, dilucide ex sanior is philosophiae & experimentiae principiis explicantur.* Lugduni Batavorum: Apud Jordanum Luchtmans, 1685.

———. *The Book of Nature, or, the Natural History of Insects.* London: n.p., 1738.

Tarde, Gabriel de. *The Laws of Imitation.* Translated from the second French edition by Elsie Clews Parson, with an introduction by Franklin H. Giddings. New York: H. Holt, 1903.

Thompson, Charles John Samuel. *The Mystic Mandrake.* London: Rider, 1934.

Thompson, D'Arcy Wentworth. *On Growth and Form.* Cambridge: Cambridge University Press, 1917.

Thorndike, Lynn. *A History of Magic and Experimental Science.* 2d ed., with corrections. New York: Macmillan, 1929.

Tissot, Samuel Auguste David. *Onanism.* New York: Garland, 1985.

Topsell, Edward. *The Historie of Foure-Footed Beastes: Describing the True and Lively Figure of Every Beast, with a Discourse of their Several Names, Conditions, Kindes & Vertues.* London: William Iaggard, 1607.

Trembley, Abraham. *Mémoires pour servir à l'histoire d'un genre de polypes d'eau douce à bras en forme de cornes.* Leiden: Jean & Herman Verbeek, 1744.

Ursprung, H., and R. Nothiger. *The Biology of Imaginal Disks.* Berlin: Springer-Verlag, 1972.

Vandermonde, D. *Essai sur la manière de perfectionner l'espèce humaine.* Paris: Chez Vincent, Imprimeur-Libraire, 1751.

Velaverde Lombrana, Julian. *Juan Caramuel: Vida y obra.* Oviedo, Spain: Pentalfa Ediciones, 1989.

Venette, Nicholas. *De la génération de l'homme ou le tableau de l'amour conjugal. Huiteme edition revue, corrigée, augmentée et enrichie de figures par l'auteur.* Cologne: Chez Claude Joly, 1702.

Vries, Ad de. *Dictionary of Symbols and Imagery.* Amsterdam: North-Holland Publishing Company, 1974.

Waddington, C. H. *Principles of Embryology.* Bristol, U.K.: George Allen and Unwin, 1956.

Walsh, James J. *Catholic Churchmen in Science.* New York: Books for Libraries Press, 1917.

Wells, Herbert George. *Patterns of Life.* Science of Life Series, vol. 2. London: Cassel, 1934.

Westermark, Edward. *The Origin and Development of Moral Ideas.* London: Macmillan, 1924–1926.

Whiteside, Béatrice. *Paracelse: l'homme, le médecin, l'alchimiste.* Including "La médecine," by Béatrice Whiteside; "Les doctrines secretes," by Serge Hutin; and "Paracelse et sa posterité," by Georges Cattaui. Paris: Table Ronde, 1966.

Wielant, Philips. *Practijke criminele; naar het eenig bekend handschrift.* Gent, Flanders: n.p., 1872.

Wiener, Norbert. *God and Golem Inc.: A Comment on Certain Points where Cybernetics Impinges on Religion.* Cambridge, Mass: MIT Press, 1964.

Wooley, Leonard. *The Sumerians.* Oxford: Clarendon Press, 1928.

Yabuti, Kiyosi. "Chinese Astronomy: Development and Limiting Factors." In *Chinese Science: Explorations of an Ancient Tradition,* edited by Shigeru Nakayama and Nathan Sivin. Cambridge, Mass.: MIT Press, 1973.

Yates, Frances Amelia. *Giordano Bruno and the Hermetic Tradition.* Chicago: University of Chicago Press, 1964.

——. *The Art of Memory.* London: Routledge and Kegan Paul, 1966.

——. *Assaigs sobre Ramon Llull.* Barcelona, Spain: Editorial Empurias, 1985.

Yoshinaga, K., D. L. Hess, A. G. Hendrickx, and L. Zamboni. "Germinal Cell Ectopism in the Strepsirhine Prosimian *Galago crassicaudatus crassicaudatus.*" *American Journal of Anatomy* 187, no. 3 (1990): 213–31.

Zamboni, L., and S. Upadhyay. "Germ Cell Differentiation in Mouse Adrenal Glands." *Journal of Experimental Zoology* 228, no. 2 (1983): 173–93.

INDEX

Boldface denotes pages containing illustrations

Index